HISTORY

OF

THE OLD COVENANT

FROM THE GERMAN OF

J. H. KURTZ, D.D.,

PROFESSOR OF THEOLOGY AT DORPAT.

VOL. II.

TRANSLATED

BY JAMES MARTIN, B.A.,

NOTTINGHAM.

PHILADELPHIA:

LINDSAY AND BLAKISTON.

1859.

CONTENTS OF VOL. II.

THIRD STAGE IN THE HISTORY OF THE FAMILY.

JACOB.

GENERAL SURVEY OF THE PATRIARCHAL AGE.

SECOND STAGE IN THE HISTORY OF THE COVENANT.

THE NATION: ITS INSTITUTIONS IN THE TIME OF MOSES.

FIRST STEP TOWARDS THE DEVELOPMENT OF THE NATION.

ISRAEL'S SOJOURN IN EGYPT;

OR THE PREPARATION OF THE PEOPLE OF THE COVENANT,

A PERIOD OF 430 YEARS.

INDEX OF SCRIPTURE PASSAGES TREATED OF IN THIS VOLUME.

THE OLD COVENANT.

REMOVAL OF THE HOUSE OF ISRAEL TO EGYPT.

§ 1. (Gen. xlvi. xlvii. 12).—The report that Joseph was still alive, and was ruler over all the land of Egypt, was like a fable to his aged father Jacob; and it was not till he saw the Egyptian waggons that he could be convinced that it was true. "It is enough," he then exclaimed, "Joseph my son is yet alive, I will go and see him before I die." Well versed as he was in the ways of God, the old man could recognise at once the *call of Jehovah* in the invitation of Joseph. He therefore went to Egypt without delay. He stopped at the border of the land of his pilgrimage, which was also the promised land, to *offer a sacrifice* to the God of his fathers; and God *appeared to him in a dream.* "Fear not," he said, "to go down to Egypt, for I will there make of thee a great nation. I will go down with thee into Egypt, I will also surely bring thee up again, and Joseph shall close thine eyes" (1). The whole *house of Jacob*, with their wives, their children and grand-children, and all their possessions (2), then went down to Egypt in Pharaoh's waggons. (3) Judah was sent forward to announce their approach to Joseph, who hastened to meet his father, "and wept on his neck a good while." He then procured from the king the formal and official sanction to his plans, and presented five of his brethren to Pharaoh, who willingly gave them the required permission to live as *strangers and immigrants* (4) in the land of Goshen (5), which was so peculiarly suited to their nomad life. As a further proof of his confidence, he instructed Joseph to give his own cattle into the charge of the most able members of his family.

At a later period Joseph introduced his aged father to the king. The hoary-headed pilgrim *blessed* the king, and replied to his friendly enquiry as to his age: "The days of the years of my pilgrimage are a hundred and thirty years; few and evil have the days of the years of my life been, and have not attained unto the days of the years of the life of my fathers in the days of their pilgrimage" (6).

(1). There seems to have been always a strong inclination in the minds of the patriarchs to turn, they probably knew not why, towards Egypt, the fairy land of wealth, of culture, and of wisdom. This bias appears in all the three, but it was only in the case of Jacob that the inclination of the heart coincided with the call of God. Abraham actually went there, but the result taught him a lesson (§ 52); Isaac was restrained by God, just as he reached the frontier (§ 71); at last Jacob turned his steps in the same direction, and Jehovah appeared to him on the border of the land, to assure him that his course was pleasing to God.

In the history of the Old Testament, so long as it evinced any life and progress, we detect a constant disposition to coalesce with heathenism; and it was not till Israel had so hardened itself, that any further development was impossible, and had sacrificed its lofty, world-wide destiny for exclusiveness of the most absolute and contracted kind, that the inclination ceased to exist. There was truth at the foundation of this disposition, viz., a consciousness on the part of Israel of its relation to the world, and a presentiment of the fact that, whilst it was to infuse new life into heathenism from the fulness of its divine inheritance, it would also require to draw supplies from the culture of heathenism, that is, of the world. But in most cases the inclination was manifested in a thoughtless way, and therefore in ungodly, perverse, and injurious efforts. We find indications of this disposition as early as the days of the patriarchs, and in their case it was associated with the same truth and the same rashness. At that time it turned exclusively to Egypt, which was then and for a long time afterwards the only representative and type of earthly power, wealth, and civilisation. The *rashness* is seen in Abraham and Isaac, the *truth* appears first in Jacob.

It was not till the days of Jacob that the promised seed attained to such maturity as to render a certain amount of intercourse with heathenism both desirable and useful.

The *first* stage in the covenant history was drawing to an end, and Israel was preparing to enter upon a second. They left Canaan as *a family*, to return to it *a people*. As *a family* they had done their work and accomplished their end, viz., to exhibit the foundations on which *national life* is based. Henceforth their task would be to show how the basis of *the world's history*, in its widest form; is to be found within *the nation*. The two epochs, the growth of the family and that of the nation, stood in the same relation to each other as two concentric circles. The force of the common centre, from which the circumference of each is generated, gives to the two circles analogous forms. And this central creative power was the divine decree, on which Israel's history rested and by which it was sustained. At the conclusion of its entire history Israel was to enter into association with heathenism, in order that its all-embracing destiny might (to a certain extent) be fulfilled by its receiving from the latter the goods of *this* world, human wisdom and culture; and, on the other hand, by its imparting to the heathen the abundance of its *spiritual* possessions, the result of all the revelations and instructions which it had received from God. And thus also at the period under review, when the first stage of its history was drawing to a close, Israel joined with Egypt, the best representative of heathenism, bringing to Egypt deliverance from its troubles, through the wisdom of God with which it was endowed, and enriching itself with the wealth, the wisdom, and the culture of that land. Thus was it prepared to enter upon a new stage of its history, a stage of far wider extent and greater importance. *Vid.* § 92, 7.

It was not merely a vague surmise in Jacob's mind, which led him to the conclusion that the time had arrived for yielding to the inclination to go to Egypt, and that this inclination was confirmed and sanctified by a call from God. All the previous leadings of God combined to make this clear and certain, even without any express permission or direction on His part now. The remarkable course of Joseph's history, no less than Joseph's dreams, which the issue had shown to be from God, and the pressure of the existing famine, prevented any other conclusion than that the invitation of Joseph was a divine call. And this opinion was expressly confirmed by the previous revelation made to Abraham, that his seed would sojourn in a foreign land four hundred years. (Gen. xv. 13 sqq.)

Still the road which Jacob took was a painful path to him. He could not forsake the land, which had been the scene of all his wanderings, the object of all his hopes, and was still the land of promise, without hesitation and anxiety, especially as he could not shut his eyes to the fact that he should never tread it again. Once already he had been obliged to leave this promised land, and did so with a heavy heart (§ 75). But Jehovah had appeared to him at Bethel then, and consoled him with the assurance that he would bring him back with abundant blessings. Nor was a similar consolation wanting here. Jehovah promised that he would go down with him into Egypt, and bring him (meaning, of course, his descendants) back again to the land of his fathers. And even in Egypt the twofold object of all His previous leadings, viz., the promised land and the promised seed, would not be forgotten. On the contrary, the final intention of the whole should be realised there; "for," said the Lord, "*there* will I make of thee a great nation."

(2). The *catalogue of the house of Israel*, which came into Egypt, as given in Gen. xlvi. 8—27, presents several points of difficulty that we must not pass over. *First*, the direct descendants from Jacob who migrated to Egypt are said in ver. 27 to have numbered *seventy* souls. They are reckoned according to their mothers, thirty-three being assigned to Leah (ver. 15), sixteen to Zilpah (ver. 18), fourteen to Rachel (ver. 22), and seven to Bilhah (ver. 25). *V. Lengerke* (i. 347 sqq.) endeavours to prove that the number 70 is merely a round and approximate number, and throws the statements of the text into such strange confusion, that he succeeds in introducing several discrepancies into a list which is otherwise straightforward and plain. He first takes Leah's descendants in hand, and finds it impossible to arrive at the number 33. If Er and Onan, who died in Canaan (ver. 12), are included, there are 34 names; and if they are omitted, the catalogue contains only 32. But it is expressly stated in vers. 8 and 26 that Jacob, the head of the family, is reckoned as one of the 70 souls, and as he is placed in ver. 8 at the head of the catalogue of the children of Leah, it can be nothing but a spirit of contradiction, that leads any one to insist upon so literal an interpretation of ver. 14 as to seek for the names of exactly 33 *sons* or *descendants* of Leah. If Jacob is to be reckoned as one of the 70, the only appropriate place in which his name could stand is at the head of the catalogue of the children of Leah, his proper and lawful wife. There is still greater confusion in *v. Lengerke's* further remark (p. 240) that "the numbers given in vers. 18, 22, and 25 are correct, but in ver. 26 the number 66 is a round and approximate number; for 33 + 16 + 14 + 7 amount to exactly 70, and according to ver. 27 this number is only arrived at by the addition of Jacob, Joseph, Ephraim, and Manasseh." This is strange. In ver. 8 Jacob is reckoned as one of the 33, and in vers. 19, 20 Joseph, Ephraim, and Manasseh form part of the 14; so that, as a matter of course, if they are deducted from the whole number, as is the case in ver. 26, there will be only 66 remaining.

Again, the statement that *the* children of Israel "which came into Egypt" were numbered (vers. 8 and 26), appears to differ in several respects from the previous history. It would be easy to offer a complete defence of the general terms employed in ver. 8, where Joseph, Ephraim, and Manasseh, who were already in Egypt, are apparently reckoned among those who had just arrived there, even if they had not been so expressly excepted in vers. 20 and 26 as to prevent any possibility of mistake; for the writer's point of view led him to regard the emigration of Joseph and his sons into Egypt as not actually completed until the whole house, of which they were members, had formally settled there. Previous to that settlement Egypt was merely a casual resting place, and Canaan their true and proper home. But we meet with *real* difficulties of another kind. Benjamin, who comes before us as a youth throughout the history of Joseph (see for example Gen. xliii. 29), and who was not more than twenty-four years old, according to the existing chronological data, had as many as ten sons (ver. 21). Reuben, who is spoken of as having only two sons when they went to Egypt the second time (chap. xlii. 37), had now four (ver. 9). Pharez, the son of Judah by Tamar, had two sons (ver. 11), a fact which seems absolutely irreconcileable with the results arrived at in vol. i. § 86. And it is very improbable, to say the least, that Jacob's two great-grandsons, the children of B'riah, the *youngest* son of Asher, were born in Canaan (v. 17), since their grandfather Asher was only forty years old at the period of the emigration, and therefore his *youngest* son B'riah must have been a mere boy. With so many circumstances leading to the same conclusion, we need not hesitate to adopt the explanation that the words of ver. 26, "all the souls that came with Jacob into Egypt," are used in so general a sense as to embrace those grandsons and great-grandsons whose birth must have fallen in the period subsequent to the emigration.

Hengstenberg (Pentateuch vol. ii. 284 sqq. trans.) has entered thoroughly into an examination of the difficulty referred to, and solves it on the ground that the grandsons and great-grandsons of Jacob, though not yet born, were in their fathers, and therefore entered Egypt with them. Objections have been raised to this interpretation from various quarters, but we must still adhere to it. *Lengerke* talks about the "orthodox *in lumbis*," but will not affirm that the objection is sufficient to set it aside. The view referred to, which sees in the father the *ensemble* of his descendants, is common to the whole of the Old Testament. We find it repeatedly in the promises of God to Abraham, Isaac, and Jacob, "I will give *thee* the land;" "in *thee* shall all nations of the earth be blessed;" "*thou* shalt be a blessing," &c.; and in the section before us there are unmistakeable examples of it: "I will *bring thee* up again," ver. 4, (evidently not the individual person of Jacob, but his descendants, who were not yet in existence, and of whom Jacob was the one representative.) Why then should not the same writer, or even another, be able to say from the same point of view that the sons of Benjamin and Pharez *went down* in their fathers to Egypt? And, "just as Joseph's sons, though born in Egypt, are reckoned among the souls who came to Egypt, because in their father they had come thither, so also may these descendants of Jacob who came to Egypt in their fathers be regarded as having come *with Jacob* thither."

The reasons already assigned serve to show that such an explanation is both admissible and necessary, and the following data heighten its *probability*. 1. In the list of the families of Israel, which was prepared in the last year of the journey through the desert (Num. xxvi.), there are no grandsons of Jacob mentioned besides those named in Gen. xlvi. "It is difficult to explain this if the arrival in Egypt spoken of in Gen. xlvi. is to be taken precisely as a *terminus ad quem*. Are we to suppose, then, that there were no children born to Jacob's sons in the land of Egypt?" 2. In chap. xlvi. 5, where there is no question of genealogy, and the individuals emigrating are described from a historical point of view, we read, not of the *grandchildren* of Jacob's sons, but merely of their children, who are described as *little ones*. 3. In the case of Hezron and Hamuel (ver. 12) the author appears desirous of intimating that they were not born in Canaan, and that he regarded them as substitutes for Er and Onan, who had died there. *Venema* has expressed the same opinion. Thus he says (i. 121): "It is probable that the sons of Pharez who were born in Egypt are mentioned, because they were substituted for the two sons of Judah who died in Canaan. The historian clearly asserts as much, and when he adds that the latter died in the land of Canaan, he plainly implies that the sons of Pharez, who were put in their place, had not been born there."

Baumgarten (i. 316, 334, 350 seq.) has taken a most decided stand in opposition to *Hengstenberg*. In his anxiety to establish the *literal* historical accuracy of the genealogy in chap. xlvi. he does violence in a most unscrupulous manner to the previous history and the chronological data afforded by it, and crowds together not merely improbabilities but impossibilities also. (See the remarks in § 86). He is of opinion that with *Hengstenberg's* explanation "the entire list loses its objective worth and its historical importance; and if such were regarded as sufficient reasons for inserting in the catalogue those who were not born till afterwards, there was no definite limit at all, and the contrast between 70 souls who entered Egypt and 600,000 who left it, on which such stress is laid in Deut. x. 22, loses all its force."

This argument proceeds upon a misunderstanding and misinterpretation of the historiographical idea and design of the document. *Baumgarten* overlooks the fact that we have here not really a historical account, but a genealogical table; and that whilst any looseness of expression would be inadmissible in the former, it is not so in the latter. Besides, it is not correct that the insertion of a few of those who were born in Egypt was an arbitrary proceeding, and that there were no essential limits to determine the selection. Not only were there such limits, but they are most clearly defined; for the only grandsons or great-grandsons of Jacob whom we find in the list are those whose descendants formed a separate family (מִשְׁפָּחוֹת) in Israel. As a general rule the *sons* of Jacob were the heads of *tribes*, and the *grandsons* the heads of *families.* The outward unity of the family of Jacob, their existence as a common household, was not disturbed by his sons; but it could not but be disturbed by his grandsons. From outward considerations this became inevitable as soon as they attained their majority; and their separate establishments formed the first step in the transition from a family to a people. Now, it was evidently the intention of the author of the book of Genesis, to trace the early history of the nation of Israel up to that point, in which the children of Israel began to lay aside their character as a family, and assume the characteristics of a people. And if we endeavour to assign some definite epoch to this change, there is none which we can fix upon but the removal to Egypt. For, as we shall afterwards show, the principal intention of that removal was to facilitate the transition from a family to a people, and to secure it against interruption. And it was just about this time that Jacob's family reached the third stage, in which the *Mishpachoth* (or families) originated. A few exceptions might be found, but they could very well be sacrificed to the general validity of the rule and the great importance of the event in question. The task of the author was to trace the history of the descendants of Jacob up to that point in which they began to form separate *Mishpachoth* (families). And thus we have a limit, both thoroughly objective and sharply defined. It was not accident and caprice, therefore, but objective historical conditions which determined the choice.

This explanation is strikingly confirmed by a comparison of our list, which describes the state of things existing when the development of the nation *began,* with that contained in Num. xxvi., which describes in a similar manner the state of the *Mishpachoth* when it was *complete.* Such a comparison establishes all the suppositions which our explanation necessarily involves. In general the names mentioned are the same. In Gen. xlvi. they are given as those of the grandsons and great-grandsons of Jacob, and in Num. xxvi. as those of the heads of separate *Mishpachoth*; and the few deviations from this rule the altered circumstances will easily and naturally explain. Thus, in Gen. xlvi. we have only two of Jacob's grandsons by Joseph mentioned, viz., Ephraim and Manasseh; whilst in Num. xxvi. we have not less than thirteen *Mishpachoth* assigned to the tribes of Ephraim and Manasseh. But so long as the two sons of Joseph had not been adopted by Jacob (and that did not take place till the end of his life, seventeen years after the emigration), they could only be regarded as Jacob's grandsons, and therefore as the founders of two *Mishpachoth.* But when once they had been adopted, and had become the heads of distinct *tribes*, the *Mishpachoth* of the tribes were necessarily traced to their sons or grandsons. On the other hand, some names are *omitted* from Num. xxvi. which we find in Gen. xlvi. among the grandsons of Jacob. This, too, may be very simply explained on the ground that probably they did not increase to a sufficient extent to be able to claim the right of forming independent *Mishpachoth*, which they would otherwise have possessed as grandsons of Jacob, or that their families became extinct. Thus, for example, ten sons of Benjamin are named in Gen. xlvi., but in Num. xxvi. and 1 Chr. viii. 1, 2, we only read of five. This diminution, however, was most probably occasioned by the punishments so frequently inflicted upon the people in the desert.

If, then, it was the design of our author to continue his history to that point of time, in which the first foundations of the *national* institutions were laid in the *Mishpachoth*, and if, as a general rule, these *Mishpachoth* commenced with the grandsons of Jacob, it was necessary that he should include *all* the sons of Benjamin as well as the rest of Jacob's grandsons in the genealogical summary with which he closes his book. The unimportant and accidental circumstance that some of these were born in Egypt, was not in itself sufficient to prevent him from completing the lists, especially as the phrase "*in lumbis*," which conveyed to his mind and to those of others in his day a sense so much at variance with modern views, would be to him both natural and ready to his hand.

And the introduction of the names of the great-grandsons of Jacob through Judah and Asher may undoubtedly be explained in a similar way. From Num. xxvi. we learn that in their case there was an exception to the general rule, that the *Mishpachoth* should be founded by Jacob's *grandsons.* With Judah's grandsons, Hezron and Hamuel, the sons of Pharez, this is very apparent. As the two sons of Judah, Er and Onan, who died in Canaan, had failed to become the founders of *Mishpachoth*, the two first-born sons of Pharez, the son of their widow Tamar, through a Levirate marriage with Judah, entered as a matter of right into the vacant places of the deceased sons. Their father Pharez also became the founder of another *Mishpachah* through the remainder of his sons; and this *Mishpachah* was called by his name. This may likewise have been the case with the grandsons of Asher, Heber, and Malchiel, who founded families of their own in addition to that of which their father B'riah was the head (see Num. xxvi. 44 seq.), but we have not the necessary genealogical data for establishing the fact.

Thus we differ from *Hengstenberg*, inasmuch as we do not consider that the ideal importance of the number 70 would be a sufficient explanation of that want of objective truth which *Baumgarten* finds in the verse before us, but trace it, as the latter also does, to an objective historical fact. We are not, however, inclined on that account to give up the importance of the number 70. We regard it as a seal impressed upon the first step in the progress of Israel towards a national existence, for the purpose of distinguishing it as the holy nation to which salvation was entrusted for all the nations of the earth. *Seven* is the covenant-number, *κατ' ἐξοχην*, *the sacred* number, and therefore the sign of separation from the world. *Ten*, again, is the mark of completeness and universality. In *seventy* we have seven multiplied by ten, and this multiplication is the symbol of the peculiar position of the people of Israel. For the two things which distinguished the nation of Israel were just its *particular* call and separation on the one hand, and its *universal* relation, as the bearer of promises, on the other. And this *universalism* was not a mere abstract idea slightly associated with the history of the people, but a concrete potential fact, which entered truly and deeply into the very first stages of that history. The nation of Israel was a blessing to the nations even before the advent of Christ. In proportion to its age and the measure of its development it was so in the person of Abraham, when he led his pilgrim-life among the people of Canaan. In a still higher degree it became so in Joseph. In the highest sense it is so in Christ.

It appears strange that in the genealogical list there are only two women mentioned among the direct descendants of Jacob; *Dinah*, his daughter, and *Serah*, the daughter of Asher. We cannot determine with certainty whether their names are inserted because they were the only female descendants of Jacob in existence at the time of the emigration to Egypt, or whether there were not rather some peculiar circumstances which led to their being singled out from the rest. In the case of Serah we might infer from Num. xxvi. 46 and 1 Chron. vii. 30 that the latter was the cause, as she had evidently attained to some kind of independence among the families of Israel after her marriage. This may also have been the case with Dinah, and the family may afterwards have lost its importance. (*Baumgarten* agrees with *Luther* in the conjecture that Dinah was Jacob's housekeeper after his wives were dead, and that this will account for the insertion of her name). We must give the preference to the first of these explanations, as most consistent with the objective correctness of the catalogue. It appears to us neither impossible nor incredible that there should have been so large an excess of male children in Jacob's family for the first two generations; on the contrary, we can see in this fact the marks of the wisdom of God, which always directed the births that took place in the chosen family. We have already seen in several instances with what difficulties the marriages of the sons were attended. It was of the greatest importance to guard against any intermarriage with the Canaanites, lest the stream of heathen corruption should break through the barriers by which this family was kept apart. But as the other branches of of the family yielded more and more to the corruption of heathenism, and as the family of Jacob himself extended, these difficulties must necessarily have gone on increasing. If, however, the immediate posterity of Jacob consisted chiefly of men, it would evidently be easier to overcome the difficulties, and there would also be less danger connected with the marriage of one of Jacob's sons or grandsons to a heathen wife, than with the marriage of a *daughter* to a heathen husband. The subordinate position of the wife would render the former of comparatively slight importance; but in the latter case the daughter would actually separate herself from the chosen family and from the Covenant with Jehovah. It was not till a later period, when the blood-relationship of the descendants of Jacob had become so distant as to present no obstacle to the contraction of marriages one with another, that the difficulties in the way of the marriage of daughters came to an end.

As a rule, however, the sons of Jacob continued to avoid contracting marriages with the women of Canaan. This is plainly implied in ver. 10, where it is expressly mentioned as an unusual occurrence that Simeon had taken a Canaanitish woman as his concubine. In addition to their relations in Syria they could have recourse to other relations, viz., the descendants of Edom and Keturah.

(3). *V. Lengerke* (ut sup. i. 347), who pronounces chap. xlvi. 1—4 an incongruous interpolation, charges the author of this passage with ignorance of the nature of the country between Beersheba and Hebron, on account of his making Pharaoh's waggons travel by this impassable road. "According to the statements of modern travellers," says *v. Lengerke*, as *e.g. Robinson* (i. 317), it seems evidently impossible that waggons can ever have been employed among the steep and rugged hills of this district, which has always been destitute of a carriage road." But *Robinson* is merely speaking of the *straight* road between the two places, by which he himself travelled; and he afterwards adds, "we are convinced that waggons for the patriarch could not have passed by that route. Still by taking a more circuitous route up the great Wady-el-Khŭlîl more to the right, (according to the map the distance would not be very much greater), they might probably reach Hebron through the valleys without great difficulty."

(4). We must not overlook the fact that, when the brothers are admitted to an audience of the king, they do not ask to be received as members of the Egyptian state, but merely request permission to settle as *foreigners and sojourners* in Egypt for an indefinite period: "to sojourn in the land are we come," (chap. xlvii. 4). In this carefully chosen expression we see not only their consciousness, that Egypt could never be the land of their home and their future history, but also their intention to retain the right of leaving Egypt whenever they pleased, and hence the subsequent oppression and detention of their descendants was an act of violence opposed alike to justice and to the original compact. There is a striking resemblance between the description of the arrival of Jacob's family and a scene which *Hengstenberg* has copied from *Wilkinson's* work on Egypt, (see Egypt and the Books of Moses, p. 40, Eng. translation). It is taken from a tomb at Beni Hassan, and represents the arrival of strangers, who have come to Egypt with presents in their hands, and with their property carried on asses. "The number 37 is written over them in hieroglyphics. All the men have beards, which was contrary to the custom of the Egyptians, although very general in the East at that period. It is usually introduced in their sculptures as a peculiarity of foreign uncivilised nations." On this *Hengstenberg* remarks, "Some believe that this painting has a direct reference to the arrival of Jacob with his family in Egypt. On the contrary, *Wilkinson* observes that the expression "captives," which appears in the inscription, makes it probable that they are some of the prisoners of whom so many were taken captive by the Egyptians during their wars in Asia. But in his more recent work, he considers this circumstance as no longer decisive, inasmuch as the contemptuous expressions common among the Egyptians in speaking of foreigners, might account for the use of this word. In fact it speaks very decidedly against their being prisoners, that they are armed. Whether this painting has a direct reference to the Israelites will, of course, ever remain problematical, but it is at any rate well worthy of notice, since it furnishes proof that emigration with women and children took place in very ancient times."

Joseph directed his brethren to introduce themselves as *shepherds*, not only in spite of the fact that shepherds were an abomination to the Egyptians, but *on that very account.* His reason for doing so is apparent. In the occupation of his brethren there was the surest guarantee that their national and religious peculiarities would not be endangered or destroyed, and that they would not be absorbed by the Egyptians. The hatred and contempt which the Egyptians cherished towards the shepherd caste, as existing monuments attest by many a characteristic sign, may be traced to the fact that agriculture, with its regular and methodical habits, was the sole support of the Egyptian state, and that the irregularities of a nomad life must have appeared to a pedantic Egyptian to be rude and barbarous in the extreme. It is interesting, however, to find traces in the Pentateuch of the different stages in the growth of that fanatical hatred, which the people of Egypt ultimately cherished towards everything foreign. When Abraham sojourned in Egypt there was no appearance of this dislike; in Joseph's time all shepherds were an abomination to the Egyptians, and it was necessary that Joseph should be naturalised by marrying the daughter of an eminent priest. But the fact that such a marriage could take place is a sign, that the hatred and antipathy towards all that was foreign, which prevailed in the time of the Exodus, had not yet reached its highest point.

Pharaoh's readiness to consent to the request of the brethren may have been dictated by political motives, as well as by a wish to gratify Joseph. He may not improbably have hoped that by the settlement of a powerful and devoted tribe in the border province he would secure a desirable *bulwark* against the devastating incursions of the Bedouin robbers of the desert, and also against the other nations of the East, from whom Egypt, with its tempting treasures, had always much to fear.

(5). For the situation of the *province of Goshen*, see *Gesenius* Thes. s. v., *Robinson* i. 76 sqq. (London Ed. 1841), *Hengstenberg* ut sup. p. 42, sqq., Eng. tr., *Ewald* ii. 52, sqq., and *Tischendorf*, de Israel. per mare rubrum transitu, Lips. 1847, p. 3, sqq. Goshen was undoubtedly the *most easterly border-land* of Egypt. Jacob sent Judah thither before the rest (Gen. xlvi. 28). There the procession halted until Joseph had obtained the king's permission (chap. xlvii. 1). And the Israelites asked for a grant of this province that they might not come too closely into contact with the Egyptians, who hated their mode of life (xlvi. 34). It is evident from Ex. xiii. 17, and 1 Chr. vii. 21, that Goshen bordered on Palestine and Arabia, and the history of the departure of the Israelites in the Book of Exodus shows that it was not far from the Red Sea. The following data help to determine the western boundary of Goshen:—It extended as far as the Nile (Ex. ii. 3; Num. xi. 5; Deut. xi. 10), and the Egyptian capital of that day was not far distant (Gen. xlv. 10, xlvi. 28, 29; Ex. ii. 5, 8), though the name of the capital is nowhere mentioned in the Pentateuch. The searching investigations of *Bochart* (sedes aulae Ægyptiacae ad Mosis tempora, opp. s. p. 1099, seq.) and *Hengstenberg* (Egypt and the Books of Moses, p. 44, 45), lead to the conclusion that it was Tanis (or Zoan), near to the mouth of the *Tanitic* arm of the Nile. This supposition is strongly confirmed by Ps. lxviii. 12, 43, where God is said to have wrought his signs in Egypt in *the field of Zoan*, *i.e.*, in the Tanitic *nomos* ; and there is an unmistakeable intimation of this in the Pentateuch, where Hebron is said to have been built seven years before Zoan of Egypt, (Num. xiii. 23). This expression, *Zoan of Egypt*, implies not merely that it was one of the oldest cities in Egypt, but that it held the highest rank, in other words, that it was the capital of Egypt. Moreover, it must not only have been well known to the Israelites, but it must also have stood in very close relation to them.* If we add to these scriptural data the statement of *Josephus*, Arch. ii. 7, 6, that Pharaoh gave up Heliopolis to Jacob and his children, " we shall probably come very near to the truth," as *K. v. Raumer* says, Beitr. Zur. bibl. Geogr., p. 1, " if we assume that the land of Goshen was the strip of cultivated land which runs from Heliopolis, on the south-west, towards the north-east, and is bounded on the east by the Arabian desert, and on the west by the eastern arms of the Nile," *i.e.*, very nearly the same ground which is now covered by the province of es-Sharkîyeh (the eastern land) ; see *Robinson*, i. 76. The only question that could arise here is whether the Tanitic arm itself, or the Pelusiac arm, which is a little further to the east, formed the western boundary. As we do not read that the Israelites crossed the Nile either when they entered Egypt or when they left, the decision of this question would depend upon the size of the Pelusiac arm, whether it was as small then as it now is (which seems very probable, from the nature and appearance of the ground, *Robinson*, i. 549), or whether it was once navigable, as some have inferred from Arrian iii. 1, 4, but without sufficient reason (*Robinson*, ut sup.).

These results are supported by the accounts which are given of the nature and fertility of the land of Goshen. From Gen. xlvi. 34 it appears to have consisted of pasture-land, and in xlvii. 6 it is described as one of the most fruitful of the provinces of Egypt. These two features are seldom found together, but in this district we have them both. Part of the land is steppe, which is only suited for pasture, whilst the rest consists of the most fertile soil, and is watered by the overflowing of the Nile. With regard to the productiveness of the province of es-Sharkîyeh, even at the present time, *Robinson* says (i. p. 78, 79):

* The author retracts this opinion afterwards ; see § 40, 2.—*Tr.*

"In the remarkable Arabic document translated by De Sacy, containing a valuation of all the provinces and villages of Egypt in the year 1376, the province of the Shurkiyeh comprises 383 towns and villages, and is valued at 1,411,875 dinars—a larger sum than is put upon any other province, with one exception. During my stay in Cairo I made many enquiries respecting this district, to which the uniform reply was that it was considered as the best province in Egypt. Wishing to obtain more definite information, I ventured to request of Lord Prudhoe, with whom the Pasha was understood to be on a very friendly footing, to obtain for me, if possible, a statement of the valuation of the provinces of Egypt. This, as he afterwards informed me, could not well be done, but he had ascertained that the province of the Shurkiyeh bears the highest valuation, and yields the largest revenue. He had himself just returned from an excursion to the lower parts of this province, and confirmed, from his own observation, the reports of its fertility. This arises from the fact that it is intersected by canals, while the surface of the land is less elevated above the level of the Nile than in the other parts of Egypt, so that it is more easily irrigated. There are here more flocks and herds than anywhere else in Egypt, and also more fishermen. The population is half migratory, composed partly of Fellahs, and partly of Arabs from the adjacent deserts, and even from Syria, who retain in part their nomadic habits, and frequently remove from one village to another. Yet there are many villages wholly deserted, where many thousands of people might at once find a habitation. Even now another million at least might be sustained in the district, and the soil is capable of higher tillage to an indefinite extent. So, too, the adjacent desert, so far as water could be applied for irrigation, might be rendered fertile, for wherever water is there is fertility."

We find another name for "the land of Goshen," in chap. xlvii. 11, viz., "the land of Raem'ses," *Sept.* Ῥαμεσσῆ. The so-called *land* of Raem'ses is generally distinguished from the *city* of Raem'ses, which was *built at a later period* (Ex. i. 11). But there is no ground for this distinction, as *Hengstenberg* in particular has shewn (Egypt and the Books of Moses, p. 49, seq., note Eng. tr.) Raem'ses is undoubtedly the name of a city in every other place in which it occurs (Ex. xii. 37; Num. xxxiii. 3, 5); and there is no reason to suppose that the city was not in existence at the time of Joseph; for Ex. i. 11 does not refer to the first building of the city, but to the fortification of it. "The land of Raem'ses" was evidently the land of Goshen, of which the chief city was Raem'ses. The question as to the city actually referred to, and its situation, will come under examination in connexion with the history of the Exodus.

6. The fact that the aged patriarch presumed to *bless* the king of Egypt, and thus, in a certain sense, to assert superiority, is to be accounted for not merely from his greater age, but also from the impulse and encouragement given to Jacob by the consciousness that he was called of God to be a blessing to the nations (Gen. xii. 2). Jacob's blessing was a return and compensation for the kindness shown by Pharaoh to the house of Israel; and we see here the type of the true relation, in which Israel was to stand to heathenism in all their future intercourse. Pharaoh offers earthly goods to the house of Israel, and Israel in return blesses him with the spiritual blessing of the house of God. We may notice, in passing, the importance of the account of Jacob's age, which is introduced at this point apparently in so accidental a manner. For, were it not for the statement here made by Jacob, we should lose the chronological thread of the patriarchal history, and that of the Old Testament in general would thereby be completely destroyed.

7. The *historical importance of the emigration of the house of Israel to Egypt* is evinced by the fact, that when the covenant was made by God with Abraham (vol. i., § 56), this was announced to him by revelation as a necessary part of the divine plan. At the same time it was expressly declared to him that the settlement in Egypt would not be permanent (chap. xv. 14), and this was repeated to Jacob in the vision at Beersheba (chap. xlvi. 4). The *design* of the emigration was made known to *Abraham:* namely, that it was necessary as a transition from pilgrimage in the promised land to the full possession of the whole. In like manner the Lord said to *Jacob* in Beersheba (ver. 3): "fear not to go down into Egypt, for *I will there make of thee a great nation.*" The two things are most intimately connected, for Israel (even if we look merely at outward circumstances), could not have obtained complete and sole possession of the land until it had become an organised nation. Canaan was already inhabited by other tribes, and they must necessarily be driven

out and the country conquered, before unlimited possession could be secured. But Israel must become a powerful people, before it could accomplish this. And whilst, on the one hand, the development of the family of Israel into a people was the condition of their taking possession of the land; so, on the other hand, was the complete and irremediable corruption of the present inhabitants to be the condition of their expulsion. This is the meaning of the words addressed to Abraham (chap. xv. 16), "for the iniquity of the Amorites is not yet full." These two indispensable prerequisites were already preparing, and during the sojourn of the Israelites in Egypt both were to progress uninterruptedly towards completion. The spiritual blessings, which the pilgrimage of the patriarchs in the midst of the Canaanites had put within their reach for two hundred years, but which had been offered in vain, were now to be taken from them. It was the just judgment of God which deprived them of the salt they had so long despised, that the corruption which existed among them might do its work the more rapidly. The Israelites, on the other hand, were led to the enjoyment of those earthly blessings which were to be found in Egypt, the fairy land of fruitfulness, that they might become a great nation with more rapidity and ease.

Two hundred and fifteen years had now elapsed since Abraham first entered Canaan. There he had completed his pilgrimage, and his remains were deposited in the family grave at Hebron. There Isaac was born and died, and there he lived and suffered. There Jacob also had fought and conquered, and his sons and grandsons, the founders of the tribes and families from which the chosen people were to spring, had all been born and brought up in that land. Thus, then, the house of Israel had lived long enough in the promised land for the home feeling, so important and necessary, as we have already shown that it was (vol. i. § 49), to be deeply and ineradicably fixed in the *national* character. It was *necessary* that the rise of the *family* should take place in Canaan; for that of the *nation* another soil was required.

The sentence of comparative barrenness had long prevailed in the chosen family; it was the curse of *nature*, which was not fitted to bring forth the promised seed. But this sentence, which had been permitted by Divine wisdom to continue in force so long, was now removed. The *mercy* of the author of the promises was unceasingly displayed, and instead of that comparative barrenness, which prevailed to such an extent that, after many decennia of apparently vain hope and patience, and unanswered faith and prayer, there was only one solitary representative of the covenant, there was now granted a productiveness of so remarkable a character, that in a few centuries there was every prospect of the fulfilment of the promise, that the seed should be as the sand which is upon the sea shore.

But Canaan, at that time, was not the land in which the promise could be fulfilled without interruptions. Israel could not possibly have grown to a great and independent nation there. And, what is quite as important, they would have been unable to maintain their national and religious peculiarities intact, amidst the temptations and attacks of a hostile principle. The elements most needed to promote their growth and bring it to perfection were not to be found there, nor would they have been educated in the school, which was best fitted to train them for their subsequent obligations.

Canaan was then in the possession of numerous tribes, who regarded the land as their own. Even Abraham had felt himself cramped in the movements of his establishment (Gen. xiii. 6); Isaac had constantly to retire before the powerful inhabitants by whom he was surrounded (Gen. xxvi.), and in the time of Jacob the difficulties must rather have increased than diminished. If, therefore, the house of Israel had remained any longer in Canaan, they would have encountered the greatest obstacles to their ever becoming a large and independent nation. If their numbers had rapidly increased, it would have been impossible for them to stand entirely aloof from the Canaanites, as they hitherto had done. In such a case, they must either have made war upon the inhabitants, in order to maintain a footing in the land (and it would not be difficult to foresee the disastrous issue if they had); or they must have scattered themselves over the neighbouring countries, and then they would have lost their national unity and degenerated into a number of separate nomad hordes; or thirdly, and of this there would be the greatest fear, they would have intermarried and mingled with the Canaanites, until they were completely absorbed by their superior numbers. But the maintenance of their religious peculiarities would have been even more difficult, than that of their national independence. The religious eclecticism of the Canaanites, their readiness to adopt the *forms* of the Israelitish religion without its spirit (of which we had an example in the case of the Sichemites), and the seductive influence, which the worship of nature exerted upon that age and would certainly have exerted upon the Israelites, if they had come into closer contact with the inhabitants of the land, would all have combined to produce a result that would have been destructive of the very foundations of Israel's destiny.

None of these dangers existed in Egypt. There they could become a great nation without any difficulties or obstructions, and without the least interference with their national and religious peculiarities. And, what was of no little importance, they had opportunities there of making many provisions for their future wants as a nation. First of all, the land of Egypt furnished them with a plentiful supply during the existing famine, and such was the fertility and extent of Goshen that there was no occasion for them to be scattered, and no inducement to the members of particular tribes to separate from the general body. There was no fear of their mixing with the Egyptians and giving up their national and religious integrity. The hatred which the Egyptians cherished towards every foreigner, and the contempt in which shepherds especially were held, furnished an indestructible safeguard against any such danger. As Goshen was just as well fitted for agriculture as it was for grazing, it naturally induced them to combine the pursuits of farming, gardening, and vine-growing with those of their earlier nomad life, and thus fostered a taste for that mode of life, which was afterwards to form an essential part of their national existence. In the midst of the science, civilisation, and industry of Egypt, Israel was in the best school for that general culture, which they would afterwards require. Their intimate acquaintance with the Egyptian modes of thought, which looked at life in all its outward manifestations and ramifications from a religious point of view, may have served to enrich in many ways even the *religious* views of the Israelites. And the *symbols* of the Egyptian worship set before them a completely developed *form* of religious life, which was the product of laws of thought that are universally inherent in the human mind, and therefore was not merely applicable to Egyptian pantheism, but could also be adopted as a welcome support to the worship of the Israelitish theism, if only it could be animated, purified, and modified by the Israelitish principle. In like manner the Egyptian constitution, with its strict rules and excellent organisation, furnished the model which, with modifications to suit the altered circumstances, was afterwards adopted in the Israelitish state. And lastly, "Egypt was the seat of the strongest worldly power, and therefore furnished the best instrumentality for the infliction of such severe sufferings as would awaken in the minds of the Israelites a longing for deliverance and a readiness to submit to their God; whilst, at the same time, it offered a splendid field for the manifestation of the power and justice and mercy of the God of Israel in the rescue of His people and the judgment of their enemies" (*Hengstenberg*, Pent. i. 362). The importance of the two elements last mentioned, and their necessary connection with the counsel of God, are apparent from the fact, that they are expressly mentioned in the revelation which was made by God to Abraham (chap. xv). Thus Israel obtained *the character of a redeemed people*, which was of such great importance in its future destiny, and *Jehovah* then showed himself to be, what he was to continue to be in a constantly increasing degree, *the Redeemer in Israel.*

(8). We reserve the inquiry respecting the *dynasties* which ruled in Egypt at the time when the children of Israel were sojourning there, and into the connection between the *Hyksos* and the Israelites, till we arrive at the period of the Exodus from Egypt, in order that we may not anticipate, or enter into separate discussions of subjects which are closely connected.

ADOPTION OF JOSEPH'S SONS.

§ 2. (Gen. xlvii. 27—xlviii. 22).—Jacob lived seventeen years in Egypt, and reached the age of 147. A short time before his death he sent for Joseph, and exacted an oath from him, that he would not bury him in Egypt, but by the side of his fathers in the promised land. Joseph then introduced his two sons, Manasseh and Ephraim, and, in virtue of the promises made to him by God, Jacob formally *adopted* and solemnly *blessed* them (1). Joseph had placed the elder son Manasseh at

Jacob's right hand, and the younger, Ephraim, at the left; but Jacob crossed his arms, and pronounced the blessing with his right hand upon Ephraim's head, and the left upon that of Manasseh. Joseph, supposing it to be an oversight, complained of his doing so; but Jacob, instead of making any alteration, explained to him that the greater blessing and the more numerous posterity would belong to the younger. The Patriarch then turned to Joseph, and, as a proof of special affection, presented him with a piece of land which he had once conquered from the Canaanites (2).

(1). We have already remarked, in the previous section, that the chosen seed had now reached the close of one of the stages of its history. The family was complete, and the basis was laid for the development of the nation. In a certain sense, too, this was a type of the absolute close of its entire history, when its course as a nation should be finished, and the basis laid for its world-wide destiny. This type, as we have seen, was chiefly displayed in the fact, that the idea of Israel's appointment, to be the medium of salvation to the nations, was here *partially* and *temporarily* realised, whilst the ultimate fulfilment would be *permanent* and *universal*. In Joseph, as the noblest product of the family life, and as the representative of his house to the heathen, Israel had become the saviour of Egypt. But it was evident that the salvation, which Israel brought to the heathen at that time, was only a passing one, and did not exhaust the *promise;* for this had spoken of salvation for *all the nations* of the earth, whereas the present fulfilment of that promise reached merely to *one* among the nations. The family life of Israel could only impart a blessing to *one people*, and that blessing was limited in force and extent. The full and unlimited blessing *for the whole world* could only be realised, when the national life of Israel was also complete. The Israelites, therefore, had not reached the goal, when the first stage of their history drew to a close. The development of the nation was now to recommence, but on a larger scale, and furnished with fresh powers and different means.

Joseph had already stept beyond the contracted limits which hedged in the chosen seed, that he might carry a blessing to the heathen. His path led him to a freer, more lofty, and we might almost say, a universal standpoint. In him Israel reached an eminence, on which the limited character of its subsequent development prevented it from standing long, and from this point it came down to the humble position assigned it, that it might afterwards attain to something infinitely higher and more glorious. Joseph's exaltation was followed by humiliation *in his sons.* He led them himself to his father, that by his blessing he might consecrate them to this. He bore them away from the posts of honour which were open to them in Egypt, that they might return to the humble shepherd-life which his brethren led. They were not to *perpetuate* the idea represented by their father, but to unite with his brethren in *originating* a new development. This act of Joseph denoted a return to a condition of exclusiveness, the transition from the first stage to the second in the history of Israel. It is a proof of Joseph's faith, gives us an insight into the plans of God, and manifests the harmony which God had determined to establish between the subjective and objective elements of that history.

Jacob's treatment of the sons of Joseph denoted two things: the restoration of the house of Joseph to the family of Israel, and the adoption of the two *grandchildren* to the position and privileges of *children.* The former was requisite, since their father Joseph had been naturalised as an Egyptian, and therefore had broken the outward ties which bound him to his family. Of the importance and effect of this we have spoken already. But, as Joseph had become the deliverer of his father's house in consequence of his leaving it, his return to it was to secure to him a larger measure of its blessings, and therefore Jacob adopted his two sons. The right to do this he founds upon the fact, that God had appeared at Bethel (vol. i. § 75) and given him the double blessings of posterity and the promised land (chap. xlviii. 3, 4). "Therefore," said he, "thy two sons, Ephraim and Manasseh, shall be mine as Reuben and Simeon." The privilege possessed by the sons of Jacob above the grandsons consisted, as we have already had occasion to remark, in the fact that the former were the founders of closely organised tribes, and the latter of merely subordinate families.

This act of Jacob's is generally regarded as a virtual exclusion of Reuben and Simeon from the rights of primogeniture, and the transference of those rights to Joseph, since the double portion was the most essential mark of the birthright (Deut. xxi. 17). But there is no ground for such an inference here. Undoubtedly Joseph did receive a double portion in his sons; but it by no means followed that he obtained the privileges of the firstborn. His sons were placed on an equal footing with those of Jacob, but Reuben's claims to the birthright were not necessarily affected in consequence. We shall enter into the question more fully in a subsequent section (chap. xlix). The only thing that makes the nature of the adoption obscure, is the fact that Jacob expressly declares upon his deathbed, that the three eldest sons have forfeited their rights, and then merely transfers to Judah the second of the two privileges of birthright (a double inheritance and the headship of the family), but says nothing at all with reference to the former.

Jacob's *blessing* is the consequence of his adoption of Joseph's sons. In addition to the formal right to found two separate tribes, he assures them also of the requisite ability, that is, he gives them the blessing of such fruitfulness, as would enable them to form and maintain such tribes. The blessing is imparted by the *imposition of hands;* for the general meaning of which see my *Mosaisches Opfer* (Mitau 1842, p. 67 sqq.) Jacob pronounces *the same blessing* on the two sons, and blesses them both *uno actu.* There is indeed a difference, but one *of degree* merely and not of kind. To the younger there is promised greater fruitfulness and power than to the elder. As there is no reason to suppose that the distinction originated in any personal predilection, we can only explain it on the ground of the prophetic foresight of the patriarch, and discover in the prediction the last expression of that παρὰ φύσιν, which predominated in the whole of the patriarchal history.

(2). When Jacob had blessed the sons of Joseph, he turned again to Joseph himself, and said: "Behold I die, but God shall be with you, and bring you again unto the land of your fathers. Moreover I give to thee one portion (שְׁכֶם אַחַד) above thy brethren, which I took out of the hand of the Amorite with my sword and with my bow." This difficult passage has been expounded in various ways, and sometimes very strangely (*vid. C. Iken* de portione una Josepho prae fratribus a patre data, in his philol. and theol. dissertations). *Calvin* and others follow the Septuagint, and suppose the passage to refer to the city of Sichem, which Jacob's sons took from the Amorites and destroyed, in consequence of the violence done to Dinah. But this explanation is irreconcileable with the use of the word אחד *one*, and it is inconceivable that Jacob should attribute to himself an event, which he so strongly lamented and abhorred (Gen. xxxiv. 30, and xlix. 5—7). Hence the שכם must in any case be an appellative, though the choice of this particular expression renders it probable that there is some allusion to Sichem, which was certainly allotted to the tribe of Ephraim. Others imagine that the reference is to the "parcel of a field" which Jacob bought from the Shechemites for a hundred pieces of silver (Gen. xxxiii. 19). This explanation apparently lies at the foundation of John iv. 5. *Iken* attempts to remove the discrepancy between the statement of chap. xxxiii., that this field was *bought*, and that of chap. xlviii., that it was *conquered*, by supposing that after the land had been purchased, it was probably taken away again by the Amorites, so that Jacob was obliged to recover it by force. He finds a positive confirmation of this opinion in a wire-drawn *Haggada* in *Jalkut Shimeon*, where Jacob and his sons are said to have returned to Sichem, and to have engaged in a fearful war with the Canaanites, in which the old patriarch Jacob performed miraculous feats of bravery, and Judah did the most extraordinary things with a kind of Berserker fury. But we cannot possibly attribute the smallest residuum of a historical tradition to so absurd a legend, which has evidently grown out of the passage before us. Besides, it appears very inappropriate, that Jacob should found his claim to the piece of land upon a forcible conquest, which is never referred to in the book of Genesis, and not upon the purchase, which is there recorded. There is a third explanation, which is given by several rabbins, and has been revived by *Tuch* (comm. p. 552), viz., that the word, לקחתי, *I took*, like the other perfects in Jacob's address, is to be regarded as a *perfectum propheticum*, and therefore that the subsequent conquest of the land by Jacob's descendants is here referred to, and that the play upon the word Shechem indicates the province which should afterwards be assigned to the descendants of Joseph. But there are difficulties connected with this explanation. It is true that, according to the ancient mode of view, Jacob might very well have attributed to himself, as the representative of the nation, such a *national transaction* as the conquest of the land by his descendants; but in this *connection* it does not appear probable. Jacob's gift is evidently referred to here, as an expression of personal favour and affection, for which there would be a much better opportunity if the land to be disposed of had been acquired by his own exertions. Moreover, it must be remembered that Jacob had already separated Joseph from his sons by adopting the latter as his own (chap. xlviii. 6), and therefore that the present was made to Joseph personally, and not as the father of Ephraim and Manasseh, who had already received their blessings (vv. 15—20). Hence we are shut up to some event in the life of Jacob, which has been passed over by the book of Genesis; and, as we can only fall back upon conjectures, that offered by *Heim* (Bibelstunden. i. 644) is perhaps the most plausible. As we learn from Gen. l. 23 that the children of Machir, the son of Manasseh, were born on Joseph's knees, *i.e.*, were adopted by him, and from Num. xxvi. 29—33 that one of these sons was named Gilead, and also from Num. xxxii. 39 sqq., and Joshua xvii. 1, that the families of the tribe of Manasseh, who sprang from Gilead, received the land of Gilead on the east of the Jordan as their possession, *Heim* supposes that the tract of land to which Jacob refers (שכם lit. the *shoulder* of land), was the hill-country of Gilead. Jacob was peculiarly interested in this district on account of his interview with Laban there (chap. xxxi. 23 sqq.), and the "heap of witness" erected by him gave him a certain claim. The Amorites may possibly have destroyed this sacred memorial, and thus Jacob may have been led to attack them, for the purpose of conquering and maintaining possession of the memorial itself and the shoulder of land on which it stood. Joseph may perhaps have bestowed the land, which was presented to him by Jacob, upon the son of Machir, who was "born upon his knees," and have named it *Gilead* in consequence. This would probably explain the abrupt introduction of the tribe of Manasseh in Num. xxxii. 39: "And the children of Machir the son of Manasseh went to Gilead and took it, and dispossessed the Amorite which was in it. And Moses gave Gilead unto Machir." Hitherto the historian had only spoken of Reuben and Gad.

JACOB'S PROPHETIC BLESSING ON HIS SONS.

§ 3. (Gen. xlix. 1—28). Jacob assembles his twelve sons around his deathbed. The germs of the future, which are wrapped up in the present, open before his prophetic glance. He says:

V. 1. "Gather yourselves together, that I may tell you
That which shall befal you in the end of the days! (1)
2. Gather yourselves together and hear, ye sons of Jacob,
Hearken unto Israel, your father!
3. *Reuben*, my first-born art thou!
My might and the first-fruits of my strength!
Pre-eminence in dignity and pre-eminence in power.
4. A fountain like water; have *no* pre-eminence!
For thou ascendedst thy father's bed,
Then defiledst thou it,—my couch he ascended!
5. *Simeon and Levi*, brethren are they!
Instruments of violence are their strokes.
6. Into their fellowship come not, my soul,
Join not in their assembly, my glory!
For in their wrath they strangled the man,
And in their wantonness lamed the ox.
7. Cursed be their wrath, for it is fierce,
And their rage, for it is cruel!
I will divide them in Jacob,
And scatter them in Israel (2).
8. *Judah* (*i.e.* praised) art thou, thy brethren praise thee,
Thy hand is on the neck of thine enemies;
The sons of thy father bow before thee.
9. A young lion is Judah.
From the prey thou risest up, my son.
He lieth down, he coucheth as a lion
And as a lioness. Who rouseth him up?
10. The sceptre shall not depart from Judah,
Nor the ruler's rod from the place between his feet,
Till he attain to rest.
And the nations obey him.
11. He binds his ass-foal to the vine,
And the young of his she-ass to the vine-branch,
He washes his clothes in wine,
His garment in the blood of the grape.

12. Dark are thine eyes with wine,
White are thy teeth with milk (3).
13. *Zebulon* (*i.e.* dwelling), on the sea shore he dwells,
He dwells on the coast of ships
And his side is at Zidon.
14. *Issachar*, an ass with strong bones,
He lieth down between the hurdles.
15. He sees that rest is good,
And that the land is pleasant.
He bends his neck to the burden,
He becomes a tributary servant.
16. *Dan* (*i. e.* judge) judges his people
As *one* of the tribes of Israel.
17. Dan is a snake in the way,
An adder in the path.
He stings the horse's heel,
And backward falls his rider.
18. *For thy help I wait, Jehovah.*
19. *Gad*, oppressors press upon him,
But he presses their heel.
20. From *Asher* come fat things, his food,
He yields the dainties of a king.
21. *Naphthali*, a hind escaped,
Speaking words of beauty.
22. Son of the fruit-tree is *Joseph*,
Son of the fruit-tree at the well,
Daughters grow up over the wall.
23. They cause him bitterness, they shoot with arrows,
They lie in wait for him, the heroes of the arrow.
24. But his bow remains firm,
Supple is the strength of his hands.
From the hands of the strong one of Jacob,
From thence, where the shepherd is, the rock of Israel.
25. From the God of thy father—and he helps thee,
From the Almighty,—he blesses thee,
Blessings of heaven from above,
Blessings of the flood, which rests beneath,
Blessings of the breast and of the womb.
26. The blessings of thy father are stronger than the blessings of the everlasting hills,
Than the loveliness of the hills of antiquity.
They come upon the head of Joseph,
On the crown of the consecrated among his brethren.
27. *Benjamin*, a rapacious wolf,
In the morning he devours the prey,
In the evening he divides the plunder (4).

(1). *Tuch*, in his Commentary (p. 561), has given a list of the numerous ancient authors who have written upon the chapter before us. Among modern expositions we may mention that of *Hävernick* (Vorlesungen über die Theol. des alten Test. p. 208 sqq.). Every prophecy is founded upon the circumstances and necessities of the period of its delivery; and it is necessary, therefore, that we should understand both the feelings of the prophet and the outward circumstances which gave occasion to the prophecy, before we can interpret the prophecy itself. The blessing of Jacob is no exception to this rule. We have now arrived at that point in the history of the chosen seed, in which the family began to expand into the people. In the *dodekad* of Jacob's sons a true basis had been laid for the future development of the nation. The law, which required the separation of Abraham from his family and the exclusion of Ishmael and Esau, was now satisfied (*vid.* vol. i. § 49). Not one of the twelve sons of Jacob had to be shut out. They were all enclosed and united by the bond of election and promise. The fulfilment of their destiny depended upon their becoming a nation and possessing the promised land. These were the two results towards which their history was leading. The germs of both were now apparent; on the one hand, in the fact that, after so long a period of comparative barrenness, they suddenly became remarkably prolific, and, on the other, in the distinct consciousness that they were strangers in Egypt, where they never could and never were intended to feel at home. The fulfilment of each of these involved the union and amalgamation of the two, for the second was dependent upon the first. And this amalgamation constituted the future of Israel. This was to be the goal, and to constitute the completion, of their history, so far, that is, as it had already struck its roots and put forth its buds. From the very nature of prophecy, then, the eye of the prophet could not look beyond this goal (*vid.* vol. i. § 7), or, at least, could only do so where the *development* of the existing germ would furnish the *basis* or the *germs* of still further expansions.

The organ of the prophecy belonging to that age was Jacob. With a heavy heart he had left the land of his pilgrimage, his trials, his adventures, and his hopes, to see it no more; but he had left it with the fullest assurance, confirmed by God, that in his descendants he should receive it as a permanent possession.

His whole soul was filled with the one thought of his return to take possession of the promised land. On this one point were all his thoughts and feelings, all his hopes and longings, concentrated. So completely was his inner life absorbed by this, that there was no room for other thoughts or feelings, and all events were viewed in their relation to this one. From the accounts we possess of his sayings and doings after the removal to Egypt, everything seems to have been merely an expression of this one deep-rooted feeling of his nature (see chap. xlvii. 29 sqq., xlviii. 3—5; 21, 22), and he could not rest till he was assured on oath that his remains should be buried in the land of his fathers. A mind thus occupied and absorbed might well urge him to prophesy. And as he draws near to death, at that moment when the fetters of the spiritual sight are often broken,* not only is he enabled to look into the future with clearer eyes, but the spirit of prophecy comes upon him from above, and in its light he sees the longings of his heart fulfilled, and the promised land in the possession of his descendants. He sees the tribes of Israel stirring and active in the full enjoyment of the rich blessings of the land, victorious over the dangers which they meet with there; each one in the situation which the elective affinity of his character and his inclinations may have led him to choose, or which the patriarchal authority of the prophet, as the medium of the divine decrees, may have assigned him by way of punishment or reward. His twelve sons are standing round his bed, the representatives and fathers of the tribes by which the land is to be taken. Before his mind there are gathered together in one living picture all the pleasing and painful events of which they have been the cause. With prophetic vision he traces the characters and dispositions of the fathers, as they are transmitted, expanded, or modified, through the history of their descendants. And aided by this insight, he allots to every one, on the authority of God, his fitting portion of that land, in which he himself has led a pilgrim life for more than a hundred years, and which now stands with all its natural diversities and with its rich and manifold productions, as vividly and distinctly before

* *Cicero* de divinatione, i. 30: *facilius evenit appropinquante morte, ut animi futura augurentur; Homer*, Il. 22, 355—360; *Plato*, Apol. i. p. 90 Bip.; *Xenophon*, Cyr. viii. 7, 21, &c; *Passavant*, Lebens-Magnetismus, Ed. 2, p. 163.

his mind as the different characters of his own sons. (See the beautiful exposition of this blessing in *Herder's* Briefe über das Stud. d. Theol. 1 Br. 5, 6, and Geist der Hebr. Poesie ii. 187—189).

That period in the future, which Jacob wishes to exhibit prophetically to his sons, is described by him as אַחֲרִית הַיָּמִים, *the end of the days.* For an explanation of this formula we refer more particularly to the excellent remarks of *M. Baumgarten* (Comm. i. p. 364 sqq. *Vid.* also *Hävernick*, p. 209 seq., and *Hengstenberg*, Balaam 175 sqq.). We must admit with *Baumgarten* and *Hävernick* (in opposition to *v. Bohlen*, *Rosenmüller*, *Hengstenberg*, and others), that in the passage before us, as well as the fifteen other passages of the Old Testament in which they occur, the words "the end of days," like the corresponding formula of the New Testament, ἐν ἐσχάταις ἡμέραις, do not merely indicate some indefinite period in the future, but the closing period, the end of days, the time of the final fulfilment, in a word, the Messianic era. For although the words in their literal signification might refer to any future times, such as were not absolutely at the *end;* yet the usage of the language was sufficiently settled to compel us to interpret them in the present instance according to their stereotyped meaning. But it is said that the blessing itself is irreconcileable with such an interpretation; that the blessing evidently refers to the time of Joshua, when the holy land was fully conquered and divided among the twelve tribes; and that the time of Joshua cannot be regarded as the end of days, *i.e.*, as the close of the history of Israel, but, on the contrary, was rather the actual commencement of that history. This objection, however, has no force, if we take a correct view of the prophecy and of the history of Israel. *Baumgarten* (ut sup.) most appropriately says:—"The true knowledge of the end must take its form from the position and the horizon of each individual. Hence for Jacob the end could be nothing else than the possession of the promised land by his seed, the people of promise. All the promises pointed to that, and beyond that nothing had been given or even hinted at." Jacob could reasonably look upon the time of Joshua as that of the completion of all things; in fact he could not do otherwise, for there was as much partiality and imperfection in his knowledge of what constituted completion, as there was in all that had been historically realised in the time of Joshua. And those very elements, which we find already fully developed and embodied in a definite form in Jacob's prophetic view of the perfection of the future, viz., the growth of his seed into a great nation and the possession of the promised land, were actually worked out and historically fulfilled in the time of Joshua. From Jacob's subjective point of view, the time when his promised posterity should have become a great nation, and taken possession of the promised land, was really the end of the days, inasmuch as their constant motion was then exchanged for rest, and wrestling and striving for possession and enjoyment. All his thoughts and hopes, his wishes and longings, were still bounded by this limited horizon. The only characteristics of the approaching end, with which he was acquainted, were the growth of his seed into a great and powerful people, and their possession of the holy land. And he knew of nothing that hindered the coming of the end, and the full and undisturbed possession and enjoyment of all the blessings it involved, but the insignificant and homeless condition of his family. Let these be once overcome, and in his view the full blessings of the promise must be enjoyed by his seed, and diffused by them throughout all the nations of the earth. This subjective view of the patriarch was imperfect, but by no means false. It was *true*, not merely because the removal of these hindrances, and the realisation of these conditions, furnished the necessary basis for the absolute completion of the Israelitish history, but also because it was in the possession of the land, the enjoyment of the blessings of that possession, and the central position which Israel then occupied among the nations of the earth, that the vocation of the seed of Abraham received its first passing fulfilment. But this fulfilment contained other germs within itself, which also required to be moulded and expanded. Jacob, however, looked at the period when the promised land should be possessed, as one of fulfilment and completion merely, and not, what it also was, as the seed of a higher development, the first stage of a still wider expansion, and therefore his view was *imperfect*. Since, then, Jacob prophesied of the time of Joshua, as though it would be the *end*, whereas it was to be only the beginning, the preparation, or an early stage of the absolute end; the prophecy of Jacob necessarily differed as much from the fulfilment in the time of Joshua, as the relative termination differs from the absolute end of all. Hence the consciousness was sure to be excited that the rest and enjoyment and possession, which are referred to as perfect in Jacob's blessing, were not fully realised in Joshua's days, and therefore that Jacob's blessing still pointed onward from the period of its first partial fulfilment to a future day, when it should be more perfectly fulfilled. As a general rule each age will see the object of its longings, and therefore the end, in the satisfaction of those wants of which it happens to be conscious. But with every essential advance in the history of the world the horizon widens, and men become conscious of new wants, new desires, new expectations, of which previously they had no suspicion. The expansion of existing germs brings new germs to light, which until then had been hidden from view. And thus every condition which seemed likely to be the end is no sooner reached, than it becomes the commencement of a new development; and this will continue till the absolute end arrives, and with it the full expansion of every germ.

This blessing was closely related to that pronounced on Jacob by his father Isaac (vol. i. § 72; vid. my Einheit der Genesis, p. 198 seq.). Jacob here communicated to his sons, in a more fully developed form, what he had already received from his father; and the many points of coincidence and, to some extent, verbal agreements, which we meet with, especially in the predictions concerning Judah and Joseph, bear witness how deeply the prophetic words of his father had been impressed upon Jacob's mind.

Hitherto we have found the blessing of promise not merely handed down to the next generation by the possessor of it for the time-being, but also expressly repeated and confirmed by Jehovah (vol. i. § 72. 1). The latter, however, was not the case with Jacob's sons; there is no intimation of their having been invested with the blessing by Jehovah. And from this time forth even the former ceased. The reason why Jacob was the last to invest his sons with the blessing of promise was, that he was the last *solitary* possessor of the covenant and the blessing. And the reason for the omission of the express investiture on the part of Jehovah in the present case, seems to have been, that now at length the way of grace entirely coincided with that of nature. So long as certain members of the family had to be excluded as natural branches, it was necessary that the divine investiture should be repeated every time; but as soon as the patriarch had been pointed out, whose entire posterity, without any exception was destined to carry forward the plans of salvation, *his* divine investiture had force and validity for all future generations.

(2). *Reuben*, the first-born, stood first in the rank of the brethren who surrounded their father's bed. According to the rules of primogeniture, the double inheritance (Gen. xxi. 17) and the headship of the family also belonged to him (1 Chr. vi. 2; Gen. xlix. 3); but he had forfeited both the rights and the honour of birthright by the commission of incest (Vol. i. § 83). He ought, as the first-born, to have been the firmest defender of the honour of the family, and it was by him that it had been violated. For that reason the crown of dignity and might, to which his birthright entitled him, was taken from his head. *Simeon* and *Levi* were the next in order, but the dignity, which Reuben had forfeited, could not be conferred upon them; for through their treachery towards the Shechemites (Vol. i. § 82) they had brought disgrace upon the house of Jacob, made his good name "to stink" among the heathen (Gen. xxxiv. 30), and acted in criminal opposition to the call of Israel, to be the channel of blessings and the medium of salvation to the heathen. They had *united* for the purpose of crime, therefore they were to be *scattered* in Israel. "This scattering of Simeon and Levi was an appropriate punishment for their alliance, which was opposed to the spirit of Israel, just as at a former period the forcible dispersion of the nations had been the consequence of their combining in opposition to the will of Jehovah" *(Baumgarten)*.

The three elder sons were thus excluded from the rights and privileges of the birthright. They were not to inhabit the heart of the land, which would otherwise have fallen to their share. Reuben's inheritance was to be outside the true holy land, and therefore was not even mentioned. Simeon and Levi were to be scattered in fragments among the rest of the tribes, and therefore to lose the advantages and independence, which only compactness and unity could secure. But, although they were deprived of the blessings of the birthright, they were not separated from the community of the chosen people, or from the call which they had received. They were not placed on the same footing as Ishmael and Esau, but still continued, as *individuals*, members of *the* family, and as *tribes*, members of *the* people, to whom the promise was given. They were, therefore, to co-operate with the rest in the duties to which the whole people had been called, and that was their blessing. But their co-operation was of a miserable kind, with very little of an independent character, and that was their curse.

(3). The earlier monographical expositions of the *blessing on Judah* have been specified by *Tuch* (Comm. p. 570). There have now to be added to the list *Hengstenberg's* Christology, sqq. : *Sack's* Apologetik ; *Hofmann's* Weissagung und Erfüllung ; and *L. Reinke's* Weissagung Jakob's üb. d. zukünftige glückliche Loos des Stammes Juda und dess. Nachkommen Schilo. The tone and substance of Jacob's discourse changed as soon as he looked at *Judah*. He was able to bestow upon the fourth son at least one part of that, which he had been obliged to refuse to the first three. The one great privilege of the first-born, the rank of chief among the tribes, with pre-eminence in power and dignity, is awarded to Judah. He is in reality, what he is in name, the *praised* among his brethren. The sons of his father bend before him, for with the courage of a lion he has fought as their leader and champion against every enemy, and having maintained their cause successfully, he holds the fruits of his victory with a lion's power. By swaying the sceptre with the force he displays, he is able not only to *enter* into rest, but to *give* rest to the tribes, at whose head he stands. The nations, whom he has conquered by the might of his arm, submit without resistance, yea willingly and cheerfully, to his peaceful government, and share in the blessings of peace and rest, into which he has entered and leads others also. The symbols of the conflict, by which the nations have been subjugated to their own advantage, are now laid aside, and he is surrounded by the emblems of peace alone. "Is he in full armour, a mighty conqueror, who has subdued the nations ? Is his garment full of the blood of the slain, his eye fired with the fierceness of battle ? No, he comes seated on the young colt of an ass, an animal of peace, and tarries in a vineyard. Doubtless he has washed his clothes in blood, but it is the blood of the grape. It is wine that makes his eyes so full of fire, and milk, the harmless food by which his teeth are whitened, has made his temper gentle and kind. The blessing to be realized in Judah's future history begins with his victorious conflict, and closes with the enjoyment of happiness and peace. His princely bearing is placed between the two. But Judah is the champion and leader of his brethren, and therefore they all share in the blessings secured by him." (*Hofmann*, Weissagung und Erfüllung i. p. 118).

The most difficult passage in the blessing of Judah is the much disputed clause "till *Shiloh* come." We have followed *Hofmann* and others in taking שִׁילֹה to be a common noun, with the meaning *rest;* and have rendered the clause: "till he (Judah) attains to rest." Most commentators, however, regard the words in question as the title of a personal Messiah, who was to spring from the tribe of Judah; though they arrive at this result in different ways. Shiloh is, of course, in this case, the subject, not the object, of the rendering: "till Shiloh (*i.e.* the Messiah) come." Thus *Delitzsch* (in his work on the prophetic theology of the Bible, p. 293) has expressed his firm conviction, "that every attempt to explain *Shiloh* as a common noun fails, and that the only correct rendering is that which treats it as a name of the Messiah, since this prediction formed an indispensable link in the historical chain, which ushered in the proclamation of salvation. For when once the patriarchal *triad* had become a *dodekad* in the family of Jacob, and thus the point of transition from the family to the people had been reached, the question necessarily arose, from which of the twelve tribes would salvation, *i.e.* the triumph of humanity, and the blessing of the nations, arise?" But *Delitzsch* himself has not adhered to this explanation.

We also admit, as will presently appear, that this prophecy forms a necessary link in the historical chain, which ushered in the proclamation of salvation; but we by no means admit that it was important that the question, from which of the twelve tribes salvation was to be expected, should receive an answer at this early age. Such a question in fact could only arise, when the idea of salvation had assumed the form of a confident expectation of a personal, individual Messiah. The organic progress of prophecy, and its close connexion in all its stages with contemporary history, prohibit us from imagining for a moment, that there was any expectation of a personal Messiah in the patriarchal age. In fact such an expectation was not only not indulged, but would have been altogether unsuitable to the character of the times. The evident intention of the whole history of that age was to develope the family into a great people; its entire tendency was to expand the unity of the patriarchs into the plurality of a nation. And this impulse, which was inherent in the patriarchal history, was not an unconscious one, but stood before the minds of the patriarchs with the greatest clearness and certainty, and was the one object of all their thoughts and hopes, and strong desires. The patriarchal history *began* with the consciousness of this their immediate destiny, as it was set before them in the clearest light by the call of Abraham. The *progress* of that history was maintained by the constant renewal, or revival of the same consciousness. Nearly every one of the numerous theophanies and Divine revelations, which occur in the history of the patriarchs, point to this end, and contain a promise that by the blessing of God it shall be attained. The earnest longing, which existed, for this expansion into a numerous people, was necessarily heightened by the delay, which arose partly from the barrenness that prevailed at first in the chosen family, and partly also from the necessity of excluding several of the actual descendants, and commencing afresh with a single patriarch. And now, just at the moment when the way was opened for this expansion, when faith in their destiny was exchanged for a sight of the first stage in its fulfilment, when the course of history was making it a reality, the consciousness must have been more vivid, and the assurance stronger, than ever it had been before. But as this was only the commencement of a coming fulfilment, and not the complete fulfilment itself, there was still so much demand for the exercise of faith and hope in connexion with that portion of their destiny, of which they were already conscious, that there was as yet no possibility of awakening the consciousness of still greater things beyond.

Since, then, prophecy, as a general rule, rests upon the age in which it is delivered, and only opens to view those features of the future, of which the germs and prototypes exist in the present, the expectations of salvation, which existed in the patriarchal age, must have been most closely related to the circumstances just referred to. An age, whose only task was to form a great nation from one single chosen man, whose movements, subjective and objective, were all concentrated upon this one result, a result longed for and looked for above all others, could only regard salvation as dependant upon the attainment of this result. The expectations of salvation, which prevailed in the whole of the patriarchal age and for some time afterwards, were summed up in the promise: " in thy seed shall all the nations of the earth be blessed." The seed of Abraham, Isaac, and Jacob, when expanded into a great and independent nation, in other words, the nation itself in its compactness and unity, appears as the bringer, the possessor, and the medium of salvation. This was, doubtless, an imperfect, undeveloped, and faulty shape for the expectation of salvation to assume, but the age to which it belonged was itself imperfect and undeveloped. The expansion of the family into a nation could certainly not in itself bring salvation, but it was the necessary condition, the preparation and first stage of its full and ultimate manifestation; and, just for that reason, in the expectations which prevailed at this time, the one was inseparably connected with the other. It was, no doubt, necessary that before this expansion into a plurality could attain its ultimate and highest end, it must by an organic process be condensed into unity, since salvation could only be exhibited in its perfect form in a personal Messiah, the noblest fruit and ἀκμή of this unfolded plurality. But before this fact could be made known in *prophecy*, it was necessary that *history* should furnish a substratum and starting point. So long, however, as the *only* thing towards which their history pointed was the multiplication of the people, the idea of a single personal Saviour could not take root at all. This could only occur after their formation into a great people was completed, and when it had become apparent that the plurality of the nation must necessarily be concentrated in a single individual; in other words, after some *one* man had arisen as the deliverer and redeemer, the leader and ruler of the whole nation. Hence the expectation of a personal Messiah would first arise and assume a definite shape on the appearance of *Moses*, *Joshua*, and *David*. Accordingly the earliest promise, which points to a personal Messiah, is found in the *Mosaic age*, and even there it stands alone and is still somewhat indefinite (Deut. xviii. 18, 19), whilst it is only the history of David which gives perfect clearness, certainty, and precision, to the announcement of a personal Saviour.—But the expectations of the patriarchal age were all fixed upon the growth of the family into a people; and as the fulfilment of their destiny seemed to be wrapt up in this, it appears impossible that in such an age salvation could have been regarded as dependant upon any individual. On the contrary, previous historical events would lead to the conclusion, that isolation would retard the desired end; for all the instances of separation and isolation that had hitherto occurred had been such as involved exclusion from the fellowship of the chosen people and from the call they had received, and rendered it necessary that the progressive development from unity to plurality should begin again.

From what we have written, it follows that we are not justified in expecting *a priori* the announcement of a personal Messiah, or rather that, so far as the history of the patriarchs in the book of Genesis affords us a glance at the progress of the ideas of salvation in that age, we are justified in not expecting such an announcement. Still this decision at the outset should not, and shall not affect in any way our exegetical inquiry into the prophecy in question. For unless an unbiassed exposition of the prophecy should lead to results in harmony with our foregone conclusion, the latter will have no objective worth, and it will be impossible to sustain it. Should a just exposition show, that the prophecy really treats of a personal Saviour, of one single individual as the medium of salvation, we shall not for a moment hesitate to accept this result, and shall willingly admit that we have been deceived in our expectations. But it will then be necessary to assume that the lives of the patriarchs must have presented some historical links of connexion with the promise of a single personal Saviour, and that unless they are to be found in the book of Genesis and have escaped our observation, the author of that book must have omitted to notice them.—Our present task will be to test the opinion, that the passage before us must necessarily be interpreted as predictive of a personal, individual Messiah.[1]

[1] The objections offered to my views by *Reinke* (l.c. p. 184 sqq.), and *Delitzsch* (Genesis p. 370), are removed by what has been said above. I fully agree with the remark made by the latter in one of his earlier writings: "History is not the measure, but the occasion of prophecy." I also agree as fully with what he now says: "We must not prescribe to prophecy, in what way it shall proceed, or decide from the history of any period, how much or how little it can prophesy, for the course of prophecy is often at variance with human logic, as can be proved from unmistakeable examples, and its telescopic vision often looks behind the hills, by which contemporary history is bounded." That the *former* is not my intention, and that I am

Our first inquiry is, whether the construction and the connection will permit of our rendering the word *Shiloh* as the subject of the sentence, which it must be if this opinion be correct. We cannot accept without reserve the confident assertion of *Hofmann* (l.c. p. 117), that "the patriarch could not have turned so completely away from Judah, and finished the sentence, which related to him, by announcing the advent of a person, who is not described as one of Judah's descendants, or even as connected in any with the posterity of Jacob." For although the words and the context undoubtedly sustain the correctness of this view, yet the connexion between Judah and Shiloh, as his descendant, might be regarded as naturally implied. But both the context and the train of thought require that we should render *Shiloh* as the *object*. In *Hofmann's* words: "The expression עַד כִּי, *until*, leads us to expect an announcement of Judah's future history, and of the result of his maintaining uninterrupted possession of his princely rank. And since, when we pass from the first half of the verse to the second, we have no reason to expect any other subject than Judah, we ought to receive proofs not only of the possibility, but also of the necessity of taking *Shiloh* to be a person and to be the subject of יָבֹא." But, as we shall presently show, no such proof can be given. On the other hand, the structure of the tenth verse will only admit of its being rendered as the object; for if we render it as the subject, we at once destroy the parallelism of thought between the two clauses

not unaware of the *latter*, will, I hope, be sufficiently attested by what I have already said. But when *Delitzsch* adds: "In the present instance it is not true that the continuous progress is interrupted, if the word *Shiloh* in the mouth of Jacob denotes the person of the Messiah, since the next great prophecy (that of Balaam, Num. xxiv. 15 sqq.) views the Messiah under the image of a star or sceptre coming out of Jacob," &c., he does not appear to have read what I have written above respecting Moses, Joshua, and David as historical links to which the idea of an individual Messiah could be attached. Whether Balaam's prophecy actually referred to this, and, if so, to what extent, are questions which cannot be discussed here. But I must confess that I cannot see the drift of *Delitzsch's* argument. It is with the meaning of Jacob's prophecy that we have to do, not with that of Balaam. I have myself shown that the foundation was laid in the time of Moses for the expectation of a personal Messiah, though I do not admit that it had been laid 400 years before. And this can never be proved by attaching Balaam's prophecy, by way of explanation, to that of Jacob. But *Delitzsch* himself does not interpret Jacob's words as predictive of a personal Messiah. And if this scholar went to the examination of the prophecy with the expectation of finding a personal Messiah, and yet did not find one, this surely favours the conclusion that *his* expectation was unfounded and *mine* correct.

עד כי־יבא שילה and ולו יקהת עמים and this parallelism is required by the arrangement of the verse. In the two clauses, "till the Messiah come," "and to him the obedience of the nations," there is no parallelism at all, but merely a progress in the thought. If, however, we regard *Shiloh* as the object, and take Judah as the subject from the previous clause, the two clauses, "till Judah come to rest," "and the obedience of the nations shall be his portion," harmonize beautifully; for the obedience of the nations, who cheerfully and without resistance submit to Judah's rule, forms a part of the rest, which Judah enjoys, after the victorious conflict just described.

The foregoing remarks apply to every interpretation, which refers the expression to a *personal Messiah.* We shall now examine them singly. One of the earliest would read שֶׁלֹּה instead of שִׁילֹה, and regards the former as equivalent to אֲשֶׁר לוֹ=שֶׁלּוֹ=שֶׁלֹּה. שֵׁבֶט is then supplied from the previous clause, and the whole passage rendered thus: "Judah shall retain the sceptre, until he come, to whom it (viz., the sceptre) belongs." The *Septuagint* rendering is based upon this view: ἕως ἐὰν ἔλθῃ τα ἀποκείμενα αὐτῷ (donec veniant quae ei reservata sunt), or, according to another reading, ᾧ ἀπόκειται (donec veniat, cui reservatum est); and most of the early versions translate the words in a similar way. The principal defenders of this view in modern times have been *Jahn* (vaticinia mess. ii. 179 sqq., Einl. i. 507 sqq.); *Sack*, christl. Apol. ii. A. S. 266 sqq.; *Larsow* (Uebers. d. Genesis); and *Herd* (mess. Weiss. ii., p. 33 sqq.). But this explanation will not bear an impartial examination; for, *first*, the favourite ellipsis is unparalleled in its harshness; *secondly*, we are compelled to act in the most arbitrary manner, by pronouncing שִׁלֹה the original reading, whereas it is found in very few MSS., and is evidently merely *scriptio defectiva* for the common reading שִׁילֹה; and *lastly*, we must declare in a dictatorial way the admissibility of the inadmissible pointing, שֶׁלֹּה for שִׁלֹה. But even supposing that this were granted, or if we determined to follow *v. Bohlen* and read שֶׁלֹּה at once, even then the sense and the connexion of the verse would compel us to protest against the interpretation. For if it were said,

"Judah shall retain the sceptre, till he come whose it is (to whom it belongs)," there would be a most inappropriate contrast drawn between Judah, who holds the sceptre, and the Messiah, to whom it belongs, from which it would follow, that the sceptre does not belong to Judah; and there would also be a not less unfounded announcement that Judah, the blessed, would one day resign, *i.e.*, lose the sceptre.—There are two things which seem to *favour* this explanation, the unanimity of the earlier translators, and an analogous passage in Ezek. xxi. 32, עַד־בֹּא אֲשֶׁר לוֹ הַמִּשְׁפָּט, which might be regarded as an exposition and paraphrase of our word שִׁלֹה *(Shiloh)*. But the two testimonies may be reduced to one, for the early translators have evidently taken the passage in Ezekiel as the foundation of their rendering of the obscure or doubtful word Shiloh, which explains their general agreement. And the proof afforded by the passage in Ezekiel also loses its worth; for whilst there is an undeniable identity of thought between the translators and Ezekiel, the original Hebrew of the passage in Genesis and the passage in Ezekiel have too little in common, to lead us for a moment to suppose that there was any reference in the latter to the former. Moreover, the two passages are totally different in other respects, for whilst Ezekiel announces *ruin* and *devastation*, which will last till he come, to whom the government belongs, the passage in Genesis would speak of *victory* and *government*, which will last till he come, to whom the government belongs.

A far more plausible interpretation is that which derives the word Shiloh from the root שָׁלָה, adopts the meaning *rest*, and, regarding this as abstract for concrete, renders it *the bringer of rest.* This view is the most prevalent of all. Among its more modern supporters are *Rosenmüller* (ad. h. l.), *Winer* (hebr. lex. s. h. v.), *Baumgarten-Crusius* (bibl. theol. p. 368), *Hengstenberg* (christol. i. 59 sqq., Engl. transl.), *Reinke* (ut supra), and many others. The supposition, that the abstract is used for the concrete, is undoubtedly admissible, and we adhere to the derivation of Shiloh from שלה in the appellative sense of "rest," or "the place in which rest is found," in spite of the opposition of *Tuch* (Comm. p. 575 sqq.), and *Delitzsch* (Comm. p. 372 sqq), who do not appear to me to have answered the arguments by which *Hengstenberg* (Christol. i. 59 transl.), and

Hofmann (Weiss. i. 116), have *defended* this derivation.—An objection might, no doubt, be offered to the rendering *tranquillator*, as שָׁלָה does not mean to *bring peace*, but to *enjoy peace* (*Gesenius*, lex. *salvus, securus, maxime de eo qui prospera fortuna secure utitur*); but שִׁילֹה might be taken as descriptive of a person, in whom the full enjoyment of rest and peace is first apparent. We should therefore decide at once in favour of this view, were it not for the two difficulties, which have been more fully explained above, (1), That Shiloh must be regarded as the object of the verb, according to the sense, the context, and the structure of the verse; and (2), That the expectation of a personal Messiah was entirely foreign to the patriarchal age.

The second objection does not affect the explanation given by *Gesenius* (lex. s. v.), who preserves the abstract signification of the word, and translates the passage: "until the rest (*sc.* of the Messianic age) come, and to him (*sc.* Judah) the obedience of the nations." But the first objection still applies, and in addition to that, the reference of the suffix in לוֹ to Judah is no longer admissible, if another subject be introduced, as the nominative of יָבֹא, in the intermediate clause. The suffix would then necessarily refer to Shiloh, the nominative of the verb, and the latter must in that case be regarded as a concrete noun. (*Vid. Hofmann, ut sup.* 116).

Some of the earlier expositors (*Jonathan, Calvin*, &c.) imagine Shiloh to mean *his* (*i.e.* Judah's) *son* or descendant. But there is no foundation whatever for the assumption that the word שִׁיל, with the meaning *son*, ever existed. (*Vid. Hengstenberg*, Christol. p. 63, 64 transl.)

Of all the explanations, which reject the Messianic reference, the only one of any importance is that which supposes *Shiloh* to be the name of the well-known city of Ephraim, where the tabernacle was erected when the Israelites entered the promised land. This opinion is supported by *Eichhorn, Ammon, Bleek* (de libri Gen. origine), *Tuch, Hitzig* (*ad* Ps. ii. 2), and others. The meaning of the passage is supposed to be that the tribe of Judah should take the first place, and be the leader of the tribes during the whole of the march through the desert, until they arrived at Shiloh. The only thing that can be said in favour of this explanation is, that in every other passage of the Old Testament, in which the word Shiloh occurs, it refers to this city of Ephraim. But every one will admit that this argument does not amount to a *positive proof;* that, at the best, it merely establishes to a certain extent the probability that there is the same reference in the passage before us. But this probability is more than counterbalanced by the number of arguments on the opposite side. First of all this explanation brings in a subject to the verb יבא, which is quite foreign to the context; for as we have already shown, Judah must be the nominative. But apart from this, there is an insupportable harshness in the *neuter* and *collective* subject thus introduced ("until they [*man*] or the *people* come to Shiloh.") It is true that this might be avoided by translating the clause: "until he (Judah) come to Shiloh;" but as it is impossible to see what Judah had to do as a tribe with this city of *Ephraim*, in contradistinction from the other tribes, there is no other resource than to fall back upon a collective subject; for although Shiloh was a spot of great importance as a resting-place or turning-point in the Israelitish history, it was not important to Judah alone, but to all the tribes in common. This explanation then loses its force unless the blessing of Jacob be regarded as a *vaticinium post eventum*, composed at a later period, say for example the time of David. For what should have led the aged patriarch to associate the glory and goal of Judah or his descendants with a place of so little importance, which is never mentioned anywhere before the time of Joshua, and probably owes both its name and its existence to the circumstance that it was there that Joshua pitched his tent, and set up the tabernacle (*Hengstenberg*, Christol. i. 80, transl.)? How bare and miserable would it have appeared, even if Shiloh were really in existence as a small town at the time, for Jacob to introduce in such high-flown terms, and in the midst of such splendid promises, the prediction that Judah would arrive at Shiloh! The assumption that the blessing was composed at some period subsequent to Joshua is overthrown by the most decisive and unanswerable objections, as we shall presently show, and in general is merely a loophole to save a foregone conclusion, that actual prophecies are impossible. But, even supposing that the blessing describes some future event, and does this under the fictitious appearance of prophecy, was there ever a period in which Shiloh was of such importance that the author, whoever he might be, could possibly regard it as the representative of the highest and most perfect glory of his people's history, a glory so brilliant that no greater could be imagined or desired? Moreover, the period to which the composition of the blessing has been assigned, that of the latest Judges and of David, was one in which the importance and glory of Shiloh had considerably declined.—And what can be made of the promise that the sceptre and dominion should be retained by Judah till the settlement in Shiloh? Was this fulfilled? or fulfilled with *such* completeness in the details as we should expect in the case of a *vaticinium post eventum*? *Bleek* (p. 19) thinks that this can be answered in the affirmative. Judah conducted his brethren till the promised land was conquered, and after that Ephraim took the lead. But *Hofmann* (p. 115) has shown that there is no foundation for the statement: "for no one would pretend that the blessing was fulfilled because the tribe of Judah took the *foremost* place in the army during the journey through the wilderness (Num. ii. 3).[1] The whole army was commanded at that time by a Levite, and after him by an Ephraimite." It was not till long after Shiloh had been fixed upon as the site for the tabernacle, *i.e.* not before but after the *terminus ad quem*, to which our prophecy points, that we meet with the first indication of Judah's actual supremacy (Judges i. 2), but then it did not continue without interruption through the period of the Judges, so that the tribe of Judah did not rise to any decided pre-eminence until David was king. How then could the blessing be applicable to Judah, if, in the midst of the splendid acquisition of power and glory, which was to distinguish this tribe from all the rest, Jacob had announced to him that the consummation of the whole would be that he would lose his supremacy as soon as Shiloh was reached? To avoid these difficulties *Tuch* translates the clause: "so long as they are assembled in Shiloh, *i.e.* for ever." But *Hofmann* has pointed out no less than five fallacies in this pretended improvement. It gives to עַד כִּי

[1] We mean of course from the point of view from which this explanation is arrived at, viz., the notion that we have here a *vaticinium post eventum*, and that the author meant to say that Judah's supremacy ceased at the encampment in Shiloh. For such a prophecy presupposes an outward harmony in matters of detail such as no one will be able to discover between the supremacy of Judah predicted here, and his position in the order of encampment and march during the journey through the wilderness.

a meaning which it does not possess, it always means "until," never "so long as;" it attributes to יָבֹא a subject which is not to be found in the context; "it makes the writer express a hope that the circumstances which prevented Israel from enjoying rest, and hindered the removal of the tabernacle to a permanent resting-place, would last for ever; and lastly, it supposes the perpetual duration of the supremacy of Judah to be dependant upon a state of things, the cessation of which is referred to by Asaph as most intimately connected with the origin of that supremacy" (Ps. lxxviii. 60, 67—72).

We now return to our own explanation. The meaning of the prophecy is that Judah shall remain in uninterrupted possession of the rank of prince among his brethren, until through conflict and victory he has reached the object, and made the fullest display, of his supremacy, in his own enjoyment of peaceful rest, and the cheerful obedience of the nations to his rule. Hence the *terminus ad quem*, which is mentioned here, does not set before us the limit or the termination of his supremacy, but rather the commencement of his secure and irresistible sway. And from this it follows quite as naturally, that the victory gained by Judah, and the blessings of peace which he secures, are shared by *his brethren* in all their fulness, because he fights as the prince and champion of his brethren; and not only so, but the blessings of this peace must necessarily be extended to all the *nations*, who now cheerfully obey him.

To what period, then, does the word "until" refer? First of all, it refers, no doubt, to that period of which the whole blessing treats, the full possession of the promised land. This is in Jacob's view the commencement of the אחרית הימים, the last time, the time of the consummation. To *him* the relative peace, which closed the strange and pilgrim life of his descendants, and the absolute peace, which is the aim and end of all the movements that originated in the call of Abraham, are one and the same. That which proves in reality to be a long continuous line, extending from the commencement of the comparative rest under Joshua to the final attainment of absolute peace under Christ, necessarily appeared from his prophetic stand-point to be merely a *single point*, since the first point covered the last as well as the intermediate line; or rather because the commencement contained the end, and only exhibited it in a typical form. The first preliminary and imperfect manifestation of the peace here promised was made in the time of Joshua; but the disturbances, to which this peace was exposed, soon proved it to be only a preliminary fulfilment of the promise. Whilst, therefore, the comparative rest enjoyed under Joshua was in one respect a fulfilment of Jacob's prophecy, in other respects it continued, on account of existing disturbances, to be still a prediction, pointing for its highest and final fulfilment to the entrance of absolute rest.

It was Judah's position and bearing, both as a prince over his brethren, and in his victorious engagements with his enemies, which secured the enjoyment of rest and peace. In proportion, then, as the rest predicted by Jacob was enjoyed in the time of Joshua, must the supremacy of Judah have been exercised *before* that time. If, therefore, the rest which was then enjoyed was true and absolute rest, the supremacy of Judah must have been manifested in its most perfect form before the days of Joshua. But if Jacob's prediction of future rest remained a prediction, as we have seen that it did, even after its first and preliminary fulfilment under Joshua, then must the prediction of Judah's supremacy have been only partially fulfilled in the period *antecedent* to Joshua, and after its first fulfilment in the lead taken by the tribe of Judah in the order of encampment and march through the wilderness,—it must still have continued a prophecy pointing onward to an ever-increasing supremacy on the part of Judah, the loftiest eminence of which would as far surpass its first appearance before the time of Joshua, as the comparative rest enjoyed in the days of the latter would be surpassed by the absolute rest secured by Christ.

Jacob's prophecy of the future rest, which Judah would enjoy in common with his brethren, whose prince, representative, and champion he was, points forward to the end. In Jacob's view, indeed, the time of Joshua was the end, for in his days all the wants of the patriarchal age, of which Jacob was conscious, were satisfied, and all the prerequisites of salvation, so far as Jacob was acquainted with them, were fully met. But there were other wants and other prerequisites, of which Jacob was not aware, and which were not supplied in the time of Joshua, and therefore, objectively considered, that time was not the end. In the prophecy of Jacob there was not only the subjective element, the product and expression of the mind of Jacob, but an objective element also, communicated to the mind of the patriarch by the illuminating influence of the Spirit of God. And hence for every succeeding stand-point this prophecy points upward to a higher form of Judah's supremacy, than the position of his tribe in the journey through the desert, and a rest superior to that produced by the occupation of the promised land.

Though we felt obliged just now to oppose the notion that *Jacob* had any thought of a *personal* Messiah, when pronouncing his blessing, yet we by no means question its *Messianic character*, as will be clearly seen from what we have already said. The announcement made by Jacob, that he was about to tell his sons what should befall them in the end of the days, indicates the Messianic character of the whole blessing, for "the end of the days" is the Messianic period. But most of all is the Messianic character apparent in the blessing pronounced on Judah, for this is unmistakeably the leading member of the whole prophecy, the centre, as it were, from which radiates all that the other blessings contain of a Messianic character, viz., the ultimate and certain enjoyment of rest and peace. It is Judah, who opens the way to repose, as the leader and champion of his brethren.

The characteristics of the Messianic idea, so far as it had yet been evolved by history and prophecy, re-appear in the sentence pronounced on Judah. For it not only announces the unparalleled blessing, which is destined for the seed of Abraham, but points out the benefits to be conferred by that seed upon other nations. The obedience of the nations, though won by conflict, is to be cheerfully rendered, and Judah's supremacy is no hard and heavy yoke, but mild and pleasant, dispensing blessings and bringing peace. The proof of this is found in the description of the pleasure of peace, to which Judah now yields himself, and the mild and gentle character which he is able to assume.

The Messianic idea is still essentially the same stage of development as in previous prophecies. This is not to be wondered at, as we are still at the same stage in the historical development as before, viz., the family history. We find the Messianic idea in the same contracted form, with salvation still concealed in the shell of earthly good and material prosperity, though in the actual kernel there are blessings of a purely spiritual character enclosed. The idea of salvation we find still as indefinite as before; as yet it has assumed no concrete shape. So much indeed is certain, that all the nations of the earth are to be blessed in Abraham's seed; but nothing further is revealed. Yet the way is paved for a further step in the progress of the prophecy, though that step is not yet taken. The new feature introduced is the designation of Judah as the chief among his brethren, who fights as their champion at their head, and secures for them rest, peace, and salvation. But, as we have already shown, this cannot have been understood by either the speaker or the hearer as meaning that the tribe of Judah was to be the sole medium of salvation, to the exclusion of the other tribes, much less that the tribe of Judah was to be shut out from the task, and the whole to be performed by a single member of that tribe. Still in the fact that, when the attainment of rest, and peace, and salvation is spoken of, Judah is named as the prince and leader of his brethren, the way is opened for the proper separation of Judah, as required by the Messianic idea. And as soon as their desires should be satisfied, and the first condition of the call of Israel fulfilled by their becoming a great people, they would be sure to learn from the results that this alone could not ensure the object for which they had been called. Thus it was soon discovered to be necessary that the plurality should be again concentrated in unity. And when such men as Moses, Joshua, and David had risen up as deliverers and redeemers, as leaders and governors of the whole nation, and by their history had furnished a substratum on which the idea of a personal Messiah could be founded, the prophecy before us necessarily led to the association of this idea with the tribe of Judah, and that with the greater facility since this tribe had risen in the meantime to a position of increasing prominence.

Delitzsch, in his latest work (Ausleg. d. Genesis p. 373 sqq.), has revived the opinion, which was first employed in the cause of rationalism, that *Shiloh* refers to the well known-city of Ephraim in this, as in every other passage of the Old Testament in which it occurs. The meaning, which he gives to it, however, is essentially the same as that which we have arrived at in another way. He says: " Judah occupied the first place in the camp, and when the Israelites were marching, Judah always led the way. This position he maintained till he came to Shiloh; for when the conquered land was divided, Judah was the first to receive his share (Josh. xv.). The division of the land of Canaan, which took place at the tabernacle, that had been set up at Shiloh, forms without doubt the boundary line between two periods in the history of Israel. Their arrival at Shiloh brought their wanderings and conflicts to a close, and formed at the same time the commencement of their settlement in full possession of the land. Shiloh was thus, as its name implied, the place of Israel's rest." But even with this explanation we cannot give in our adhesion to the opinion; for many of the objections, already offered to it in its rationalistic form, are equally applicable to it in its present shape. So accidental an event, as the selection of Shiloh, rather than any other town, as a temporary resting-place for the tabernacle, could not have been a subject for prophecy. We admit that the settlement at Shiloh was a boundary line in the history of Israel, and that as such it might very well be a subject for prophecy. But the settlement itself, the acquisition of a resting-place, was all that was essential; the choice of Shiloh in preference to any other place was something unessential and accidental, with which prophecy had no concern. Not that we would for a moment dispute the fact that the form in which the idea of a prophecy is expressed often coincides in a remarkable way with the (accidental) form, in which the prediction is fulfilled. But we most firmly deny, *that the sons of Jacob could have looked upon this insignificant town* (even if it then existed), *as the end of their dying father's prophecies.* Still we are certainly inclined to recognise a connexion between the Shiloh, in which the tabernacle was placed, and the Shiloh referred to in Jacob's prophecy; only, we regard the former as dependent upon the latter, as *M. Baumgarten* does, and not the latter upon the former, which is *Delitzsch's* opinion. For it appears to us a very probable thing, that the Israelites gave the name of *Shiloh* to the place in which they rested for the first time, and set up the sanctuary after their victorious conflict with the Canaanites, and that they did so with a conscious reference to the blessing of the patriarch, and as a sign and testimony that his prophecy had here received its preliminary fulfilment. Moreover, we can readily conceive that, in the fulness of their first delight at the enjoyment of rest, they might look upon this as the complete and adequate fulfilment of the prophecy, and overlook the troubles that were still before them.

(4). The "grammatico-historical method of exposition," as the rationalistic exegesis is called, starts from the concession, that Jacob's blessing is descriptive of circumstances, which had no existence till after his descendants had taken possession of Canaan. It professes "to leave the dispute as to the *possible* or *impossible* composition of the piece by Jacob to those, whose special interest it is to cultivate without effect this barren soil." But yet regarding it as *above all things* certain that a real prophecy is thoroughly incredible, it denies that it was written by either Jacob or Moses, and then proceeds "in a *conclusive* (? !) way to determine the date of the composition on historical grounds." (*Tuch* comm. p. 554 seq.). But the safety of the "conclusive" method, to which this "grammatico-historical" criticism lays claim, is not confirmed by the many different and discordant results to which it leads. *Heinrichs*, for example, in his *commentatio de auctore atque aetate* cap. Gen. xlix. (Göttingen 1790), and *Friedrich* in "*der Segen Jakobs, eine Weissagung des Proph. Nathan* (Breslau 1811), confine themselves to the blessing pronounced on Judah, and pretend that they have demonstrated that it was written in the time of David; *Tuch*, who considers the blessing of Levi the safest criterion, considers it indisputable, that it was composed in the time of Samuel; whilst *Ewald* (Gesch. i. 80), appeals to the blessing on Dan as sufficient to establish the fact that it was written in the latter half of the period of the Judges, most likely during the life of Samson.

It so happens, however, that the *data* which we possess for fixing the time of its composition are so numerous, so decisive, and so favourable, that there is scarcely any disputed passage in the Old Testament, whose authenticity is as certain as that of Jacob's blessing. For (1), its style is not at all that of a *vaticinium post eventum;* (2), it can be proved that there was no one period *post eventum vaticinii, i.e.*, after the conquest of the promised land by Joshua, in which all the different expressions could have been written; (3), the blessing itself contains positive data, which compel us to assign it to a prae-Mosaic age;

D 2

and (4), the matter and the form are perfectly in harmony with the views and expectations of the patriarch, and there is nothing which we might not expect him to say, always supposing that he was enabled to look into the future by a prophetic inspiration.

We have already shown how completely this prophetic picture harmonizes with the historical background, on which it is drawn, how perfectly the substance of it tallies with the patriarch's state of mind, his views, his desires, and his expectations at the time. And as, on the one hand, there is nothing to hinder our receiving the song as an actual prophecy, and recognising the historical frame in which it is set; so on the other, are we led by a careful and *unprejudiced* examination to the inevitable conclusion that the blessing is not a *vaticinium post eventum* either in whole or in part, and that there is a total absence of the characteristic marks of such pretended prophecies. A *real* prophecy looks from the present into the future, or rather it sees the future in the present. The germs and preformations of the future, which are already discernible in the present, and all the imperfections and wants, of which there is an existing consciousness, are viewed by it in the light of God, not merely as germs and deficiencies, but in that state of perfect development, towards which they are striving and at which they must *of necessity* arrive. At the same time the various *phases*, through which the maturity of these germs and the satisfaction of these wants will be actually attained, and the outward *forms*, which they will eventually assume, are not made known even by this real prophecy, inasmuch as the conditions of both of these will be determined by the course of history, and therefore there is as yet no existing substratum or point of contact for such a prophecy. Hence, however definite a prophecy may be in relation to the *idea*, and however keen and clear its gaze, yet in respect of the outward forms, in which the idea will appear, it is always general and indefinite. Still more, if we compare the prophecy with the details of its fulfilment, we shall generally notice an apparent want of congruity between them. The cause of this will be found partly in the fact that in the prophecy we have but a single field of view in which everything is represented in its perfect form, whereas in the actual fulfilment there are successive stages, attended by many oscillations and by retrograde as well as progressive movements ; and partly also from the fact that, in order to give expression to the idea, with which alone it is concerned, it clothes it in a certain drapery, which is intended *for no other purpose than this*, and therefore very frequently is not found to harmonize exactly with the outward form eventually assumed. This is not the case with *pretended* prophecies. They clothe in the garb of prophecy events which have actually occurred. However great, then, the anxiety to avoid every thing that could betray their real character, they cannot so far overlook the concrete phenomena which lies before them, as to assume the features of a true prophecy in sufficient measure to hide the fraud. And where they are the result of an ingenious illusion, of a character not absolutely evil, where there is therefore not a distinct consciousness of any intention to deceive, there is sure to be all the less ability or disposition to disguise.

If, now, we apply this test to the prophecy before us, we shall be constrained to confess, that it bears the marks of a real prophecy, and not of a *vaticinii post eventum.* It is true, the opposite has been asserted with the greatest confidence. The special details given in the blessing, and their peculiar harmony with the fulfilment, are appealed to as removing all doubt that we have here only a masked copy of the present, and not a real prediction of the future. But, notwithstanding these, the blessing in whole and in part is expressed in such general terms, its descriptions are so free from any sharply defined sketches, any concrete forms, and any reference to such accidental circumstances, as are only of importance to the age itself, and are so little in harmony with the *external, accidental* circumstances of the period, of which it is descriptive, that the idea of a *vaticinii post eventum* is thoroughly inadmissible. We have clearly a case before us, in which the prophecy is *too definite* in certain respects, to be merely the product of natural intuition or subjective anticipation, and yet is too *indefinite* in its general character and in some of its details to have been written after the event. Rationalistic criticism, therefore, as it has no third to fall back upon, naturally attempts through thick and thin to prove one of these two.

We may get an idea of the indefinite and general manner, in which throughout the whole blessing concrete forms and special incidents are referred to, from the blessings pronounced on Judah and Joseph. Each of these occupies almost as much space as all the others put together. We see that the author was desirous of giving a much fuller description of their prospects, that he entered *con amore* upon this description, and wished to dwell as long as possible upon the picture of their lot, and of their superiority to all their brethren. If now he had taken his materials and his colours from the past or the present, his description would surely be full of references to special details, and rich in concrete forms. Yet how indefinite the two blessings actually are! We find only general ideas and references to lion-like courage and strength for battle, to victory and dominion, to fulness of blessings and pre-eminence of rank, all of which resemble the external events only so far as was absolutely necessary to produce the impression required. Who is there that would for a moment assert that these blessings can only have been copied from events which had actually occurred? The whole blessing is acknowledged to point to the completion of the conquest of the promised land and the distribution of that land among the twelve tribes; and how little do we find, in either of these sections, that is characteristic of the period referred to! If we did not know it beforehand, who would be able to discover a reference to the provinces allotted to the two tribes in the promise to Judah of an abundant supply of wine and milk, and to Joseph of dew and rain, or to recognise in these the distinguishing characteristics of each of those provinces? It is only in the prediction of Judah's supremacy that it could possibly be maintained, that the general idea assumes a concrete, external form;—but even here there is so little outward resemblance to the circumstances, which really existed at the supposed date of its composition, that it is still necessary to assume that the subsequent glory of this tribe was anticipated by the author, a fact which may be assumed in the case of an actual prophecy, but not where the prophecy is merely feigned.

We shall now pass on to the other blessings. The writer says nothing about the circumstances and possessions of the tribe of *Reuben*. How inexplicable is this in the case of a *vaticinium post eventum!* However insigificant the tribe may have been, and though its province may not have been within the limits of Canaan proper, yet the same may be said of Gad and the half-tribe of Manasseh, and of these the author has something to say.

It is not difficult to understand why *Jacob* should only speak of Reuben himself, and say nothing about his tribe except that it would not take the lead ; but we cannot conceive how a contemporary of Samson, or Samuel, or David, could so entirely overlook the tribe, as to mention the founder alone, or how he could record the curse pronounced on Reuben, without shewing how, where, or by what means the effect of the curse was manifest in the history of the tribe.

Simeon and *Levi* receive precisely the same blessing. There is apparently no difference whatever in their lot. They are both to be scattered in Israel. Now Jacob might express himself in this way, but not a writer who saw how completely different were the modes of their dispersion. *Tuch* is right in saying that "Simeon received his inheritance in the midst of the tribe of *Judah*," but he goes further than he has any right to go when he adds "but without any continuous boundaries" *(vid. Keil's* commentary on Joshua p. 419, translation Clark's For. Theol. Lib.). Again, how different was this distribution of Simeon from that of Levi! So different, that a later writer could not possibly have employed the same words to describe them both.

To the tribe of *Zebulon* there is promised a dwelling-place on the sea-shore and near to the Phœnician city of Sidon. Here certainly there is something, which offers apparently no little support to the views of our opponents, and if all the blessings referred to the future in the same manner, there would be some ground for the notion of a *vaticinium post eventum.* But if the minuteness and precision, with which the blessings are here described, appear to furnish an argument to our opponents, they are immediately deprived of it by the want of congruity between the prophecy and its fulfilment. "If the prophecy of Jacob had been written *post eventum*, there would certainly have been greater geographical accuracy, and the description of the boundary towards Sidon would have belonged to *Asher* (Josh. xix. 28) *rather than to Zebulon" (Baumgarten).* So far as it is possible to determine the boundaries of the tribe of Zebulon from the book of Joshua (chap. xix. 10—16), they did not touch the sea at all *(Keil's* commentary on Joshua p. 422 sqq., Martin's translation). If, then, the blessing pronounced on Zebulon cannot have been a description taken from existing circumstances, since it is only partially in harmony with the circumstances of Joshua's days, there must certainly have been something in Zebulon himself, the founder of the tribe, which led Jacob to place him by the sea, and which furnished a substratum and a starting point for the prophecy. The fact, that we do not know what the reason was, is no argument against its existence.

Issachar is represented as a strong but lazy nomad, who enjoys the fruits of peace in his fertile and genial inheritance in a state of careless repose, and who puts up with many an inconvenience rather than disturb his comfortable rest by a firm and warlike bearing. But from what we know of the condition of this tribe in the period of the Judges, the prophecy is by no means so completely in harmony with it, as we should expect it to be if taken from the facts; for it was "just this tribe of Issachar, together with that of Zebulon, which acquired such renown for heroic bravery (Judg. v. 14, 15, 18), whereas Reuben, Dan, and Asher remained inactive." If the author lived, as is supposed, at a later age, he must have been aware of this, and it is pure imagination to say that this heroic courage gave place to cowardice in the second half of the period of the Judges. But the agreement between the blessing and its fulfilment is to be found, not in any single outward event, occurring at a particular period of time, but in the general characteristics of the history of the tribe. And here, as in all the other sections, the whole of the history of the tribe subsequently to the conquest of the land is compressed into one single field of view.

With reference to the prediction concerning *Dan*, *Ewald* says (p. 81): "This clearly points to the times of Samson and to his administration of the office of judge; for then the small tribe of Dan could take its place by the side of any other tribe, however great it might be, possessing as it did in Samson a judge and leader, of whom it could be proud, whose success for a time at least was great, and under whom, though small and oppressed, it boldly resisted the pride of the Philistines, as a snake craftily conquers a powerful rider. And the greater the certainty that this attitude of the tribe under Samson was transient and without important results, the stronger is the evidence that such a description must have been written during Samson's brief and successful career." The argument is plausible enough, but it is nothing more. For the miserable and despicable state of Judah in the time of Samson, the cowardice and want of common-sense which were manifested by it at that period (Judg. xv. 9 sqq.), when contrasted with the proud picture of the lion-like courage, the conquest, the leadership, and the supremacy of Judah, as set forth in this blessing, are totally irreconcileable with *Ewald's* opinions. Moreover, his views with regard to Dan and Samson are founded upon a misapprehension of the true characteristics of the office of judge, which Samson filled. For even though all the miraculous and wonderful accounts of Samson's deeds were really myths, as *Ewald* says, yet so much would certainly be left as a historical *residuum*, that Samson was distinguished from all the previous judges, by the fact that through his own fault he was isolated, not only from the general body of the tribes, but even from his own, that he was left to fight alone on account of the torn and heartless state of the times, and therefore, that the most gigantic exertions and the most striking success on his part were nearly if not totally barren of permanent results. We can hardly imagine a contemporary ascribing so unreservedly to the whole tribe, what was not merely achieved by a single member of that tribe, but by one who was left alone and forsaken by all the rest. Still, it cannot but appear strange that just this and no other tribe should be selected for the office of judge, and that it should be done in such a manner (for how came the patriarch to be so specific in this instance?); and this fact would furnish an almost unanswerable argument in favour of *Ewald's* views, were it not that the name of the tribe affords a sufficient explanation of so striking a phenomenon. Wherever it is possible, the blessings are founded upon an explanation of the name, and the favourite motto of the patriarchal age "*nomen habet omen*" was a sufficient starting-point for Jacob's prediction that Dan, the judge, should judge his people.

There is nothing special and concrete in the blessing of *Gad*, a triple play upon the name is all that we find in the prophecy concerning this tribe. *Asher* is promised a rich and fertile territory in such general terms, that there is no indication of a *vaticinium post eventum*. The blessing on *Naphtali* and that on *Benjamin* have none of the characteristic marks, which we should look for in a description drawn from existing events. And the fact that the tribe of Joseph is only referred to in its united form, that no particular reference is made to the powerful tribe of Ephraim, and that nothing is said about the geographical separation of Manasseh, which might have been described as "divided in Jacob" with even greater justice than that of Simeon, can hardly be reconciled with the assumption of a *vaticinium post eventum.*

The views of our opponents are not merely at variance with the individual blessings, but also with the introductory clause, supposing, that is, that the words "in the last days" are to be taken as descriptive of the final era, the time of consummation, as we showed above that they necessarily must be. For a contemporary of Samson, or Samuel, or David would not have been very likely to speak of *his* age as the time of perfection, when there were still so many perceptible wants and deficiencies, and so many germs and unfinished beginnings.

There is a decisive proof of the pre-Mosaic origin of the blessing in the address to *Levi. V. Bohlen* is perfectly right when he maintains (p. 453) that Levi cannot have been a priestly tribe at the time when this song was composed; but he jumps to a wrong conclusion when he infers from this that the tribe of Levi cannot have obtained exclusive possession of the priesthood till after the time of Moses; for if there is one thing connected with the early history of Israel, which is indisputably established, it is the fact that the priesthood was conferred upon the Levites by Moses himself (*Tuch* p. 557). But this address does not contain one syllable about the priesthood, nor is there the slightest hint, or reference, from which it could be inferred that the author knew that it had been bestowed upon Levi. *Tuch* further adds, it is true, that "the scattering in Israel, to which our author refers, proceeded from Levi's priestly vocation." But this is evidently *eis*egesis, not exegesis; *the* scattering, "*to which our author refers*," is merely the consequence of the curse, which is here pronounced upon Levi; it is a fit punishment for that perverse union for perverse ends, in which he had sinfully taken part. This curse was changed into a blessing when the sinful combination and ungodly zeal for which the patriarch had merited dispersion as a curse were cancelled by the proper association and godly zeal, for which the tribe of Levi merited dispersion as a blessing and a favour (Ex. xxxii. 27—29). The outward form remained the same, but the reason of it, and therefore its real nature, were entirely changed. If the author had already known the tribe of Levi as a priestly tribe, he could not, with his religious, Israelitish mind, have passed over the priesthood in silence, when it must have appeared to him as the essence and guiding star of the whole constitution. He could not possibly have described the dispersion as a curse, when that dispersion was known to result from the priesthood, for by doing this he would pronounce the priesthood a curse likewise. The force of this argument *Tuch* endeavours to evade by remarking that "we find ourselves in the midst of circumstances, in which the national sanctuary united the people with but a slender bond, when the Levites wandered almost houseless through the land, and acted as priests for any one who would pay them (Judg. xvii. 7—12, xviii. 4, 19 seq. cf. ver. 30), and when the descendants of Aaron drew upon themselves the contempt and indignation of the people by their behaviour at the tabernacle (1 Sam. ii. 12—17)." But how unhistorical it is to take the case of a single vagrant belonging to the tribe of Levi (for all the passages quoted from the book of Judges refer to the same individual) and to infer from this that the whole tribe consisted of such vagrants; and how unwarrantable to take the example of a single pair of boys belonging to the priestly family, who were spoiled by their father, and who drew upon themselves the indignation of the people on account of their crimes and acts of violence, and to conclude from this, that the whole tribe to which they belonged, was equally corrupt, and therefore equally despised. The priestly tribe may possibly have lost their rank, their influence, their incomes, etc., during the confusion which prevailed in the period of the Judges, partly on account of the circumstances of the times, and partly by their own fault. But in any case, they had not done so to anything like the extent which *Tuch* supposes. And a theocratic man, so truly religious and thoroughly patriotic, as the author of this song undoubtedly was, could not possibly at this, or any other time, have regarded it as an unmitigated curse to belong to the priesthood of Israel. In fact, the history of that vagrant Levite in the book of Judges shows how highly even this worthless man was esteemed on account of his connexion with the priestly tribe. Micah kept him "as one of his sons" (Judg. xvii. 11), and the Danites, who were wandering northwards, considered it so great an advantage to have him with them, that, when he refused to go of his own accord, they employed force rather than go without him. If therefore these two things are firmly established, (1) That Levi was not a priestly tribe when the sentence on Levi was written, and (2) That the priesthood was conferred upon the tribe as early as the time of Moses, the prae-Mosaic origin of the blessing is certain, and in that case we have approached so nearly to the date assigned it in the present passage, that no one who admits these two premises will hesitate to adopt the conclusion that it really belongs to Jacob, by whom it is expressly said to have been composed.

It is equally impossible to point out any other period between Joshua and David in which this blessing can have been composed; and within those limits the assumption of a *vaticinium post eventum* must necessarily be confined. For, whether we assume with *Tuch* (in order that the blessing on Levi may appear fulfilled, in however partial or distorted a form), that it was written in the time of Samuel, or with *Ewald* (on account of the blessing on Dan), that it belongs to that of Samson, the blessing on Judah, which was certainly written at the same time, is perfectly irreconcileable with either hypothesis. For how does the glory, which the author heaps in such splendid colours and high-flown expressions upon Judah's head, correspond to the miserable, cowardly, and senseless conduct of the tribe of Judah in the time of Samson (Judges xv. 9 sqq.), or to the thorough insignificance of that tribe in the life-time of Samuel? During the whole of Samuel's career, and even up to the time of David's independent appearance, this tribe is scarcely ever incidentally referred to (1 Sam. xi. 8, xv. 4). And even in the passage in which it appears, its comparative insignificance is very apparent. In the war against the Ammonites, described in 1 Sam. xi., out of 300,000 Israelites only 30,000 belonged to Judah; and in the army which Saul led against the Amalekites, out of 200,000 infantry only 10,000 were of the tribe of Judah (1 Sam. xv. 4). How do the boasted princely rank of Judah, and the imperishable supremacy and rule, attributed to him in this blessing, square with the fact that it was not till the time of David, and only by his instrumentality, that this princely rank was attained? Are we to suppose that the mere outward precedence in the camp and in the order of march through the desert can really have been regarded by the author as fully answering to the supremacy which he so highly extols, and exhaustive of the description in all its fulness? But apart from the sentence pronounced on Judah, and looking more at the blessing as a whole, is it conceivable that a discourse, which is so full of exultation at the prosperous condition of the tribes, which cannot find words or figures adequate to give expression to the abundance of blessings and power, to the conquest and peace secured by almost every tribe, can possibly have been written in the latter half of the period of the Judges, and composed with the intention of describing both the circumstances existing at the time and those belonging to the immediate past? No truly, the torn, and mournful, and down-trodden period of the Judges, of which our opponents generally draw a darker picture than we can admit to be correct,—that period, in which Israel was again and again oppressed and enslaved by the Gentiles, whilst reproach after reproach was heaped upon the people of God on account of their frequent apostasy, *cannot possibly* have been the time at which so exalted a description of the condition of Israel as our blessing contains was written down in the form of a prophecy, supposed to have been fulfilled in the age in which it was composed.

Perhaps, however, all that is necessary to avoid these insuperable difficulties is to fix a somewhat later date for the composition, the time of David or Solomon, for example, as is done by *Heinrichs.* The blessing on Judah would then remain in full force, and all its gorgeous pictures be realised in David's splendid victories and the pomp or magnificence of Solomon's peaceful reign. But *incidit in Scyllam, qui vult vitare Charybdin.* As the blessing on Judah overthrew the former hypothesis, so now does the sentence on Levi rise up with fatal testimony. For from the time of David the priestly tribe was in possession of the highest rank and the greatest favour, and therefore with this assumption there vanishes the opportunity, so warmly contended for, and firmly defended, of bringing the curse pronounced on Levi into apparent harmony with the pretended date of composition.

If, then, the tone of exultation pervading the whole blessing, and the blessing pronounced on *Judah* especially, preclude us from tracing the origin of the song to the period of the Judges, whilst on the other hand the sentence on *Levi* hinders us from assigning it to the times of David and Solomon, and if it is only within these limits that it can be supposed to have arisen as a *vaticinium post eventum*, then such a supposition falls at once to the ground as inadmissible and worthless, and we are brought back to the conclusion that the blessing owes its origin to the prae-Mosaic age, and that there is nothing to hinder any one from admitting its authenticity and its claim to the character of a genuine prophecy, except the rationalistic *placet:* "there are no real prophecies at all."

Other objections to the authenticity of the blessing, such as that "so sublime, imaginative, and lively a style of poetry could not be expected from an old man at the point of death," or that "it is impossible to conceive how such a blessing pronounced by Jacob can have been handed down word for word to the time of the author or compiler of the Pentateuch," with more of the same description, no longer merit any notice, and *Hengstenberg*, in our opinion, has paid them too much honour by his reply. *Hävernick* has founded an argument in favour of its prae-Mosaic origin upon the peculiar character of the *poetry* itself (Introd. to Pentateuch, p. 228. Clark's For. Theol. Lib.).

REPLY TO HENGSTENBERG'S OBJECTIONS TO THE FOREGOING REMARKS.

Since the above was written, the passage before us has been most elaborately expounded by *Hengstenberg* in the second edition of his Christology (i. 47—90 translation), and as my mode of treating the subject is keenly criticized and warmly opposed, I am induced to add the following supplementary remarks. *Hengstenberg's* work has made me more than ever convinced of the correctness of my views, and the fallacy of those advocated by him; and his retractions, so far from improving his theory, have rather tended to deteriorate it. But the author has written in so confident a tone, made his assertions with such unbending determination, and heaped up such an overwhelming abundance of supposed proofs, that any reader who does not examine his arguments with the most critical care, is likely to be dazzled and carried away by them. I will begin with the objections brought against me by *Hengstenberg* from the most general points of view.

(1). P. 69. "The most superficial objections have been considered sufficient by *Hofmann*, *Kurtz*, and others, to induce them to disregard the *consensus* of the whole Christian Church. We cannot, indeed, but be astonished at this." I leave the reader to judge whether my reasons are superficial or not. *I* do not think them superficial. But I am more concerned about the charge that I have set at nought the common consent of the whole Christian Church. I attach as much importance to the assurance that I am supported by the common consent of the whole Christian Church, even in matters of exegesis, as my honoured opponent, perhaps rather more, and I believe that my writings will bear comparison in this respect with those of *Hengstenberg*. Take, for example, his subtle and trifling remarks on the signs and wonders in Egypt, especially on the last plague. In this and many other instances, on grounds to which I will not apply the appropriate epithet, he has disregarded not only the *consensus* of the whole Christian Church, but that of all sound grammatical and historical interpretation, at which *I* was not the only one or the first to feel astonishment. No one indeed will deny, and least of all *Hengstenberg* himself, that even a christian-minded commentator may and must deviate in many cases from the traditional exegesis. The *consensus* of the whole Christian Church has understood Ps. xxii. 16 to refer to a piercing of the hands and the feet; but *Hengstenberg* in his later writings has disregarded this *consensus.* Many persons, who have thus felt themselves deprived of one of the most cherished, most important, and most convincing predictions of the sufferings of Christ, have probably been as much surprised at this, as *Hengstenberg* himself at my interpretation of Gen xlix. 10. And yet he is undoubtedly in the right.

But let us look more closely at the common consent of the Christian Church in reference to Gen xlix. 10. It is true, the early Christian Church without exception referred this passage to a personal Messiah, and so did the ancient synagogue, *but* on the ground of a decidedly false rendering of the word in question, and one which *Hengstenberg* is no less confident in pronouncing false than I am, viz., the rendering given by the *Septuagint* and *Vulgate*. It is absurd for a man to boast of the consensus of the Church, when he has pronounced the basis on which it rests erroneous, in other words has declared the consensus itself to be without foundation.

(2). *Hengstenberg* constantly speaks of my views as non-Messianic, reckons me without reserve as one of the opponents of the Messianic interpretation, and therefore places me in the same category with the rationalistic commentators *Tuch*, *Gesenius*, and *Knobel*. This is very unjust. I have opposed the opinion that the passage refers to a personal Messiah, but I have expressly and most firmly defended its Messianic character and importance. *Hengstenberg* himself is of opinion that the prophecies concerning the seed of the woman (Gen. iii. 15) and the seed of Abraham (Gen. xii. 3) do not refer to a personal individual Messiah, and yet he calls them Messianic!

(3). At p. 71 *Hengstenberg* says, " a suspicion with reference to the non-Messianic (he *means* the non-personal) interpretations is naturally suggested by their variety and multiplicity, as well as by the fact that the opponents of the Messianic explanation never agree among themselves, but that on the contrary one of their interpretations is invariably overthrown by another. *Such is, in every case, a sure indication of error.*" This is excellent. *Hengstenberg* himself has already disposed of two Messianic interpretations; *Sack* propounds a third, and others have been given by different commentators. If the variety be "*in every case*" a sure indication of error, it must be so here. On which side again has there been the greatest diversity, or the most frequent change of opinion? The non-personal interpretations are *three* in number, (1) till rest comes, (2) till he (or one) comes to rest or to the place of rest, (3) till he (or one) comes to Shiloh. Of the personal interpretations there are four. (1), *ἕως ἄν ἔλθῃ τὰ ἀποκείμενα αὐτῷ* or *ἕως ἀν ἔλθῃ ᾧ ἀπόκειται*; (2), Donec veniat qui mittendus est; (3), Donec veniat filius ejus; (4), Till the hero (*alias:* rest, *i.e.*, the bringer of rest, *alias:* the man of rest) comes. It is to be observed here, however, that the division of the expositors into two classes, those who refer the passage to a personal Messiah and the non-Messianic, is a very wrong one, even from an exegetical point of view. The principal exegetical difference relates to the question whether *Shiloh* is the subject or the object. And here there are *five* different explanations on *Hengstenberg's* side, and only *two* on ours (and these two, as we shall presently show, are, exegetically considered, one and the same). Thus *Hengstenberg* has pronounced sentence on his own interpretation. Nevertheless *we* are magnanimous enough to cancel it for the present as undeserved.

(4). P. 71. "It is possible in every case to trace out some interest, *apart from* the merits of the question, which has led to the objections against the Messianic interpretation. . . . . *Hofmann* and his followers do not in the least conceal the fact that they are guided by the principle of a concatenation of prophecy with history."—How far the latter is correct, at least in my case, we will enquire by and by. For the present I shall simply say, that it is untrue that I have any interest *apart from the merits of the question.*

(5). At p. 67, *Hengstenberg* says: "The entire relation of the Pentateuch to the sacred literature of later times, and the circumstance that the former constituted the foundation of the latter, and contained, in the germ, all that was afterwards more fully developed, entitle us to expect to find some expression of the Messianic idea in the books of Moses. The more prominent the place occupied in the later books by the announcement of a personal Messiah, the more difficult will it be to one who has acquired correct fundamental views regarding the Pentateuch, to conceive that this announcement should be wanting in it—especially the announcement of the Messiah in his kingly office. . . . . But there cannot be any doubt, that the promise of a personal Messiah in his kingly office, if it be found in the Old Testament at all, *must* exist in the passage which we are now considering." That is to say, the Pentateuch prepares the ground in every direction, therefore the Messianic idea *must* have taken root in it, and everything that we find subsequently expanded, must have existed here in the germ. Who is there that will dispute this, if he believe in the history of the plan of salvation at all? But imperceptibly the Messianic idea is exchanged for "the announcement of the Messiah in his kingly office," the germ, that is, for the full grown tree. We, too, are of opinion that the foundation of the Messianic idea *must* be laid in the Pentateuch, but we do not consider that we are justified in maintaining *a priori* that it must have existed in the Pentateuch in this or that expanded form. *Hengstenberg* decides that, since the Messianic idea appears in subsequent books as an announcement of a personal kingly Messiah, it must be found in the Pentateuch, not merely in the germ, but in its fully developed form. But what are we to say, then, of the announcement of a suffering Messiah, which also appears in the later books? According to *Hengstenberg's Hermeneutics*, this also *must* be found in the Pentateuch. Let him point us, then, to such a prophecy in the books of Moses. No doubt the antecedents are already there, the soil is prepared in which this idea shall strike its roots, namely in the institution of sacrifice, but the application of the idea of sacrifice, and its expansion into the concrete announcement of a personal suffering Messiah belong to a later age.

We do maintain, however (not *a priori* as *Hengstenberg* does, but *a posteriori*), that the idea of a personal Messiah is to be found in the Pentateuch. But in spite of *Hengstenberg's* decision that it exists in Gen. xlix. 10, and nowhere else, we take the liberty of looking for it, not there but in Num. xxiv. 17 (see vol. iii., § 57. 1), and Deut. xviii. 18 (see vol. iii., § 60. 3).

(6.) In commenting upon the remark made by me, to the effect that the historical conditions and preparations requisite to the development of the Messianic idea did not exist in the time of Jacob, but that they are to be found first of all in the time of Moses, and afterwards in a more perfect form in that of David, *Hengstenberg* writes with the greatest indignation (p. 70): "Do you mean to teach God wisdom? we might ask, in answer to such argumentation. To chain prophecy to history, in such a manner as this, is in reality nothing short of destroying it. How much soever people may choose to varnish it, this is but another form of naturalism, against the influence of which no one is secure; for it is in the atmosphere of our day. Men who occupy so narrowminded and trifling a ground of argument as this, who would rather shape history, than heartily surrender themselves to it, and find out, meditate upon, and follow the footsteps of God in it, will be compelled to erase the promise in Gen. xii. 3: 'In thee shall all the families of the earth be blessed,' yea, even the words, 'I will make of thee a great nation,' with which the promise begins—for *that* also violates the natural order."

I admire the zeal which is apparent in these words, for it is zeal in a holy cause, though it arises from prejudice, misunderstanding, and error. But before I proceed to prove this, I will point out to what extent there are errors in my own line of argument. First, I have done wrong in putting the historical proofs of my opinion before the exegetical, for by so doing I have undoubtedly made it appear that I regarded the former as the more important, though I guarded against such a mistake by the most explicit declarations. Exegesis ought to do its work, free and unconstrained, without the fetters either of tradition or of its own system; the results of the exegesis should then be linked on to the system, and the latter should be shaped, completed, or rectified according to the former. In the present instance, the results of exegesis, at which I have arrived with the greatest care and conscientiousness, are completely in harmony with the historical data and the expectations founded upon them. Hence my error was merely one of form. I have only to put the exegetical enquiry before the historical, and everything will be in order. Then again, I have to confess that my historical researches have perhaps been conducted in a more confident tone, than human speculations in general ought to assume, and that this may appear to have been peculiarly unjustifiable in the present case, as no exegetical foundation had yet been laid. But even this is a mere error of form, and I have only to alter the expressions, not the matter.

Let us look, however, at the charge of *naturalism*. Even if I looked upon the history of Israel as purely natural, a purely human development, a concatenation of history and prophecy, regarding these as props and conditions, the one of the other, I ought not to be regarded as the precursor of naturalism. Would it, for example, be naturalism, if I were to maintain that the point of time at which God became incarnate in Christ was affected by the natural development of heathenism, that God performed this, the greatest miracle in the history of the world, just at the time when all the conditions requisite for the cheerful acceptance of salvation on the part of the heathen, and all that could promote the diffusion of the gospel through the earth, were to be found in the political and social state of the Gentile world?

But I can see in the history of Israel, in which, with which, and about which prophecy is occupied, not merely a natural, human development, but on the contrary a product of nature and grace, of human freedom and the sovereignty of God. If, then, I look at divine prophecy in its relation to the history of Israel, that is, to a history which was the result of the most special guidance and constant active interference on the part of God, how can this be condemned as a naturalistic degradation of prophecy? Do not the traces of God's mercy and wisdom in the history of salvation come first and most clearly to light, do they not appear in their most wonderful and attractive form, when we see how divine prophecy was introduced as a living and organic part of history, and on the other hand how the course of history was so directed by God, and his operations therein were of such a kind, as to be constantly opening the way and preparing a place for new and more glorious forms of prophecy? I fall in the dust and worship when I thus discover how the living God was ever moving in history and prophecy, how the mercy and wisdom of God, through his adorable condescension, adapted themselves in both of these to existing wants and circumstances. Is this naturalism? Is this shaping history and destroying prophecy? To my mind, prophecy first acquires its full value, when I can see what God has done in history to prepare a fitting place for prophecy. The incarnation of God in the fulness of time loses nothing of its adorable worth, but rather gains the more, from the fact that it required a historical preparation of 4000 years.

For my own part I am conscious of having "heartily surrendered myself to history," and of having "meditated upon and followed the footsteps of God therein." I have doubtless done so in great weakness and with much liability to error, and shall therefore be always delighted to learn not merely of *Hofmann*, but of *Hengstenberg* also. There may be many an error in the work I have written; but no one can charge me with want of hearty devotion or thoughtful research. Again, there is as much injustice as bitterness in the accusation brought against me, of giving way to the desire to *teach God wisdom*. Might I not, with equal justice, or rather injustice, bring the same charge against *Hengstenberg*, for saying at p. 67, that God *must* have caused the announcement of a personal Messiah and of his kingly character to be made in the Pentateuch, or for similar remarks which might be found in a hundred other passages of his writings? But what shall I say, when *Hengstenberg* is so carried away by his zeal as to maintain that, with the views which I hold on the relation between history and prophecy, I shall be compelled to erase the promise in Gen. xii. 3, " In thee shall all the families of the earth be blessed," as well as the prediction, " I will make of thee a great nation," since they also violate the natural order ? Such arguments as these bear upon the face of them convincing proof that the writer either could not, or would not, understand his opponent.

The charge of naturalism, of destroying prophecy, &c., would only be justifiable, if I looked upon prophecy as *Ewald* does, as a natural product of the human mind, and supposed it to be attributable to an elevated and enlightened spirit, skilled in anticipating the future, as the result of the study of the history of the past. But *Hengstenberg* knows, or ought to know, that these are not my views. Prophecy, in my opinion, is an objective communication of divine knowledge to man, but one that is vitally associated with the circumstances of the age in which it is made, which supplies its wants and enters as an organic element into the general course of affairs. The dependence of prophecy upon history, as I understand it, is no other than this, that God does not scatter the seeds of prophecy, until by his guidance of history, he has brought the soil to such a state, that as soon as those seeds are scattered, they will strike their roots and bring forth fruit. The seeds of prophecy do *not* resemble the grains of wheat, which the Egyptians placed in the hands of their mummies, to lie there perhaps for thousands of years, before they fell into a genial soil, where they could unfold the blessing that was in them. They bear a far greater resemblance to the sowing of the husbandman, who scatters every kind of seed at the proper season, and either seeks a fitting soil, or *makes* it so by cultivation.

Hengstenberg has left the field of scientific discussion, and made a very cutting appeal to my conscience. I am far from denying that any one has a right to do this. But before bringing against another charges so sweeping as those of naturalism, of shaping history, destroying prophecy, and sacrilegiously wishing to teach God wisdom, charges which, as *Hengstenberg* might well have known, would go to my heart like a two-edged sword, it is a duty to weigh the terms employed with greater care than *Hengstenberg*, in his excessive zeal, appears to have exercised. I desire no mercy, even from *Hengstenberg*, but I desire justice and truth, and these I do not meet with. Nor can I avoid acknowledging that I look upon *Hengstenberg* as having even less right than others to speak upon such subjects in a way like this, for, were he measured by his own standard, he would hardly escape the same, or rather, I believe, far greater condemnation. *I* shall not call it naturalism that we find him so often depriving miracles of their miraculous character, nor shall I say that he is a destroyer of prophecy, though so frequently he dissipates the concrete substance of a prophecy into shadowy ideas. I will not speak of him as shaping history, when he explains away everything in it that displeases him, nor will I charge him with wishing to be wiser than God, when he so completely sets at nought all the laws of exegesis, in his interpretation of the miracles wrought by God for Israel, as to bring out exactly what *he* would have done if he had been in the place of God.[1] As I have said I *neither will nor can* bring such severe and unjust charges against him; but I say with confidence and without reserve, that if *Hengstenberg* were measured by the same standard by which he has measured me, there are none of these charges which he would be able to rebut or evade.

(7). *Hengstenberg* had formerly translated the passage under review: "till rest, *i.e.* the bringer of rest, shall come," and had endeavoured to prove from such examples as קיטר, כידור, שילח that שילה might be an abstract noun. But it is very clear that this explanation is not a true one, even apart from the context, the structure, and the parallelism of the verse. It might indeed be possible to defend the use of an abstract for a concrete noun; but as שלה does not mean *to bring rest*, but *to enjoy rest*, שילה (Shiloh) cannot indicate one who brings rest, but one who enjoys it, and this is a predicate which can hardly be applied to the Messiah, who came not to enjoy rest himself, but to impart it to others (Gen. xii. 3). *Hengstenberg* has, therefore, done right in dropping this explanation, but he has done wrong in substituting for it one which is even weaker and more untenable. He now interprets *Shiloh* as a personal appellative, or (what he appears to regard as the same thing) a proper name, and translates it *man of rest.* He has been led to make this modification,

[1] See my treatise on Jephthah's Sacrifice in the luther. Zeitschrift, 1853.

partly by the discovery that such forms as שִׁילֹה = שִׁילוֹן cannot possibly be *abstract* nouns, and partly by the fact that in every other passage, Shiloh is the *proper name* of the town in which the tabernacle was first set up after the conquest of the Holy Land. "An interpretation," he says at p. 74, "which dissevers the connexion betwixt Shiloh and Shiloh, betwixt Shiloh and Solomon, betwixt Shiloh and the Prince of Peace, betwixt Shiloh and him "whose right it is," must for that very reason be self-condemned." But this town of Shiloh is just the Achilles' heel in *Hengstenberg's* explanation of the passage as referring to a personal Messiah, and, to say the least, it is not a prudent thing to run, with the heel exposed, upon the adversary's sword. If once we decide that the passage alludes to the town of Shiloh, then all reference to a personal Messiah is hopelessly gone; for we shall have no other resource open to us than to say that the word Shiloh is the *object* of the passage, indicating the point at which they were to arrive. But how unsuitable does the conjecture, expressed by *Baumgarten* and myself, that the town of Shiloh owes its name to this prophecy, appear in *Hengstenberg's* mouth! For such a thought is just as much at variance with his interpretation, as it is in harmony with ours. Shiloh, he says, is a proper name, the name of the Messiah, and its appellative signification is *man of rest.* Then, Joshua named the town where he first erected the tabernacle "*man of rest,*" because Jacob had called the personal Messiah the man of rest! What an absurd idea! For what had the town of Shiloh to do with the personal Messiah, the future king of Israel? What a ridiculous name for a town: *man of rest!* Can we conceive of the Jews returning from the Babylonian captivity and calling Jerusalem "*Messiah,*" in commemoration of the rebuilding of the temple?!! If not, it is just as *inconceivable* that Joshua should have given the name of *Shiloh* to the town where he erected the tabernacle, if Shiloh was then an appellative noun, or, as *Hengstenberg* says, a proper name of the personal *Messiah.*

Hengstenberg's new interpretation has thus left all the weak points of his former explanation unaltered (we shall discuss them presently), and has merely added fresh impossibilities. He has even retained the weak point already referred to, viz. the derivation of שִׁילֹה from שָׁלָה, which means *salvus, securus fuit, maxime de eo qui prospera fortuna secure utitur (Gesenius* thes.), and the inference that *Shiloh* can only mean a *man of rest*, in the sense of one who enjoys rest, not of a man who brings rest and peace.

(8). The first and most essential question to be asked in connexion with the interpretation of this passage, the question, in fact, upon which everything else depends, is not whether the passage speaks of a personal Messiah or no, but whether שילה is to be translated as the subject ("till Shiloh come"), or as an object ("till he come to Shiloh"). To the latter rendering, which I gave in my first edition in the abstract form ("till he comes to rest"), but in the present edition in the concrete shape ("till he arrives at the place of rest," *i.e.* at the place where rest shall be made apparent), *Hengstenberg* offers the following objections: (1) *Shiloh*, from its very form, cannot be an abstract or appellative noun, but must necessarily be either a concrete adjective or a proper name, and (2) if *Shiloh* were either of the two former, the object, to which they were to come, would necessarily have been introduced with a preposition.

I do not consider the comparison of שילה with כידור, שילוח, קימוש, קיטר absolutely inadmissible, although *Hengstenberg* has adopted *Tuch's* arguments against such a comparison. Nor can I adopt the opinion of *Delitzsch*, that where there are already so many synonymes for the one word *rest* (מְנוּחָה, שָׁלוֹם, שַׁלְוָה, שֶׁלֶו), it would be impossible that the form שִׁילֹה should have the same meaning. Does the fact that there are four words in a language with the same meaning, rest, establish the impossibility of our meeting with a fifth? Still I see no objection on the other hand to the derivation of *Shiloh* from an original form שִׁילוֹן, which is advocated by *Hengstenberg* and *Tuch*. The existence of such a form is rendered very probable by the *nomen gentile* שִׁילֹנִי, which we meet with in 1 Kings xi. 29, xii. 15.

But so much may be admitted without our being, therefore, unable to interpret שילה as an abstract noun. *Ewald*, at least, informs us, that adjectives *and* abstract nouns are formed by the terminations *an* and *on* (Lehrbuch § 163 b.). The adjective signification he regards as the primary one, and states that at present there is no distinction in the terminations, but that it is *certain* that *an* was originally the form of the adjective, *on* that of the abstract noun.

We have already pointed out an instance, in which *Hengstenberg* has condemned himself whilst attempting to rectify his opinion. We have another proof of this in the case before us; but here also we must reverse the sentence as an unjust one. In the excess of his zeal, for example, in attempting to overthrow our explanation, he has adopted *Tuch's* assertion, that "it is quite impossible to give the word the signification of an appellative noun, since it is only in proper names, where the signification of the derivative suffix is of the less consequence, that *on* is shortened into *oh*." This reasoning suits *Tuch's* interpretation very well, for in his opinion *Shiloh* is the name of the well-known town, in this and every other passage of the Old Testament, in which it occurs. But instead of sustaining *Hengstenberg's* view, that Shiloh means a *man of rest* in the passage before us, it is directly opposed to it. Is *Shiloh*, then, simply a proper name in this connexion? Is the word *Messiah* a proper name? Are such terms as the king, the ruler, the conqueror, &c., proper names? Undoubtedly these and other similar words may all become proper names, but they only become so when they are associated with particular individuals. *Victor* is primarily an appellative noun, but it becomes a proper name by becoming the name of a person; *Shiloh* is an appellative noun, but it becomes a proper noun by being used as the name of a town; so with the name Solomon, &c. If Jacob, then, predicted the coming of a *man of rest*, did he mean that "*man of rest*" was to be his name? Certainly not; he surely meant that he would *be* a man of rest, and did not intend to say whether that would be his *name* or not. If he had, he would have predicted something, which was *not* fulfilled, for in Luke ii. 21, we do not read that "when eight days were accomplished for the circumcising of the child, his name was called *Shiloh*." It is evident therefore that if the word Shiloh in Gen. xlix. 10 refers to a person at all, it must be an appellative noun descriptive of that person, and not his proper name. The identification of an adjective and a proper name is a self-delusion, which in our case, at least, has not succeeded in imposing upon others also.

With regard to the assertion itself, it is certainly true that only one exception can be found to the rule, that the abbreviation of the ending *on* into *oh* took place in none but proper names. But the fact that there is at least *one* exception (אֲבַדֹּה

= death, hell, in Prov. xxvii. 20), is a proof that the rule is not an absolute one. And *Ewald* has shown that there are other analogous instances of the softening down of the final consonant *n: e.g.* אֶתְנָה (Hos. ii. 14) for אתנן and שִׁרְיָה (Job xli. 18) for *Shirjân* (1 Kings xxii. 34) and *Shirjôn* (1 Sam. xvii. 38); *vid.* Lehrbuch § 163 seq.

What, then, is the meaning of *Shiloh?* Two things are certain, that Shiloh is derived from the root שָׁלָה, and that Shiloh was the name of a *town*. Either of these is sufficient to establish the appellative signification of the word, from which undoubtedly the name of the town originally sprang. And from these two data, even apart from the laws which regulated the formation of the language, we may argue conclusively that the original notion expressed by the word is either *rest* in the abstract, or, what I decidedly prefer, *the place of rest, i.e.*, the place in which one rests, or where rest is first enjoyed. But we must defer the consideration of the question, whether, in the passage before us, Shiloh is the proper name of the well-known town, or still retains the appellative signification which was kept in view when the town was named.

(9). It is not necessary to offer proofs that the verb בּוֹא is often followed by an accusative without a preposition, to indicate the object arrived at. We find it in various connexions, both with proper and appellative nouns, *e.g.*, to come to Shiloh, to come to Jerusalem, to come to the town, to come to the gate, to arrive at wisdom (Prov. ii. 19), to come to the sabbath (2 Kings xi. 9, בּוֹא הַשַּׁבָּת *i.e.*, for the purpose of performing the priestly duties of that day), to come to the feast (Lam. i. 4 בּוֹא מוֹעֵד). But the arguments of our opponents assume that the *objects* can only stand without a preposition when it is a *concrete*, not when it is an *abstract* noun. And if the two expressions, "to come to the sabbath," and "to come to the feast," are not allowed to be cases in point, I must candidly confess that I know of no other instance in which בּוֹא is connected with an *abstract* noun without a preposition, and that in every other case we find it with *places* or *persons*. Still even if we must admit, that the ordinary rules of the language required a preposition with abstract nouns, this would not prove that poetry may not have emancipated itself from this law, seeing that it always adopts so very different, and so much bolder a style.

However I do not require this admission. I have already stated that I also regard the word Shiloh not as an abstract, but as a concrete noun, with the meaning *place of rest*. But I have not yet been able to determine, whether it is to be taken as an appellative or a proper noun. Should further investigation establish the former, I have no doubt that it will be just as possible to do without the preposition in the phrase, "to come to the *place of rest*" as in the other phrases "to come to the *town* (Jer. xxii. 24), or "to the *gate*" (Gen. xxiii. 10, 18 ; Ps. c. 4).

(10). There is nothing in the rules of the language, therefore, to prevent our rendering the passage עַד כִּי־יָבֹא שִׁילֹה : "till he come to the place of rest (town of rest)." *Shiloh may be* the *object*, and there is nothing to prove that it *must* be the *subject*. This we have already demonstrated, and therefore all that we have to do here is to adduce still further evidence, and to answer *Hengstenberg's* objections. I said above that the parallelism of the verse leads us to consider *Shiloh* as the object. I have probably laid too much stress upon this argument, but I must still maintain so much at least, that in my view the parallelism is unmistakeably clear ; and *Hengstenberg* admits that the parallelism is "somewhat concealed" by his interpretation, inasmuch as, instead of (?) : "till the bringer of peace comes, *and he, to whom* belongs the obedience of the nations"—we have in the second member, "*and to him* belongs the obedience of the nations."

The context and the train of thought in the blessing on Judah speak much more decidedly and, as I think, with absolute proof in favour of my interpretation, and in opposition to *Hengstenberg's*. The following reasons may be assigned : (1), We should expect the word "*until*," to introduce some information as to the course of Judah, and what would be the result of his uninterrupted possession of the post of leader ? (2), What could induce the patriarch, when describing the blessings that awaited Judah, to look so far away from Judah himself, as to place the climax of the blessing in the announcement of a person, who is not said to have been connected with Judah in any way whatever ? For it is nowhere stated that the person, supposed to be indicated by Shiloh, will be the descendant of Judah, nor is this by any means necessarily implied. (3), But even granting that the supposed person, Shiloh, can or rather must be regarded as descending from Judah, and that the word Shiloh describes the person of the Messiah according to his kingly office and his peace-bringing rule, then Jacob will have prophesied that Judah should rule until the ruler sprang from Judah, *i.e.*, that Judah should rule till Judah ruled. There is no sense in this. (4), If the word Shiloh really denoted the Messiah, *i.e.*, a particular, well-defined personality, there would be every reason to expect that the article would be prefixed, and that thus the expression would be somewhat less general. (5), The first half of the tenth verse speaks only of Judah, and according to *Hengstenberg* another subject, viz., Shiloh, is introduced into the second half. Be it so; but what are we to make of the next verse (11) which commences, "he binds his colt to the vine," &c., "he washes his garments in milk," &c.? Who is the *he* in this case? Judah or Shiloh? According to the laws of exegesis *Hengstenberg* ought to reply, Shiloh. But how does the description given in ver. 11 apply to the Messiah? This verse is most clearly descriptive of Judah's inheritance in the Holy Land, a province rich in wine and milk. Hence *Hengstenberg* says without the least reserve (p. 74): "What is here assigned to Judah, belongs to him only as a part of the whole, as a fellow-heir of the country flowing with milk and honey." The subject is *Judah*, then, not Shiloh? But what is to be done with the "*he*" in ver. 11, which can only refer to Shiloh? (6), The train of thought in the whole of Jacob's address to Judah (ver. 8—12) requires that we should render Shiloh as an object, and precludes our taking it as the subject of this sentence. How beautifully and smoothly does thought link itself to thought with our interpretation! What life there is in the whole section; and how natural is every part! Judah, the praised one, is the conqueror of his enemies, the champion of his brethren. By his victorious, lion-like power, and his inalienable supremacy, Judah passes on from conflict to victory, from war to peace, and the nations gladly obey the conqueror. This peaceful and happy condition is still farther pictured in vers. 11, 12, by a description of the abundant blessings to be enjoyed in the land, into which Judah enters as the leader of the rest. What man is there, with any feeling for the proper order and consecutiveness of thought, who will not grant, that with this interpretation the connexion and the train of thought are as natural, and free from violence, as they are intelligible and easy? And what does *Hengstenberg* say? "We further remark, that verses 11 and 12, which ancient and modern commentators (*e.g.*, *Kurtz*) have attempted to bring into *artificial* connexion with ver. 10, simply finish the picture of Judah's happiness by a description of the luxurious fulness of his rich territory" (p. 74). Indeed! Then the connexion, which I have pointed out, is *artificial*, and it is sufficient that *Hengstenberg says so* without waiting to prove it. But when we ask what natural, simple, and unforced connexion he suggests instead, we receive for answer, *none*. Now, undoubtedly, a connexion which has no existence at all, cannot be called an *artificial* connexion. But if there is any place, in which an expositor must necessarily find out the connexion between two consecutive sentences, it is just here between ver. 10 and those which follow. For as there is no subject named in ver. 11 seq., the subject must be sought in the verses immediately preceding, and, therefore, there must be a connexion between the two, which it is the duty of the expositor to point out.

This is the exegetical ground on which I have based my view. I will not maintain that all these arguments are absolute proofs: on the contrary the only ones to which I attribute such force as this are Nos. 3, 5, and 6; though I do not regard the others as unimportant. Yet all that *Hengstenberg* has to say in reply to the whole of these multifarious arguments is found in the bare and unsupported assertion, that I have attempted to bring ver. 11 into artificial connexion with ver. 10.

If, now, we further consider the fact, that *Shiloh* is the name of a *town*, and that a town cannot possibly have been named the "man of rest" or have been called by the personal name of the Messiah, I think I shall have adduced all the exegetical proof that can be required of the impossibility of *Hengstenberg's* opinions, whether new or old.

The word Shiloh occurs forty-one times in the Old Testament as the name of a town. What then is more natural than to suppose that in the forty-second passage, that is, the passage before us, either this town is expressly designated, or there is some essential connexion between the Shiloh mentioned here and the name of the town? Everything depends upon the question, whether the town was in existence in Jacob's time, or rather, whether it was then called *Shiloh*. For if so, there would be no doubt that Jacob's prophecy had reference to the town, and we should have to adopt the rendering of *Tuch*, *Delitzsch*, *Diestel*, and others: *till he come to Shiloh.* But if not, then the name of the town had some reference to Jacob's prophecy. *Shiloh*, therefore, will in that case have been used by Jacob as an appellative noun, meaning the *place of rest*, and will subsequently have become a proper name by being transferred to the town as the "town of rest."

I still give a decided preference to the latter explanation. All that I have said *in opposition* to the former appears to me as convincing as ever. Moreover, I am now of opinion that I can support my view, that the name of the town was changed with direct reference to Gen. xlix. 10, by biblical data (for which I am indebted to *Hengstenberg* himself, p. 81). In the first passage, in which the word *Shiloh* occurs as the name of the town, viz., Josh. xvi. 6, we find it written *Taanath-Shiloh*, and shortly afterwards it is mentioned in a connexion which points unmistakeably to Gen. xlix. 10. In Josh. xviii. 1, we read that "the whole congregation assembled together *at Shiloh*, and set up *the tabernacle* of the congregation there, *and the land was subdued before them.*" With this we should compare Josh. xxi. 44: "And the Lord *gave them rest* round about, according to all that he sware unto their fathers; and *there stood not a man of all their enemies before them*," &c., and Josh. xxii. 4: "And now the Lord your God *hath given rest unto your brethren*, as he promised them; therefore, now return ye, and get you unto your tents," &c. From these passages we perceive that Israel regarded the erection of the tabernacle at Shiloh as a boundary-line in its history, marking the termination of its previous wanderings and homeless condition, and the commencement of its quiet and peaceful possession of the land, which was promised to the fathers. And they had good reason for so doing, for the permanent erection of the tabernacle, the setting up and taking down of which had hitherto served as an invariable signal of the encampment and the departure of the Israelites during the journey through the desert, naturally served as a sign and guarantee of the termination of their wanderings and the attainment of a settled rest. What Jacob had foretold in his blessing to the fathers, was now fulfilled (at least in a preliminary form). And whilst it was perfectly natural that the blessing of the patriarch should be remembered on that occasion, the passages referred to distinctly intimate that it was so remembered. When the tabernacle was set up at Shiloh in the place of its rest, all Israel had also arrived at its resting-place. If the town had actually been called Shiloh before, it was not till now that it became fully and truly what its name indicated, *a place of rest.* To judge from appearance, however, this was *not* its name previously, but it was so called *for the first time on the occasion referred to*, in commemoration of the important manner in which their previous history had been brought to a close. It is true that the absence of any reference to a city of Shiloh in the earlier history, is not a proof that no such town existed, or that it bore some other name, but it gives a certain amount of probability to the assumption. More than this, the fact that, when the town is first mentioned, we find another name, *Taanah*, by the side of the name Shiloh, and that this name subsequently vanished, confirms the conclusion, to which we were brought by the other data mentioned above. There is not the least improbability, therefore, in the opinion, which we have been led to form, that the town was formerly called Taanah, but that it received the name of Shiloh, after the erection of the tabernacle, with especial reference to Jacob's prophecy.

Hengstenberg agrees with me in this, except that with the greatest *naiveté*, he adopts the most incredible notion, that the town was named the *man of rest* or Messiah. But in a note on p. 81 (transl.), by the use of the word *vielleicht* (perhaps) he suggests the possibility that the name Taanath-Shiloh, in Josh. xvi. 6, " may not be a combination of the earlier and later names, but the full form of the original name, of which the latter, Shiloh, is only an abbreviation. From the well-ascertained and common signification of the word אנה, we are entitled to translate Taanath-Shiloh: the *futurity*, or *the appearance of Shiloh. Shiloh shall come:* such was the watchword at that time. The word *Taanah* would then correspond to the יבא of the fundamental passage."—*Hengstenberg* has certainly acted with great prudence, in leaving a backdoor open, when setting up his impossible theory; only it is unfortunate that it should lead to a תֹהוּ וָבֹהוּ. For (1) There is something very beautiful and edifying in the assurance that "'*Shiloh will come*,' was the *watchword* of the time;" but unfortunately this assertion is a mere piece of imagination, as there is not the slightest or most remote trace of such a *watchword* in the whole of the book of Joshua. The *actual* watchword is given most clearly and unmistakeably in Josh. xviii. 1, xxi. 44, xxii. 4: "*Jehovah has given rest to Israel*," and *this* watchword was incorporated in the new name, that was given to the town. (2). It is just as fatal an objection, that the "well-ascertained and common" signification of אנה does not admit of the explanation: the future, or the appearance of Shiloh. תאנה, in Jer. ii. 24, is generally admitted to mean sexual connection, *coitus*. And even if we assume that this is merely a derivative meaning, and that the primary meaning (the one applicable here), is a *meeting*, or *combination*, it will not be easy to extract from this the idea of the "future, or appearance, of the Messiah." The verb אנה is not used in the Kal. In the Piel, Pual, and Hithpael it has the meaning to light upon, to happen *accidentally*, and the notion of that which is accidental always appears as essentially connected with the verb. The meaning of the Kal, from which *Taanah* is derived, is given by both *Gesenius* (p. 123), and *Fürst* (Handwörterbuch 112), as, "to be a suitable, convenient, proper time; to meet or fit exactly." This does not in any way suit the explanation "future of the Messiah." If this be the true meaning of the Kal, the proper interpretation of Taanath-Shiloh would be not "Shiloh's future," but "Shiloh's present." The name could only be intended to say: what Shiloh signifies, has now come to pass. And this would harmonise with my views very well, but not with those of *Hengstenberg*.

(11). "Up to the time of their arrival in Shiloh," says *Hengstenberg*, p. 72, "Judah was never in possession of the sceptre, or lawgiver; and this reason would alone be sufficient to overthrow the opinion, which we are now combating" (viz., that advocated by *Tuch*, *Delitzsch*, &c.) "We have already proved that, by these terms, *royal* power and dominion are designated, and that, for this reason, the *beginning* of the fulfilment cannot be sought for in any period *previous* to the time of David." This argument is equally applicable to the views I entertain. I will not enter into a controversy with *Hengstenberg* on account of his having translated מְחֹקֵק *lawgiver*, though

I regard this rendering as decidedly erroneous, and feel myself compelled by Num. xxi. 18 (where the word is used in just the same connexion and from the same point of view as in Gen. xlix. 10), to render it the *ruler's staff.* But it is an assertion altogether without foundation to say that *Shebeth* and *M'chokek* can only refer to *royal* supremacy. The context in the case before us shows, that they must both of them be interpreted as referring to the lead taken by the tribe of Judah. *Shebeth* occurs in Judg. v. 14, as the staff of the head of the tribe of Zebulon, and *M'chokek* in Num. xxi. 18, as the ruler's staff held by the nobles of the nation. And in neither of these passages can it denote really *royal* insignia.—*Hengstenberg* then continues (p. 72, 73): "But even if we were to come down to the mere *leadership* of Judah, we could demonstrate that this did not belong to him. His marching in front of the others cannot, even in the remotest degree, be considered as a leadership. Moses, who belonged to another tribe, had been solemnly called by God to the chief command. Nor was *Joshua* of the tribe of Judah." But in spite of all this, the fact, that when Jacob said, "the sceptre shall not depart from Judah, nor the ruler's staff from the place between his feet," he merely thought of the lead to be taken by the tribe, may be inferred (1), from the passage itself, for he promises the sceptre and the staff to the tribe, and not to one particular member of it; and (2), with still greater certainty from the relation between the words addressed to Judah, and those previously addressed to *Reuben*, *Simeon*, and *Levi.*—*Reuben*, the first-born, had forfeited the pre-eminence in might and dignity, which properly belonged to him, on account of his wickedness. And for a similar reason, the pre-eminence in might and dignity, which naturally belonged to the first-born, could not be transferred to either Simeon or Levi. The patriarch's eye then fell upon *Judah*, and he at once exclaimed: "It is thou my son, the children of thy father bend before thee." Thus Judah was assured of the pre-eminence in dignity and power, which had been taken away from Reuben. And what was this pre-eminence in power and dignity, or the bending of the other children before this one, but the leadership and rule? But, replies *Hengstenberg*, in the journey through the desert and during the conquest of the promised land, Judah was not the leader. In making this remark, however, (1), He overlooks the fact, that the words of Jacob speak of the leadership of one tribe among the rest. Moses and Joshua were what they were, not on account of their belonging to this or that tribe, but by virtue of an extraordinary call on the part of Jehovah. Judah was still the first of the tribes, notwithstanding that neither Moses nor Joshua belonged to that tribe. "In every numbering of the people, Judah appears as the most important and populous of the tribes, and whenever the camp broke up, Judah led the way. When the land was divided, it was Judah again which received its inheritance in Gilgal before any other tribe." *(Delitzsch.)* (Compare also *Hengstenberg's* remarks at p. 76 seq.). Moreover, the blessing pronounced upon Judah by Moses was based upon the fact, that Judah was the acknowledged leader of the tribes. *Hengstenberg* himself says with reference to this (p. 79): "The whole announcement (of Moses concerning Judah) is based upon the supposition that Judah is the fore-champion of Israel; and this supposition refers us back to Gen. xlix. This is especially apparent in the words: 'bring him to his people,' on which light is thrown only by Gen. xlix. It is for his people that Judah engages in foreign wars, and the Lord, fulfilling the words: 'from the prey, my son, thou goest up,' brings him safely to his people." Is not this a leadership, or chieftainship?—(2), We have also to observe, that our interpretation, at all events, does not compel us to regard the fulfilment of the prophecy contained in Gen. xlix. 10, as completed, exhausted, and therefore terminated by the erection of the tabernacle at Shiloh. The conquest of the land by Joshua ushered in the period in which the Israelites were to dwell in peace and quiet in a land of their own. But when it became apparent that the repose already secured was to be mixed up with, and even exchanged for, disquiet and trouble, it was also apparent that Jacob's prophecy was not yet absolutely fulfilled, that it had only received a provisional fulfilment, and was now entering upon a new stage, which would lead to a later fulfilment in a higher sense and wider form. We are, therefore, justified in appealing to the progressive development of the chieftainship of Judah during the subsequent history, as first exhibited in Judges i. 2, xx. 18 *(vid. Hengstenberg*, p. 81), and continually advancing till the time of David, and then till that of Christ.

(12). Having thus exhibited the *exegetical* proofs, that Jacob did not announce a personal bringer of rest, but merely a period and a place of Messianic rest, and having defended these proofs against all attacks, it will now be perfectly in keeping and very proper, that we should show how completely this exegetical result answers to the historical data furnished by that age, and in how vital, harmonious, and organic a manner history and prophecy are blended together. I have already entered fully, and as I think conclusively, into this subject. All therefore that I have to do, is to refer to what I have said before. But it is necessary here to test the arguments by which *Hengstenberg*, in his description of the connexion between this prophecy and history, has attempted to establish his views and overthrow mine. We read, for example, on p. 67 : "The promises which were first given to Jacob's parents and then transferred to Jacob himself, included two things: *first*, a numerous progeny and the possession of Canaan ; and, *secondly*, the blessing which should come through his descendants upon all nations. How, then, could it be expected that Jacob, in transferring these blessings to his sons, and while in spirit seeing them already in possession of the promised land, and describing the places of abode which they should occupy, should have entirely lost sight of the second object which was much the more important, and just as often repeated?" There are two statements here which are not true: (1), It is *not true* that the second portion of the prophecy is *as often* repeated as the first. It is only in those promises in which *Jehovah himself* pronounces the blessing directly, formally, and solemnly upon the three patriarchs (on Abraham, Gen. xii. 3 and 18, xxii. 17 seq. ; on Isaac, xxvi. 4; on Jacob, xxviii. 14), that the spiritual blessing is mentioned in connexion with the temporal. In Gen. xii. 6, xiii. 16, xv. 5, 18, xvii. 4—8 and 16, we have a whole series of promises made by God to the patriarchs, in which the temporal blessings alone are referred to. (2), It is *not true* (at least according to our interpretation), that Jacob has altogether passed over the blessing which was to flow through his descendants to all the nations of the earth. It is expressed in ver. 10, "and to him shall the willing obedience (the cheerful submission) of the people be." No doubt the reference made by *Jacob*, when blessing his sons, to benefits of a spiritual kind, is less distinct than in Gen. xii. 3, xxvi. 4, xxviii. 14, where *Jehovah* himself bestows and describes the blessing. But this is equally applicable to Gen. xxvii. 29, where Isaac bestows the blessing upon Jacob. The relation between these striking variations in the patriarchal blessing has already been examined and put in the proper light (*vid.* Vol.i. § 72. 4, and my *Einheit der Genesis*, p. 94, 95). We see here the difference between the *objective* proclamation of the blessing on the part of God, and the *subjective* apprehension of that blessing on the part of the patriarchs. On this point I need not repeat what I have already written.

Hengstenberg continues (p. 67), "Is it not probable that, as formerly from among the sons of Abraham and Isaac, so now from among the sons of Jacob, *he* should be pointed out who should become the depository of this promise, which was acquiring more and more of a definite shape?" We reply (1), It is *not true* that this blessing had acquired *more and more of a definite shape* from the time of Abraham's call to that of Jacob's death. On the contrary, the whole of the descriptions and repetitions referred to above, which extended over the entire patriarchal age, did not open it a hair's-breadth wider, and nowhere, I say nowhere, did it receive a more definite shape till Gen. xlix. This is a fact of great significance, that the blessing, however often it was repeated, was not extended or more clearly defined during the whole of the patriarchal age. And for that reason we have at least *no a priori ground* for expecting, that under Jacob, who stood upon the same footing, under the same influences, with the same hopes, this blessing would make such enormous progress in the attainment of a more definite shape. (2), It shows an utter want of insight into the nature of the progress observable in the patriarchal age, when *Hengstenberg*, in so unreserved a manner, desires and expects, that because a distinction had been made between Isaac and Ishmael, and between Jacob and Esau, the blessing being transmitted to the one to the exclusion of the other, therefore the same distinction should be made by the blessing of Jacob among his twelve sons. Did Judah, then, stand in exactly the same relation to his eleven brethren as Isaac to Ishmael, or Jacob to Esau? Did the selection of Judah from the twelve amount to a *rejection* of the rest, a severance from the tree of the history of salvation? (3), I have maintained that there is some progress apparent in Jacob's blessing, viz., in the elevation of Judah *above* his brethren, but I cannot possibly class this elevation with the distinction made between Isaac and Ishmael, or between Jacob and Esau.

Again, at p. 68, we read: "If we do not admit the reference in this passage to the Messiah, then a very large department of the future, which was notoriously accessible to Jacob, is left untouched by his announcement."—This sentence is left without any proof. But an *ipse dixit* is not admissible in the field of science. Let *Hengstenberg* demonstrate to us, therefore, that the expectation of a personal Messiah was a "department of the future which was notoriously accessible to Jacob!"—Till then, I shall very properly continue to doubt it. Still, the Spirit of God, by whose inspiration Jacob prophesied, was not necessarily restricted to that department of the future which was notoriously accessible to Jacob; and therefore the Spirit of God may have opened up to him for the first time a department of the future which had not been accessible before. Let us assume, then, for the moment, that *Hengstenberg* has given a correct interpretation of Gen. xlix. 10. In that case the expectation of a personal Messiah would be set forth in this passage in a manner so clear and intelligible, so definite and free from ambiguity, that the anticipation of a personal Messiah must henceforth have pointed out a department of the future notoriously accessible to every Israelite, and therefore most certainly to *Moses.* It is an indisputable fact, however, that in his blessing on the twelve tribes, which is completely parallel and analogous to Jacob's blessing on his sons, Moses does not make the slightest reference to a personal Messiah. Hence, if *Hengstenberg's* exegesis of Gen. xlix. 10 be the correct one, there is an entire department of the future which was accessible to Moses, and yet which is not in any way referred to in his announcement. It is evident, therefore, that either *Hengstenberg's* mode of arguing is inadmissible, or his assertion that, after Jacob's prophecy, the expectation of a personal Messiah was a department of the future notoriously accessible to every Israelite, is incorrect.

"If," he proceeds (p. 68), "the reference of the passage to a personal Messiah be explained away, we should certainly be at a loss to discover, where the fundamental prophecy of the Messiah can possibly be found. We should then, in the first place, be thrown upon the Messianic Psalms—especially Ps. ii. and cx. But as it is the office of prophecy alone to make known to the congregation truths absolutely new, it would subvert the whole relation of Psalm-poetry to prophecy if, in these Psalms, we were to seek for the origin of the expectations of a personal Messiah. They are unintelligible unless we recognise in Shiloh the first name of the Messiah."—Is this proof? Is there any one holding our views, who would think of appealing to Ps. ii. and cx. as the primary prophecy, the source and starting point of the expectation of a personal Messiah? Have we not 2 Sam. vii.? And why should not this be regarded as the primary prophecy on which Ps. ii. and cx. are based?

Lastly, on p. 70 he says: "But the historical point of connexion for the announcement of a personal Messiah, which here at once, like a flash of lightning, illuminates the darkness, is by no means so completely wanting as is commonly asserted. . . . . . All the blessings of salvation, which the congregation possessed at the time when Jacob's blessing was uttered, had come to them through single individuals. . . . Why should not Abraham be as fit a type of the Messiah as Moses, Joshua, and David? . . . . Or why not Joseph, who, according to Gen. xlvii. 2, 'nourished his father and his brethren, and all his father's household,' and whom the grateful Egyptians called 'the Saviour of the world.'"—This is evidently the most plausible, or rather the only plausible argument which *Hengstenberg* has employed in opposition to my interpretation. And yet it is mere plausibility, which vanishes as soon as any one takes the trouble to examine my arguments more closely. I have said, for example, that in Jacob's time the Messianic expectation was still bound up with *the* promise and expectation, that the unity of the family would be expanded into the plurality of a nation. The entrance of salvation could not be regarded as dependent upon the selection and singling out of any individual. On the contrary, from the nature of their previous historical experience, this could only be regarded as deferring the end desired. For whilst, on the one hand, the multiplication of the family into a great nation, and the possession of a land of their own, had been made prominent in all the promises, as the first and for the present the only conditions of the entrance of salvation, on the other hand, when any had hitherto been singled out, it had always involved the exclusion of others from the chosen community and the necessity for a fresh commencement. It was not till the unity of the family had been expanded into the plurality of the nation, and it had been historically demonstrated that it was not only advantageous but necessary, that this plurality should be recondensed into the unity of *one* helping, saving, and governing individual, that the true foundation was laid, on which the expectation of a personal Messiah could be based.

(13). On p. 76 sqq. *Hengstenberg* traces the blessing on Judah through the entire history of Israel, for the purpose of showing that this prophecy was made prominent in every period of the Old Testament, and particularly that the Shiloh passage was understood by the biblical writers and prophets in the sȧme way in which he has interpreted it. But we have still only arguments in which confident assertions are used as substitutes for proof. Thus in p. 83 he says: "*There cannot be a doubt* that David gave his son the name Solomon, because he hoped that he would be a type of the Shiloh" predicted by Jacob. We cannot be required to examine these arguments one by one, and treat them as they deserve. I will merely notice two points more. On p. 79 *Hengstenberg* mentions the blessing of Moses. He very properly maintains that this is connected with the blessing of Jacob, and that it carries it forward. How then, we ask, are we to explain the fact that Moses' blessing on Judah does not contain the slightest trace of the expectation of a personal Messiah, if that of Jacob had already announced this expectation in so clear and unmistakeable a manner, and had placed it on so firm and indestructible a foundation? My answer to this question may be found in Vol. i. § 98. 2. But what is *Hengstenberg's* reply from his standpoint? The most charitable supposition, which I gladly adopt, is that he makes no reply. For if the answer is to be found in p. 79, where he says, "even the remarkable brevity of this utterance (Moses' blessing on Judah) points back to the blessing of Jacob; and with this brevity the length of the blessing upon Levi, of whom too little had been said by Jacob, corresponds,"—I must say that I have seldom met with anything more flimsy. For why is the blessing on Joseph so long in both instances, if length and brevity alternated in the two blessings? —In conclusion, I will again refer to Ezek. xxi. 32. It is time that the words עַד־בֹּא אֲשֶׁר לוֹ הַמִּשְׁפָּט should cease to be taken as the rule by which to render and explain the word *Shiloh*, in

Gen. xlix. 10, especially after the theory, that שִׁילֹה is but another form of שֶׁלֹּה = שֶׁלּוֹ = אֲשֶׁר לוֹ, has been most properly given up as utterly fallacious. Moreover, we should altogether abstain from attributing to the prophet Ezekiel such a play upon words, as *Hengstenberg* imputes to him when he says (p. 86): "the words אֲשֶׁר לוֹ הַמִּשְׁפָּט, which Ezekiel puts in the place of Shiloh, on the ground of Ps. lxxii., allude to the letters of the latter word which form the initials (?) of the words in Ezekiel. That שׁ is the main letter in אשר is shown by the common abbreviation of it into שֶׁ, and that the י in שילה is unessential, is proved by the circumstance, that the name of the place is often written שלה." If the passage in Ezekiel bore any conscious reference to Gen. xlix. 10, and this I no longer dispute, it is not to be regarded as an explanation or confirmation of it, but simply as a free allusion to the passage, which the prophet has enriched with the fulness of his own more expanded views in relation to the coming Messiah.

DEATH OF JACOB AND JOSEPH.

§ 4. (Gen. xlix. 28—l. 26).—When the patriarch had thus looked forward with prophetic eye; had seen his descendants in possession of the land of his pilgrimage; and had announced in prophetic words the vision he had seen: he concluded by uttering with renewed earnestness the last wish of his life, that he might be buried there, in the land of his reminiscences and hopes, and in the *family grave* of his fathers. The execution of this wish, of which Joseph had already given him an assurance on oath, he now pressed most urgently upon all his sons. His account with life was closed, and he died at the age of 147 years. (1). Joseph had the body *embalmed* by his *physicians* in the Egyptian mode, and after the usual period of mourning, obtained *Pharaoh's permission*, and went with all his brethren and their households to convey the corpse to its place of destination.

The Israelites were accompanied *to the borders of the promised land*, by a solemn and numerously attended *funeral procession* of Egyptian courtiers and officers of state. There they remained for seven days mourning together; after which the Egyptians *departed*, and left the members of the family to bury the corpse in the cave of Machpelah. (2). The guilty conscience of Joseph's brethren now began to trouble them again, and they became uneasy, lest Joseph should perhaps have only deferred their well-merited punishment till their aged father's death. But the noble-minded deliverer and protector of his family anticipated their fears, and dispelled them with words of comfort: "Ye thought evil against me; but God meant it unto good to bring to pass, as it is this day, to save much people alive." Joseph lived sufficiently long to witness the commencement of the fulfilment of his father's blessing, for he saw his *grandchildren* and *great-grandchildren*; and as his end approached, looking with faith at the promises of the future, he took an oath of the children of Israel, that whenever these promises were fulfilled, they would carry his bones with them to the promised land. He died at the age of 110 years. His body was embalmed and placed in a mummy-case for preservation. (3).

(1). On chap. xlix. 33: "And when Jacob had made an end of commanding his sons, he gathered up his feet into the bed, and yielded up the ghost," *Calvin* correctly observes: non est supervacua locutio, nempe qua exprimere vult Moses placidam sancti viri mortem, ac si dixisset, sanctum senem tranquillo animi statu membra direxisse quo volebat, qualiter sani et vegeti se ad somnum componere solent;" and *M. Baumgarten* adds: "Jacob is the only one of the Old Testament patriarchs, whom we are able to accompany to his very last hour. And here we see how the Old Testament death-bed was surrounded by brightness and peace, the fear of death being swallowed up in the certain hope of the rest that remaineth for the people of God."—On the family-vault and the interest attaching to it in the minds of the patriarchs, see Vol. i. § 66.

(2). For the *Egyptian customs* referred to here, consult *Hengstenberg's* Egypt and the Books of Moses, p. 66 sqq. (translation). The fact that Joseph is said to have possessed a large number of *physicians*, may be explained from *Herodotus* (ii. 84), where we read that there were special physicians in Egypt for every disease. On the different modes in which the *mummies* were prepared, see *Herodotus* (ii. 86—88) and *Diodorus* (i. 91). Compare also *Friedreich* on the Bible (ii. 199 sqq.). The difference between the account given here, that Joseph's physicians embalmed his father, and the statement of *Diodorus* (l. c.) to the effect that there was a regularly organised, hereditary guild appointed for that purpose, and that the different departments were assigned to different individuals, may easily be explained, if we take into consideration the different periods to which the two accounts refer. *Hengstenberg* is certainly correct in saying (p. 67) that "it is quite natural to suppose, that in the most ancient times this operation was performed by those to whom any one entrusted it; but that afterwards, when the embalming was executed more according to the rules of art, a distinct class of operators gradually arose." There is a striking coincidence between the statement made here, that the whole period of mourning, evidently including the forty days of embalming, extended to seventy days, and the account given by *Diodorus* (i. 72, 91). *Hengstenberg* (p. 68) has shown that there is no discrepancy between *Herodotus*, ii. 85, and *Diodorus*, i. 72, 91. The extravagances of the funeral rites of the Egyptians are depicted in both these passages, and their monuments show the intensity and solemnity of their lamentations (*vid. Wilkinson*, i. 256). Joseph appeals to the courtiers to intercede for him, and obtain *Pharaoh's permission* to bury the corpse in Canaan. The reason why Joseph did not lay his own request before the king, has been correctly explained by *Hengstenberg* (ut supra) on the ground that, according to Egyptian customs, Joseph allowed his hair and beard to grow during the term of mourning (*Herod.* ii. 36), and that no one was permitted to enter the presence of the king in this unseemly condition (Gen. xli. 14). Moreover, the request had reference to Joseph himself, for as a matter of course, the minister of a well organised state could not leave the country without the knowledge and consent of the king. The rest of the brethren required no royal permission to bury the body in Canaan and accompany it thither. The fact that so numerous and influential a body of the Egyptians, viz. the elders of the house of Pharaoh (*i.e.* the officers of the court), and the elders of the land of Egypt (the state officials), accompanied the procession, most likely with an armed guard, shows how highly Joseph was esteemed and beloved by both the court and the king. "The custom of *funeral-processions*," says *Rossellini*, ii.3, p. 395. "existed in every province of Egypt and in every age of its history. We have seen representations of them in the oldest graves of Elethyas; there are similar ones in those of Saqqarah and Gizzeh, and others also exactly like them in the tombs of Thebes, which belong to the 18th, 19th, and 20th dynasties." To this, *Hengstenberg* adds, "When we look at the representations of processions for the dead upon the monuments, we can fancy we see the funeral train of Jacob (*vid. Taylor*, p. 182)."—As *the threshing-floor Atad* (גֹּרֶן הָאָטָד : the buck-thorn threshing-floor), at which the Egyptians turned back after seven days' mourning, is on the other side, *i.e.* the east, of the Jordan, the procession did not take the nearest road, by Gaza and through the territory of the Philistines, but went by a long circuitous route round the Dead Sea, and so crossed the Jordan and entered Canaan on the eastern side. The reason of this may be attributable to political circumstances, with which we are unacquainted. So large a procession, attended by an armed guard, would probably have met with difficulties from the contentious Philistines. It is a remarkable coincidence, however, that Jacob's corpse should have taken, or have been compelled to take, the same road, which his descendants were afterwards obliged to follow in their journey to the promised land. We should not be surprised to find some critic detecting in this an unmistakeable proof, that the road, by which the legend states that the body of Jacob was carried, was first taken from the journey of the Israelites. For our part, however, we do not hesitate to express our opinion most freely, that we discover in this similarity of route one of those events, unintentional and therefore apparently accidental, that abound in history in general, but particularly in sacred history, and from the stand-point of the observer are proofs of the *prophetic character* with which the biblical history is always secretly pervaded. *Tuch* (p. 593), with his usual delight at the discovery and imputation of crudities, says that the Egyptian escort is described in the Saga as stopping short before reaching the Jordan, because "the foreign attendants could not be allowed to tread the holy promised land;" and so important does he consider the discovery, that he has had the words printed in italics. But where do we find, in any part of the Old Testament, the least trace of so harsh and trivial an idea? And how particularly crude and absurd would such a notion have been, at a time when the "holy promised land" was entirely in the possession and occupation of foreigners. But *Tuch* himself assigns the true and perfectly satisfactory reason for the departure of the Egyptians, when he says: "the actual interment of the corpse was a matter for the family alone." This sufficiently explains, why the Egyptians only accompanied them to the frontier of Canaan. Had so numerous an escort gone further, it might have excited political disturbances in Canaan. From the very nature of the case, too, an escort only goes, as a rule, to the line which separates their own from a foreign land. But in this instance the procession had hitherto passed only through a desert, in which there were none but nomad-hordes, and therefore the boundary of Canaan, at which the escort stopped, might be regarded in a certain sense as the boundary of Egypt, especially when we consider, that it was their intention to pay the greatest honour to the funeral procession, by going as far as they possibly could. No one will consider it an improbable thing, that the place where the Egyptians encamped, by the *floor of Atad*, may have received the name "meadow of the Egyptians," אָבֵל מִצְרַיִם from the fact that this splendid procession sojourned there for seven days; and it will hardly be regarded as a crime, either against the grammar or the lexicon, that the author should have laid stress upon the paronomasia between this name and אֵבֶל מִצְרַיִם "the mourning of the Egyptians." We have no means of determining the site of the threshingfloor of Atad with exactness. *Jerome* identifies it with Beth-Hogla, two miles from the Jordan on the road to Jericho, *i.e.* to the *west* of the Jordan (*vid.* Onomast. art. Area Atad), but this is at variance with the evident meaning of the text.

(3). In v. 23 we read that the children of Machir, the son of Manasseh, were born *on Joseph's lap*. From chap. xxx. 3 it is evident that this can only mean, that they were adopted by him; and as that would not lay the foundation of a new tribe, the tribes of Israel having been fixed once for all, it could only involve the transfer of Joseph's special rights and property to these children of Machir, *vid.* § 2. The body of Joseph was placed in a *wooden sarcophagus.* The Egyptian coffins were generally constructed of sycamore wood, and were made to resemble the human body. (See *Herodotus* ii. 86). *M. Baumgarten* has most truly observed: "the last instructions, which Joseph gave to his brethren, and made them swear that they would fulfil, are peculiarly important. Joseph remained an Egyptian to the day of his death, and was, therefore, separated from his brethren. If, then, before his death, he expressed his certain hope that they would one day return to Canaan, and his wish to be associated with that return, his former separation must have given the greater force to such a desire. From that time forward the coffin with Joseph's remains became an eloquent witness of the fact that Israel was only a temporary sojourner in the land of Egypt, and continued to turn its face towards Canaan, the promised land."

The intercourse between Joseph and his brethren terminated with their anxiety on account of the injury, which they were conscious of having inflicted upon him, and with Joseph's declaration of his forgiving love, by which he removed all doubt as to the unalterable nature of the reconciliation that had taken place, and the perpetuity of his affection for them. Henceforth the brethren were able to give themselves up to the full enjoyment of the rich provision he had made for them, without any lingering fear lest they might one day be punished for their fault, by one whom they had so deeply injured, in fact without a thought that such a thing was any longer possible. The touching history of Joseph is now lying in all its completeness before us, and we have therefore a fitting opportunity for surveying it as a whole.

All the teachers of the Christian Church, who regard the Old Testament history as the result of God's special and supernatural direction, have recognised in Joseph a distinct type of Christ (*e.g. Sack*, Apologetik 2. A. p. 340 seq.). "In the person of Joseph," says *Luther*, "God foreshadowed both Christ and his entire kingdom in the most brilliant manner in a bodily form. He received his name on account of his perpetually growing and increasing, heaping up and accumulating, for *Joseph* means *one who adds.* And the crowning point of the figure is this: as Joseph was treated by his brethren, so was Christ treated by his brethren, *i.e.*, by the Jews." Following this rule, there are some who have discovered the most striking agreement between Joseph's call and the events of his life on the one hand, and those of Christ on the other, even in the most trifling, and apparently the most accidental circumstances (*vid. e.g. Vitringa* observv. ss. l. vi. c. 21; *Heim*, Bibelstunden i. 540 sqq. and others without number). There is, in our opinion, just ground for regarding Joseph as a type. But in this, as in other instances, the true historical relation between the type and the antitype has been reversed. The proper method would have been, first of all, to determine *the fact*, that the position, the calling, and the task of Joseph bore the same relation to the lower stage of development, at which the kingdom of God had then arrived, as was borne by those of Christ to the fulness of time, or the time of fulness, and also to decide *how*, *why*, and *to what extent* such a resemblance existed. When this had been done, then would have been the time to show that the resemblance, which can be traced between the events and results of their lives, was necessary and essential; whereas otherwise it could only be regarded as accidental, and therefore unimportant, or else as purely imaginary. And in this way it would be shown, that the dissimilarities, which would otherwise appear sufficient to outweigh and destroy the resemblance, were equally necessary and essential. Instead of this, expositors have contented themselves with a merely external comparison of particular phenomena, and thus have lost themselves in strange and arbitrary conjectures, and grasped a baseless and visionary result.

There are two things to be considered in the history of Joseph, his relation to heathenism, and his relation to his own people. He brought salvation to the heathen, and to his brethren also. We have already shown, in § 1 and 2, both *how* and *why* Joseph's peculiar position as the deliverer of Egypt, the representative of the whole heathen world, was in itself a prophetic event; an event, which was the result of the deepest impulses at work in his history, and which, although merely transient and imperfect, on account of the imperfection of the age of Joseph himself, and of the circumstances, was for that very reason prophetic. But the salvation, which was to proceed from the house of Israel, was not merely salvation *for the Gentiles*, but first of all salvation *for the house of Israel* itself. And in this respect also, the moving principle of the history of Israel was typically exhibited in the person and life of Joseph. The reason and the cause of this prototypical manifestation of Israel's vocation, precisely at that time, and in the person of Joseph, are one and the same. We have already explained, that the patriarchal epoch formed the first complete and definite stage of the kingdom of God in Israel; and that this stage bore the same relation to the whole of the Old Testament history, as the smaller of two concentric circles bears to the larger. The common centre will generate in both *the same forms;* but in the smaller circumference these forms are on a smaller and less perfect scale, in the larger they reach their fullest development. So we do find in Joseph the noblest blossom of the patriarchal life, the embodiment of all the true worth that it possessed; but in Christ we see the perfect blossom, the entire fulness of the whole of the Old Testament dispensation.

The opposition which Christ and Joseph both met with from their own people, the hatred, contempt, and persecution, to which both were exposed, on the part of those to whom they were bringing salvation, were not accidental. They sprang from the same soil, and were the fruit of the same perverse and hostile disposition, the same evils, which are so exuberant in the whole of the Old Testament history, but which appeared in a concentrated and more fully developed form, just at those epochs in which salvation itself was manifested in a similar way. The soil, from which they sprang, was the perversity and selfishness of human nature; and these had to be overcome by the devotion and self-sacrifice, in which alone salvation comes to view. In other words, it was that natural enmity of the heart, which consciously or unconsciously resists the ways of grace, but which has to be subdued by the power of the love that comes to meet it. This selfishness and enmity were manifest, not only in the rude and profane minds of an Ishmael and an Esau, whose hearts were hardened into perfect insensibility, and in whose case they were not subdued by the grace of God; but also in the expressions of self-will, of weak faith or of unbelief, to which Abraham, Isaac, and Jacob, gave utterance; though in their case, after a conflict, of less or greater violence, between grace and nature, submissive faith and resisting unbelief, they were entirely subdued. They showed themselves more decidedly in Jacob's sons, since with them the selfishness of nature was no longer under the immediate and express control of God, but had to submit to one who was himself a recipient, as well as a mediator of the divine mercy, one who was naturally their equal, but, according to the hidden and marvellous wisdom of God, was destined to be their deliverer and redeemer. Yet even in this instance the power of forgiving love, displayed by Joseph, triumphed over the obstinacy of selfishness in the hearts of his brethren.

This then being the leading principle, on which the course of salvation in the kingdom of God depends, that its victory over the evils existing in human nature shall be gained by godlike love, submission, and self-sacrifice, it is a fundamental law of the whole of the sacred history, till its ultimate completion, that the way of salvation leads through abasement to exaltation, through serving to ruling, through sacrifice to possession, through suffering to glory. And this fundamental law, of which the highest and most perfect manifestation is seen in the life of the Redeemer, was first displayed in a definite and concrete form in the life of Joseph.

The typical character of the life of Joseph, then, consists in this, that he, the first temporary deliverer of Israel, who brought the first stage of its history to a close, like the perfect Saviour of Israel, in whom its entire history terminated, was slighted, despised, persecuted, and betrayed by "his own;" that, like Him, he passed through abasement, service, and suffering, to exaltation and glory, and also that, like him, he succeeded at length in softening their hardened hearts by the fulness of his forgiving love, and in raising his own to the enjoyment of the benefits which he had secured for them. If, in addition to this, there is often a striking resemblance between particular incidents and the accidental circumstances, we cannot lay any very great stress upon this, though we regard it as a mark of that prophetic spirit, by which the history was directed and controlled.

GENERAL SURVEY OF THE PATRIARCHAL AGE.

REVELATION, RELIGION, AND GENERAL CULTURE IN THE TIME OF THE PATRIARCHS.

§ 5. We have already seen (Vol. i. § 12. 13), that in order to determine to what extent the consciousness of God was developed under the Old Testament economy, it is essentially necessary to make a twofold distinction in the process of divine revelation; that is to say, it is necessary to distinguish the preservation and government of the world in general, from the more special operations connected with the introduction and working out of the plan of salvation. We have also seen that this distinction was exhibited to the religious consciousness of the chosen people, in the two names by which God was known, *Elohim* and *Jehovah*. The only questions remaining for discussion at present are, whether there was any distinct apprehension in the patriarchal age, of the difference between these two manifestations of God? and if so, whether it was expressed by the two different names of God at that early age? Some have thought that a negative answer to these questions is rendered necessary by Ex. vi. 3; but this is not the case. For, on the one hand, the explanation of the passage on which this answer is founded is an erroneous one (1), and on the other, whatever opinion may be entertained respecting the composition of the book of Genesis (Vol. i. § 20. 2), such a reply is decidedly at variance with the contents of that book (2).

(1). On the ground of Ex. vi. 3 (where *Elohim* says to Moses: "I am Jehovah, and I appeared unto Abraham, unto Isaac, and unto Jacob אֵל שַׁדַּי, but by my name יְהוָה was I not known to them"), it has been very confidently maintained by modern critics, that the name יְהוָה was not in existence before the time of Moses, but was first introduced by him in connexion with his peculiar instruction respecting the nature of God. But in my work *Einheit der Genesis* (p. xxii.—xxxii.) I have proved at length that this is an erroneous explanation of the passage. We shall therefore content ourselves with giving the correct explanation here, and for a fuller discussion of the question refer to the work just named, *vid.* also *Keil*, in the *Luth. Zeitschrift* 1851, i. p. 225 seq.; *Hofmann, Schriftbeweis*, i. 82 seq.; *Delitzsch, Auslegung der Genesis*, p. 26).

For a correct understanding of these words of God, it is indispensable that we should first determine whether the word נוֹדַעְתִּי, "I was known," is to be regarded as emphatic or not. The whole tenor and connexion of the passage, its peculiar mode of construction and expression, the remarkable importance of its contents, and the great solemnity with which the words were uttered, compel us to take the word as emphatic, and to seek the meaning of the solemn address of God in this word alone; and doing so, it is necessary to take into account all the depth and fulness of meaning of which the verb ידע is capable. Now it is well known, how deep and comprehensive a meaning this verb is capable of, and, where it is used emphatically, must necessarily have. In such a case it denotes a thorough insight into, and grasp of any object, even in its inmost essence. *Perception* in its primary and peculiar sense is by no means merely a superficial knowledge, which only touches the shell, and is content with the external and accidental appearance of an object; on the contrary it is the reception of an object into one's own spiritual life as the result of actual personal experience. It presupposes a close and intimate communion between the subject and the object, the perceiving mind and the object of perception. Hence it appears to us to be by no means a forced explanation, but a very natural one, and one which suits the words as well as the circumstances, and does full justice to the history contained in the book of Genesis, as well as to the expression itself, if we suppose the meaning intended to be conveyed to be this: that the Israelites were to be made fully conscious, that they would immediately receive such a glorious manifestation of the operations of God, as even their celebrated ancestors had not been permitted to see. The latter had never witnessed, known, or experienced the whole extent of the fulness and glory of the divine operations, expressed by the name *Jehovah*, but these were now shortly to be displayed. *El-Shaddai* is the Almighty God, who, by his creative omnipotence, prepared the natural conditions and vital agencies required for the development of salvation, and hence the word sets forth one view of the Elohistic existence of God, on which it was necessary that peculiar stress should be laid (see *Einheit der Genesis*, p. 124). Jehovah, on the other hand, is the God engaged in the development of salvation, who enters into it himself, manifests himself in it and with it, and therefore conducts it with absolute certainty to the desired result. *Jehovah* had already ruled and worked in the history of the patriarchs. Their history commenced with *Jehovah*. It was by *Jehovah* that Abraham was *chosen* and *called;* and He *appointed* him to be the father of the chosen people, the channel of blessings to the nations. But to accomplish this result, Jehovah had to become Elohim, El-Shaddai, that as creator he might produce the *promised* seed from an unfruitful body, and make of it a numerous *people*. And therefore *that which was actually accomplished in the patriarchal age, that which the patriarchs (not merely hoped for and believed, but) saw and experienced as a fact fulfilled, was the work, not of Jehovah, but of* El-Shaddai. All that *Jehovah* had performed, in connexion with the patriarchal history, *was limited* to *the election and call of individuals*, to *the communication of directions and promises, and the fostering of faith in the directions and promises given.* Hitherto, there had been no embodiment in *fact;* there had been merely the introduction of an idea, which was to be realized and embodied for the first time at Sinai. *Hence the patriarchs could only grasp the operations of Jehovah in faith and hope;* they could not *see* them; they did not *feel and know them as something actually accomplished and fulfilled.* This was reserved for their descendants, to whom Moses was sent with the message that it was now about to happen. This then, and this alone, is the meaning of the words of God: "They have *known* me, my nature, and my operations, as El-Shaddai, but not as Jehovah; *you*, however, shall soon know me as Jehovah also."

(2). It is a fact that the name *Jehovah* occurs all through the book of Genesis, quite as frequently as the name Elohim, not only in the objective narration of the author, but also in the mouth of God and of the patriarchs. Various suggestions have been made, for reconciling this fact with the words of God in Ex. vi. 3. *De Wette*, *Tuch*, *Stähelin*, *Lengerke*, and many others suppose the meaning of these words to be, that the name Jehovah was not in existence before the time of Moses; and on this supposition they deny the unity of Genesis, and assume that such passages of that book, as do not contain the name Jehovah, form together a complete work (the so-called groundwork), whose author intentionally and consistently avoided using that name in consequence of the statement made in Ex. vi. 3. A subsequent interpolator or finisher extended this groundwork, and, overlooking the statement contained in that passage, either used the two names promiscuously in his additions, or with special reference to their different significations. On the other hand *Hävernick*, *Hengstenberg*, *Drechsler*, *Keil*, and many others, oppose this interpretation of the verse in Exodus, and defend the unity of the book of Genesis. The interchange of the names of God in that book, they explain entirely on the ground of the different notions conveyed by the two names. *Ebrard (das Alter des Jehovahnamens: hist. theol. Zeitschrift v. Niedner*, 1849. iv.), and *Delitzsch*, in his exposition of Genesis, endeavour to find a *via media* between the two, but seek it in opposite directions. For whilst *Ebrard* adopts *Tuch's* explanation of Ex. vi. 3, and yet wishes to maintain the unity of Genesis, *Delitzsch* gives up the unity of the book of Genesis, but yet adopts *Hengstenberg's* explanation of the passage in Exodus (Vol. i. § 20. 2).

We have already given our opinion as to the meaning of Ex. vi. 3; and all that we have still to do, is to say whether we give in our adhesion to the views of *Hengstenberg* or of *Delitzsch*. But this question has little connexion with our present topic, and, therefore, we shall defer the discussion of it to a more fitting occasion (see, in the meantime, Vol. i. § 20. 2). The only point of importance here is whether the name Jehovah, and the consciousness of the difference in the manifestations of God which that name expresses, were in existence at so early a period as the patriarchal age. If we admit the unity of the book of Genesis, this question must of course be answered in the affirmative. But we are convinced, and that we have now to prove, that it can and must be answered in the affirmative, even *if the correctness of the supplementary hypothesis be assumed.* As proofs of this we mention the following: (1). In Ex. vi. 3, it is not expressly said that the name Jehovah was *unknown* before the time of Moses, but simply that in the patriarchal age God had not revealed the fulness and depths of his nature, to which that name particularly referred. The author of the ground-work, however, from the peculiar nature of his legal and priestly standpoint, was chiefly desirous of making it as clear as possible to his readers, and of keeping the fact constantly before their minds, that the Sinaitic covenant and legislation had introduced into the sacred history a fresh and incomparably superior element of divine revelation, and that this element alone expressed all that was included in the name Jehovah. For this reason he purposely avoided the use of that name, in connexion with the earlier history. But he had no intention of saying, that the name Jehovah was entirely unknown in the patriarchal age; for (apart from other reasons), we have an absolute proof of this in the fact that in Jacob's blessing, which indisputably belongs to him, he puts that name into the mouth of the patriarch (ver. 18). And this he could very well do, without at all departing from his original purpose, since Jacob was carried by the spirit of prophecy into the heart of the Jehovistic times. If, then, this blessing was actually pronounced by the patriarch, and handed down by tradition in the form in which the author has recorded it, as we think we have unanswerably demonstrated (§ 3. 3), the evidence afforded by the occurrence of this name is all the more important.—(2). The supposed finisher of the work cannot have intended, that Ex. vi. 3 should be understood in the way in which *Tuch* and the rest explain it; for in that case he would have placed himself in conscious and evident opposition to the ground-work, which it was his design to extend. We cannot imagine this a possible thing, especially when we consider, that a slight alteration of the expression contained in the ground-work would have been sufficient to remove the discrepancy, which is supposed to be so apparent; and if the critics are correct, he has frequently made such alterations when there was far less to be gained.—(3). If it be undeniable, that the later author represented the name Jehovah, as already known and current in the patriarchal age, his historical representation is in our estimation authoritative, for we regard him as a writer who was filled and directed by the Spirit of God, just as thoroughly as the author of the ground-work.—(4). It is *a priori* both a natural and probable supposition that the name Jehovah was in existence in the patriarchal age. For if the patriarchs were conscious of the special call, which they had received, of the peculiarity of their position, and of the extraordinary relation in which God stood to them (and even the ground-work teaches as much as this), there must have been some definite terms, which expressed this consciousness, especially when we consider that it was the source and guiding star of the whole course of their lives.

§ 6. *Miracle* and *prophecy* are the two indispensable accompaniments, vehicles, and messengers of revelation (see Vol. i. § 4). In each there is a manifestation to man of the fulness of the godhead; in the former of the power of God, in the latter of his wisdom. And through each the divine fulness enters into a covenant association with the history of humanity, co-operates in its development, and ensures its safe arrival at its destined end. That end is the incarnation of God and the consequent entrance of the whole fulness of the divine essence, in a living and personal form, into an intimate and abiding union with man. We have already shown in Vol. i. § 50, how the first advances towards this end were manifested in elementary forms, as it were; how, for example, there was as yet no miraculous power given to man, whilst the gift of prophecy was but seldom possessed, and that only in particular, culminating points of history (1).—The *substance* of patriarchal revelation, and its *results* in patriarchal history, have already appeared, as we followed the course of that history in the former parts of this work. The sum of the whole is, that the *will* of God was revealed in the selection, the call, and the appointment of Abraham and his seed, to be the instruments through whom salvation should be introduced and completed; the *knowledge* of God in the announcement of this call to those who were intrusted with it; and lastly, the *power* of God in the creative production of the promised seed from an unfruitful body, in the separation of that seed from the natural branches, and in the protection and guidance of those who had been chosen.

(1). It is a striking fact, that *in the whole of the patriarchal history, and in the primeval history anterior to it, we do not meet with a single miracle performed by a man.* Not even by an Enoch, who had this testimony that he walked with God, nor an Abraham, with whom God talked as a friend with his friend; in fact, none of the fathers of the old world were workers of miracles. Where any miracles occur, they are performed solely and exclusively by God himself. We have in this fact a decisive argument against every mythical explanation of the patriarchal history, and a strong proof of the historical credibility of this portion of sacred history, as *Sack* has already shown (Apologetik Ed. ii. p. 174). With what a dense *nimbus* of miracles would any legendary tale have enveloped the heads of the celebrated founders of the race! They would assuredly have been made to surpass in this respect an Elijah and an Elisha, who were far less celebrated, and whose forms were not so obscured by the haze of a distant antiquity. The same may be said of the gift of prophecy, for, though not perhaps altogether wanting, there is an analogy in its infrequent and exceptional appearance. Abraham is, no doubt, called a prophet in Gen. xx. 7; but evidently in so general and indefinite a sense (Vol. i. § 63. 3), that we cannot for a moment think of that specific gift of prophecy, which we meet with at a later period as an essential co-efficient in the development of the nation's history. We do not find the least trace of a prophetic utterance on the part of Abraham. Isaac and Jacob both prophesy, as Shem had done before them in an exactly similar way, but each of them prophesies only once in his life, and in a manner perfectly unique. Prophecy does not appear in their case as a continuous endowment to all, and this is the main point of importance. It was not an office with which they were entrusted. In all three the paternal authority to bless and curse was the principal thing; the prophecy was a subordinate matter. The supernatural force of this paternal authority assimilated itself both to the authority of God, of which it was the symbol and the medium, and also to the foreknowledge of God; it brought them down to itself, as it were, on this particular occasion (see Vol. i. § 72. 1). If we take a comparative survey of the further course of the sacred history, we find that Moses was the first to work a miracle, and that from that time forward, there was a visible increase in the number of miracles performed by men, through several stages of the history: again they appear less frequently, and for a period cease altogether, till at length the miracle appears in its most absolute form in the incarnation of Christ. The gift of prophecy passes through essentially the same phases. On the other hand we find that visions of God, which are almost the only form of revelation in the patriarchal history, gradually decrease in the subsequent history, in proportion to the increase in the number of prophets and workers of miracles. In the visions of God the divine power and knowledge did not enter *into* human nature, but moved *by the side of*, and *in connexion with*, the agency of man. But in the gift of prophecy, and the power to work miracles, they entered *into* human nature and became subservient to it. In the impartation of these gifts to man, there was an advance towards the incarnation of God. This absence of miraculous powers and of the gift of prophecy in the patriarchal age, and the frequency with which God appeared, are therefore to be easily explained, as parts of God's regular plan for gradually revealing and communicating himself to the people of the covenant. On the other hand, it was no less conditioned by the regular and gradual development of the people of the covenant themselves, and especially by the fact that as yet the history of the patriarchs was a *family*-history, and they had not become a numerous and organised people. It was an essential element in the gifts of miracles and prophecy, that the performer of miracles did not work them primarily *for himself*, but *for others*, and that the prophet did not proclaim the message from God *for himself*, but *for those around*. Now Abraham, Isaac, and Jacob were the solitary recipients of the divine call; God was related to them as a friend to a friend, and all the blessing, the protection, and the light, which he had to impart to them, were necessarily imparted directly to themselves, since there was no third person in existence who could mediate between the two. It was very different when the seed of Abraham had become a numerous people. Individuals could then be raised up, endowed with divine power and wisdom, to be the channels of power and light from God to the rest of the people. In fact, *it was necessary* that such persons should rise up, to be the typical representatives of the perfect mediatorship of the God-man, to whom the whole history of the covenant pointed, and at the same time to prepare the way for his coming, so that when he appeared, it might be not as a *deus ex machina*, but as the ripe fruit, the complete and mature result of the entire history.

§ 7. The *religion* and *worship* of the patriarchs were modified and determined by the nature and extent of the revelation, which had been transmitted to them by their ancestors, or communicated directly to themselves. As the accounts of primeval times, which are preserved in the book of Genesis, must, if historically true, have been handed down by tradition, and as this tradition must have been restricted to the family of the patriarchs, we must necessarily assume that this family possessed an acquaintance with the religious views embodied in those accounts. Hence we must presuppose a knowledge on their part of the unity, the personality, and the holiness of God, the almighty Creator of the heavens and the earth, of the image of God, in which man was created, of the corruption into which he had fallen through sin, and of the hope of a future victory to be gained by humanity over the principle of evil. These views were now to receive a fresh vitality, to be deepened, expanded, and rendered more definite, by the revelations of which they were to be the personal recipients. The peculiar intimacy with God, which they enjoyed, the call they received, the promises given to them, and the guidance of God, which fitted them for their vocation, all confirmed and enlarged their knowledge of God and of salvation, and awakened the faith which was reckoned to them for righteousness, the obedience which cheerfully followed the leadings of God, and the hope, which grasped the promised salvation as something already possessed, and rested upon it amidst all the privations they had to endure. The truth and purity of the religious knowledge of the patriarchs are great and marvellous when contrasted with heathenism, which was so deeply sunk in mere nature-worship. But when looked at from an objective point of view, however thoroughly it was fitted to the progressive character of the sacred history, it appears faulty, imperfect, and one-sided; for it does not present a single religious notion, in a form sufficiently complete and definite to express fully the objective truth, and even heathenism often surpassed it in the greater richness and comprehensiveness of its religious views, although they were perverted to pantheism, and therefore issued in its own destruction (1). In its comparative poverty, yet absolute purity, the patriarchal worship resembled the patriarchal religion. It was always sufficient to meet the necessities of the moment, but it was destitute of any systematic and complete organisation; it had no established, binding rules, and was not attached to any particular persons, places, or times (2).

(1). The patriarchal *consciousness of God* did not comprehend the doctrine, which was the crowning point of its full developinent, viz., the Christian *doctrine of the Trinity;* whilst heathenism had prematurely grasped this truth. But for that very reason the conceptions of the latter were false and distorted, and in a pantheistic *Trimurti* the truth was so caricatured, as to preclude the possibility of any return or advance towards a true and purified form of belief. The doctrine of the Trinity could not be conceived of, comprehended, and preserved in its fulness and purity, until it had appeared as a fact in history, that is, until the Logos had become man in Christ, working out redemption as the incarnate God, and the Spirit had been poured out upon all flesh on the ground of this complete redemption. A premature revelation of this sublime mystery in the Divine nature (that is a revelation, for which no sufficient preparation had been made in history, and which had not assumed a concrete shape in these two facts, the incarnation and the outpouring of the Spirit) would have been all the more injurious, since the spirit of the age and of the world at that time tended towards a pantheistic and polytheistic perversion of the idea of God, and Israel was in danger of being drawn away by its attractions on account of the elective affinity of its natural inclinations. In contradistinction to this perversion, and as a safeguard against it, it was necessary that the idea of the unity of God should be ineradicably implanted in the consciousness of the people of the covenant, and that the basis should thus be laid for the manifestation and appropriation of His true tri-unity. But as these two facts, the incarnation of the Logos, and the outpouring of the Spirit, set forth the predetermined end, and the highest perfection of the covenant-history, and as the whole of that history from its very commencement was constantly urged forward towards this point by the vital principle, with which it was imbued, a corresponding intellectual culture must also have existed throughout the Old Testament, so as to pave the way for the announcement of this doctrine, and therefore the germs of the doctrine itself must have been deposited even in the patriarchal history. We have already pointed out in *Fr. Delitzsch's* words, how the two names of God, *Jehovah* and *Elohim*, contained the undeveloped and unconscious germs of the perfect doctrine of God (Vol. i. § 13. 1), and we have also shown that the appearance of God in the *Maleach Jehovah* (the angel of the Lord) was a typical precursor of his incarnation. Moreover, the description of the vivifying and fructifying action of the Spirit of God in creation, contained in Gen. i. 2, was adapted to prepare the way for the revelation of the triune nature of God. Yet we do not find in the patriarchal history the least indication of any development of this doctrine. The patriarchs had no definite conception of any hypostatic plurality in the God, who appeared in the *Maleach Jehovah*, and the recognition of the personality of the Spirit of God was still at so great a distance, that there is not the slightest reference to it in the patriarchal history.

There was something in the *exclusiveness* of the call given in the time of the patriarchs, in the fact, that is, that Jehovah was solely and exclusively the God of Israel, which must have rendered the views entertained, respecting the nature and operations of God, one-sided, rude, and contracted, though there was a wholesome counteracting influence in the universality of the promise. But this very rudeness and partiality were necessary and salutary, for they opposed a powerful barrier to the threatened amalgamation with heathenism. It was only out of a mature and self-sufficient exclusiveness, that true universality could be produced. The doctrine of salvation, also, had not yet advanced beyond the very earliest rudiments, as we may learn from the fact, that the idea of a personal incarnate Messiah, without which that doctrine could never become perfectly definite and clear, or be in any way richly developed, was not yet understood even in its first principles. It was just the same with the *doctrine of eternal life* as with that of the Trinity, and the revelation of the former in contradistinction to the false and distorted belief in immortality, which prevailed at that time in the heathen world, (*vid. Hengstenberg* Beitr. iii. 565 sqq.), was wisely delayed by the providence of God. In this case also we find, not error, but imperfection. The doctrine of *divine retribution* in general was not wanting, but it had not yet led to a knowledge of retribution *hereafter*. In the living consciousness of the retribution, which takes place in this life, the true basis was laid for the belief in retribution in the life to come. Still the old Israelitish notion of death was, that it was followed, not by annihilation or by the cessation of the individual life, but by a departure into *Sheol*. (שְׁאֹל is not a derivative of שָׁאַל, to ask, with the meaning, "the ever-craving, that which demands all life for itself," as *Hengstenberg*, on the Psalms, still maintains; but is to be regarded as derived from שָׁעַל = *cavum esse*, as *Gesenius*, *Fürst*, *Böttcher*, and others assume. On the etymology of the word, *Gesenius* says, *s.v.*: "The true etymology of the word seems to be, that *Sheol* signifies a hollow and subterraneous place; just as the German *Hölle*, hell, is originally the same with *Höhle*, a hollow. For the thorough discussion of this question see *Böttcher*, *de inferis rebusque post mortem futuris*, vol. i., Dresden 1845, p. 64—78, where the frequent softening of ע into א is clearly shown. The imperfection of this view consisted in the fact, that the hopes of the future could not pass with any clear consciousness beyond this *Sheol*, that *Sheol* itself was not what it is described in the New Testament as being, a *middle place* and an intermediate state (*vid.* Matt. xii. 40; Luke xvi. 22 sqq.; 1 Pet. iii. 19, iv. 6; Phil. ii. 10), from which the righteous would pass to the blessedness of everlasting life, but was regarded as a state in which the development of life would for ever *terminate*. There was another respect in which the early notion of *Sheol* was imperfect; but the imperfection in this case was conditioned and demanded by the objective imperfection of the actual reality. It was this, that *Sheol* was supposed to be a thoroughly gloomy place of abode, which had only negative advantages over this earthly life, inasmuch as it afforded to those who were oppressed by the pains and sorrows of life, or by the burden of a weary and decrepit old age, the rest they longed for, and oblivion of earthly care and toil (Gen. xxv. 8, xxxv. 29); whilst it was actually destitute of the rich blessings of our earthly existence, since it condemned to an inactive vegetation and the loss of all the pleasures of life (Ps. vi. 6, xxx. 10, xxxi. 18, lxxxviii. 13, xciv. 17, cxv. 17). The latter notion was the necessary effect, produced by the consciousness that death was the wages of sin (Gen. ii. 17, iii. 19), and therefore a sentence and a punishment, and the absence of any clear consciousness of redemption and of its influence upon our future state of existence. Yet there were certain provisions connected with the patriarchal age for the further development of these eschatological elements. There was a source of comfort in the fact that death was regarded as being gathered to their fathers (Gen. xlix. 33), and though the gloominess of the prophet was not entirely removed in consequence, it was certainly considerably diminished. Here was at least *one* element of positive happiness, connected with the life after death, which opened and prepared the way for the New Testament doctrine of a separation of the righteous from the wicked, and a happy meeting of the former with one another and with the Lord (Luke xvi. 22 sqq.; Phil. i. 23, &c.). There is a more distinct reference to an *everlasting* life, superior to the barren and gloomy shade-life of Sheol, and stretching beyond it, in the account of Enoch's translation to God, "in which it is of especial importance to remark, that his walk with God is intentionally and expressly placed in a causal connexion with his being taken by God. And this passage also bears an enigmatical character, tending to produce the impression, that the original revelation was meant to spread a veil of secrecy over this doctrine, the blessed influence of which presupposed conditions, that were not then in existence," *Hengstenberg*, Psalms vol. iii. p. lxxxvii. translation). In the work just quoted, *Hengstenberg* calls attention to another element of great importance in the development of the doctrine of eternal life, viz., the belief that death was not the natural and necessary concomitant of human existence, but the wages of sin. " With this view of death, faith in an everlasting life could not but break forth, as soon as the hope of redemption, and of the restoration of that which was lost in Adam, had taken root. As death entered into the world by sin, it could not but be removed by the redemption, which restored to man the happy state of paradise" (see Is. xi.)—On the Old Testament doctrine and its gradual expansion consult, particularly, *Hengstenberg*, Beitr. iii. p. 559—593 ; *Dess*, comment. on the Psalms iv. 2, p. 314—326 ; *H. A. Hahn*, de spe immortalit. sub. vet. test. gradatim exculta, Breslau 1846 ; *Oehler*, vet. test. sententia de rebus post mortem futuris, Stuttg. 1846 ; *Hävernick* Theol. d. A. T., p. 105 sqq. ; *Hofmann*, Schriftbeweis i. 500 sqq.

(2). On the *worship* of the pre-Mosaic times see *C. Iken's* two dissertations *de institutis et cærimoniis legis mosaicæ ante Mosem* (in his diss. theolog. vol. ii. 1770).—The fact that so many of the forms of worship, and of the manners and customs, which are mentioned in the pre-Mosaic age, re-appear in the legislation of Moses, has been regarded by modern criticism as so inexplicable a phenomenon, that it can only be accounted for on the ground that the author transferred the full-blown " Levitism" of his own age in a thoroughly unhistorical manner, into his (mythical) description of earlier times. But for our part, all that we find thoroughly unhistorical is this discovery of modern criticism itself ; for nothing appears to us more *natural*, than that the forms of worship, and the manners and customs which had already taken deep root among the people, should be adopted and sanctioned by the legislation of Moses, inasmuch as they were not at variance with the principles of that legislation, but, on the contrary, were completely adapted to its requirements. On the other hand, nothing appears to us more *unnatural*, than to suppose that there were no forms of worship in the pre-Mosaic times at all, or, if there were any, that they were entirely ignored by the Mosaic legislation. Besides, it must also be observed, that whilst there are many points in which the forms in question resemble each other, there are many others in which they diverge and differ. And this *dictum* of criticism appears the more absurd, inasmuch as the forms adopted in the early history of the Israelitish people were of so general, simple, and inartificial a kind, that their adoption was just as natural and intelligible as the absence of them would be unnatural and inexplicable (see my *Einheit der Genesis*, p. xlix. seq., &c.).—We shall content ourselves at present with merely mentioning the forms of worship existing in the patriarchal age, and shall reserve any discussion of their meaning till we come to treat of the Mosaic legislation.—The most general expression, descriptive of the patriarchal worship, is the frequently recurring phrase קָרָא בְּשֵׁם יְהוָֹה (Gen. xii. 8, xiii. 4, xxvi. 25, xxxiii. 20); which means "to call, to address by the name of Jehovah," and always implies the adoration of Jehovah (Ps. lxxix. 6, cxvi. 17; Is. xii. 4). *Luther's* translation: "to preach the name of the Lord," is very correctly criticized by *M. Baumgarten* (i. 1, p. 172), as follows: "The different epochs in the divine economy are confounded by those who suppose, that the patriarchs ever thought of, or aimed at, the conversion of the heathen. Missionary work was by no means the task of the Old Testament. When Abraham built altars, and praised the name of the Lord, this was the expression of his own personal feelings, and his service as the father of his race." The more special forms of worship, which we meet with, are prayer (chap. xxiv. 63), altars and sacrifice (the former principally upon hills and high places, chap. xii. 8, xxii. 2, for the hills were already regarded as natural symbols of exaltation, from the humility of their earthly condition to one more heavenly and divine), purification (chap. xxxv. 2), vows (xxviii. 20 sqq.), tithes (xiv. 20, xxviii. 20), and circumcision. This exhausts the forms of worship, to which any reference is made.

There are still two points, however, about which a great deal has been written on both sides, and on which we must give our opinion as briefly as possible, viz., on the observance of the Sabbath, and the existence of any priestly institution in the pre-Mosaic times. With reference to the *Sabbath*, see *Iken*, p. 26 sqq. The week of seven days is the earliest measure of time amongst all nations (*vid. G. H. Schubert* Lehrb. d. Sternkunde, Erlangen 1847, p. 204 sqq.), and *Philo* justly designates the weekly cycle as *πάνδημον καὶ τοῦ κόσμου γενέσιον* (de opif. mundi). We need not discuss the question here, whether the universal agreement in this respect is to be explained on the ground of the agreement between such a division and the four phases of the moon, or the number of the planets, or from the symbolical dignity of the number seven, or whether it should rather be referred to a universal revelation made before the dispersion of the people, in which case we should have to seek the record of it in Gen. ii. 2. At all events the division by weeks was known in the patriarchal age: we find it in fact as early as the history of the flood, and we have a proof of its symbolical or religious meaning in its connexion with the marriage festival, chap. xxix. 27, 28, and also with the rite of circumcision, chap. xvii. 12. Hence it is not in itself an improbable thing, that there may have been some kind of festival connected with the seventh day, as early as the days of the patriarchs. At the same time, it must be confessed that we cannot bring any proof of the existence of a Sabbatic festival in the ante-Sinaitic period. Neither the divine determination in Gen. ii. 3, to sanctify the seventh day, nor the peculiar form in which this is first enjoined in the law: "*remember* the seventh day to keep it holy," nor the event, which prepared the way for the legal proclamation of the Sabbath, viz., the fact that no manna fell upon the seventh day (Ex. xvi. 22 sqq.), can be appealed to as yielding decisive testimony in the affirmative. But, on the other hand, we cannot quote these passages as proofs of the contrary as *Hengstenberg* has done (The Lord's day, p. 7 sqq., Engl. transl.).

According to the Talmud and the Rabbins, the *priestly* rights belonged exclusively to the *first-born* before the giving of the law, and this opinion is shared by *Jerome*, *Selden*, *Bochart*, &c. But it has been warmly opposed by *Outram* and *Spencer*, and especially by *Vitringa* (*de synag. vet.* ii. 2, and *observ.* ss. ii. 2, 3). And their objections are certainly just, for the arguments adduced in favour of any peculiar priesthood are quite untenable. That Esau's raiment, mentioned in Gen. xxvii. 15, was priestly raiment, is an absurd fiction. That Jacob's blessing in Gen. xlix. 3 included the priesthood among the privileges of the birthright, is a notion founded entirely upon *Luther's* false rendering. That the young men, whom Moses sent to offer sacrifice (Ex. xxiv. 5), were all eldest sons, is a gratuitous assumption; and the substitution of the tribe of Levi for all the firstborn of the congregation does not prove anything, since *Vitringa* is certainly right in saying (p. 272): *illos Deo consecratos esse ad ministerium sacrum non ad sacerdotium, s. non ut sacerdotes sed ut sacrificia.* The natural and historical order of events was certainly this, that the priestly functions were usually discharged by the fathers and heads of the families; and therefore, if the firstborn inherited any priestly rights, it was simply on account of his becoming the head of the family. See *Buddei hist. eccl. ed.* iv. Vol. i. p. 311 sqq.

§ 8. The general *culture* of the patriarchs was undoubtedly affected by their nomadic mode of life. But nothing can be more unwarrantable, than to attribute to the patriarchs all the rudeness and hopeless degradation of ordinary nomad-hordes, who determinately fence themselves against any influence from the civilization by which they may be surrounded. Their wandering mode of life in the holy land was the necessary consequence of their being foreigners without a home. Their pilgrimage was *forced upon them*, and the period of its cessation was the constant object of their hopes and desires. Hence we find that, so far as it was possible, they did participate in the benefits resulting from the culture and civilization of the more settled tribes, with whom they came in contact. (1)—The external *constitution* of the patriarchal commonwealth partook of the characteristics of a family. The head of the family concentrated the whole authority and jurisdiction in his own person; he even possessed the power of life and death, controlled only by certain fixed traditions (Gen. xxxviii. 24). The position of the woman was a subordinate one, as it always was before the time of Christ, her claim to equal rights being nowhere fully recognised. Hence polygamy was regarded as perfectly justifiable. But we find no trace among the patriarchs of such degradation of the woman, as is found wherever she is regarded as nothing but a slave of the man, affording him the means of perpetuating his race and gratifying his lusts. On the contrary, we find many a proof of the esteem and love which she received as a wife, and of the personal rights which she possessed as the mistress of the house. (2) We also find the *inviolable* purity of the marriage bed maintained with such severity that adultery was punished with death (Gen. xxxviii. 24), and in the case of the patriarchs it was rendered peculiarly important from their consciousness of a divine call and of the destiny of the family. The strongest incitement to polygamy arose from the desire to maintain and enlarge the family, and this was also the cause of the peculiar institution of the Levirate marriage (see Vol. i., § 86. 2).

(1.) *Hengstenberg* (Beitr. ii. 431 seq.) has made an excellent collection of proofs that general culture was both sought after and possessed: "In the case of the patriarchs it is very apparent, that their wandering mode of life was forced upon them by the fact that they were sojourners in a land, the whole of which was held in possession by its original occupants. We find no marks of the rudeness of nomad tribes. Both mentally and morally they were on a level with civilized nations. They shared in the advantages, conveniences, and luxuries enjoyed by more favoured nations. Jacob possessed a signet-ring; Joseph wore a richly ornamented dress; Abraham paid for the field he bought, in coin; the sons of Jacob also took money with them to purchase corn; and Abraham's servant presented Rebekah with a gold ring and armlets. Wherever it was possible, the nomadic life was immediately relinquished. Lot settled in Sodom, occupied a house there, and entered too readily into the habits of the town. When Abraham went down to Egypt, instead of doing what nomads by profession and inclination have been in the habit of doing for thousands of years, namely taking up his abode in the pasture lands on the border, he went direct to the court of the king

(Gen. xii. 10 sqq.). He afterwards settled in Hebron as a home, and was there the prince of God in the midst of the Hittites (Gen. xxiii.). Isaac lived in the capital of the Philistines, and occupied a house opposite to the palace (Gen. xxvi. 8). He also sowed a field (ver. 12). Jacob built himself *a house* after his return from Mesopotamia (chap. xxxiii. 17).—Joseph's dream of the sheaves of his brethren bowing down to his sheaf is also an important illustration of the point in question (cf. Vol. i., § 84. 1).

(2). There are many proofs that the person of the woman was highly esteemed. The history of Sarah shows, that in several respects she had the right to exercise her own authority in the sphere of domestic life. The consent of the bride was asked on the occasion of her marriage (chap. xxiv. 58). The husband showed the most devoted affection to his wife (chap. xxiv. 67, xxix. 20). The multiplication of wives does not appear to have been entirely dependent upon the caprice of the husband, but was generally founded upon, and defended by the barrenness of the lawful wife (chap. xvi. 2 sqq., xxx. 3, 4, 9). And when any plan was decided upon, which was intended to alter the general condition of the family, the wife was asked to give her consent. Thus, for example, when Jacob fled from Mesopotamia, he explained his reasons to his wives, that he might obtain their approbation (chap. xxxi. 4 sqq.).

SECOND STAGE

IN THE

HISTORY OF THE COVENANT.

THE NATION:

FORM ASSUMED IN THE TIME OF MOSES.

EXTENT, CHARACTER, AND IMPORTANCE OF THIS STAGE,

IN THE

HISTORY OF THE ANCIENT COVENANT.

---

§ 9. The first stage in the covenant history, which displayed itself in the form of a *family*, was brought to an end by the death of Jacob, unity being an essential element in the idea of a family. With the death of Jacob, the last solitary representative and father of the whole tribe, this unity of the one family was resolved into a plurality of families; and thus the way was opened for their becoming a nation. We have now reached the commencement, therefore, of the *second* stage in the covenant-history, in which we shall see the family expand into a nation. But the growth of any nation is directly and primarily determined by the *people* themselves, that is, by the mass of individuals and families who are united together in a higher, independent commonwealth, by virtue of a common ancestry, a common language, a common religion, and a general uniformity of character. Such an association, of course, necessarily requires a *constitution*, by which the individuals are held together. This again involves another indispensable condition, viz., a prosperous population, in independent possession of a *land* of their own, and one that is suited to the character of the inhabitants. But at the commencement of the stage before us, we find none of these conditions fulfilled; though by the decree and promise of God they existed potentially in the *dodekad* of the families, and gradually attained to the requisite fulfilment. The *first step*, then, towards the future nation is to be found in such an organisation of the people, as formed the substratum of all further development. This was the embryo-state of the nation. Egypt (§ 1. 7) was the womb, as it were, in which the germs of the promised people were deposited, that it might guard them and nourish them by its natural powers, till they had grown into a great nation. As soon as the embryo had reached maturity, *i.e.*, as soon as the people had become so strong as to require and demand an independent existence, an impulse from within urged them to seek that independence, and did not rest till it was secured. The *exodus* from Egypt represents the natural birth of the people, and the Egyptian oppression resembles those labour-pains without which, in this earthly state, no life can be brought into existence. The wonders of God in Egypt, the strong arm of the Lord, which was stretched out to help and save, were the instruments of divine surgery by which the natural force of the mature embryo, then striving for independent existence, was enabled to attain its end. By the *exodus* Israel gained an independent position, and stood upon an equal footing with other nations, in fact, became a nation *like* all the rest. The *first step* in the development of the national existence, viz., the *preparation of the people of the covenant*, had now attained its object. *Moses*, the man of God, was the instrument of the divine assistance; being called by God, and furnished with divine power to be the saviour of Israel.

But Israel was *not* to be *merely* a nation, like the other nations, resting on no other basis than that of natural life. According to its vocation and its destiny it was to be *the nation of God*, the holy nation, the chosen race, the possessor and messenger of salvation for all the nations of the earth. And thus the nation entered upon the *second* stage in its history. *Moses*, the deliverer of the people by the power of God, led them to the majestic altar of the Lord, that altar which He, the creator of the heavens and the earth, had erected for himself among the rocks of Sinai, with their heads lifted towards heaven; and there they were set apart as a *holy* nation. If the *exodus* from

Egypt was the natural birth of the nation, the conclusion of the covenant at Sinai was the religious consecration of the new-born infant; its *regeneration* to a higher life. But as God never demands without giving, so he never gives without demanding. And therefore, when Israel entered upon the *privileges* of the covenant-nation, and obtained possession of the gifts and goods, the promises and hopes of the covenant, it necessarily undertook the *duties* of a covenant-nation, and submitted to the commandments, the restrictions, and the sacrifices which such a relation involved. The conclusion of the covenant was therefore accompanied by the *giving of a law*, which defined the privileges and prescribed the duties of the covenant-nation. This law also conferred upon Israel *a constitution*, suited to its vocation and its future destiny, by which its internal organisation was completed, its external distinctions defined, and its safety ensured. The events attendant upon the legislation and the conclusion of the covenant ushered in the *second step* in the onward progress of the nation, namely, the determination of the peculiar constitution, which was henceforth to regulate the course and development of the history of Israel, in other words, *the establishment of the theocracy*. The mediator of the covenant and the agent in the foundation of the theocracy was *Moses*, the man of God (1).

But the development of the nation was not yet complete. In the first step of this stage in its history, Israel had received its natural freedom and independence ; in the second, its sacred dedication and covenant. One thing was still wanting, however, which was an essential pre-requisite to the actual realization of the whole of these, viz. a *country* suited to its natural and spiritual character, its position, and its destiny. In the *third step* of its national history this want was satisfied, and it *obtained possession of the land*, which the providence of God had selected as the arena on which the covenant-history was to run its course, and which the mercy of God had already promised to the fathers. The divine hero, by whom Israel was led through conflict and victory to the possession of this treasure, was *Joshua*, who continued and completed the work which Moses had begun.

The condition and possessions of Israel now embraced all that was requisite, to sustain and exhibit a national existence *devoted to God*, by the side of the other nations, which were *at enmity against God.* Country and people, laws and promises, constitution and worship were given; and they contained the germs of all their future development. This brings us to the commencement of the *fourth step* in the history of the covenant, which we find in the existence of a nation entrusted with the task of *working out its peculiar nationality.* Hitherto the operations and gifts *of God* had stood in the foreground. But the time had now arrived, when the works of Israel in performance of the covenant were to stand prominently forward; when Israel might, and should have shown, that the gifts, and leadings, and revelations of God, which it had hitherto received, it could now use and apply for itself; and when it should have taught the way in which this could be done. Again and again, however, it forsook the path of the covenant; and God had continually to interfere, and by punishment and chastening to save and heal. Surrounding nations were employed to execute his sentences, and *Judges* were afterwards sent as his messengers of salvation.

(1). The second step of this stage was indisputably the most important and eventful. We must, therefore, examine it with especial care. In doing so we shall divide it into two parts. The first will contain an account of the *historical foundations*, on which the theocracy was based, and the *circumstances* amidst which the legislation, that established it, was completed. The second will consist of a systematic analysis of the *legislation* itself.

The sources from which our knowledge of the first two steps must be derived are the last four books of Pentateuch. As critical and exegetical *aids* we recommend especially the works already mentioned (in Vol. i. § 14—20), of *Hävernick*, *Ranke*, *Hengstenberg*, *Welte*, *Keil*, *Rosenmüller*, and *M. Baumgarten*. In addition to historical works of a more general character, the following monographs deserve particular notice: *Warburton's divine legation of Moses* ; *Fr. Hauff*, über Mose's welthistorische Bedeutung (Studien der evangelischen Geistlichkeit. Würtemberg vi. 2 p. 3 sqq.) ; *E. Osiander*, Blicke auf Moses (Christoterpe, 1837 p. 77 sqq.) ; *Patr. Fairbairn's* Typology of Scripture, vol. ii., the Mosaic period, Edinburgh, 1847.

SCENE OF THE HISTORY.

Compare the aids mentioned in Vol. i., § 15. 2; also *Léon de Laborde et Linant, voyage de l'Arabie pétrée*, Paris, 1830, and *Léon de Laborde, Commentaire géographique sur l'Exode et les Nombres*, Paris and Leipzic, 1841-4, as well as the works named in Vol. iii., § 2 and 23.

§ 10. An immense tract of desert stretches along the north of Africa, commencing at the coast on the north-west, and running not only through Africa, but into Asia as far as the steppes of the Euphrates. The only interruption which it meets with is from the *Nile*, whose fertilising waters flow completely across the desert, and have produced a fruitful oasis, which bears the name of *Egypt*, and is one of the most ancient and important of all the civilized lands, that have figured in the history of the human race. By far the larger part of this desert, towards the west, consists of low land, and is known by the name of the *Sahara*. The portion immediately bordering upon Egypt is called the *Libyan desert*. On the other side of the Nile, at the point where the sand regains its supremacy, the *Arabian desert* commences, and stretches thence to the Euphrates. This eastern division, which is much smaller than the other, is hilly, and is intersected or bounded by mountain ranges, which vary in extent, and on which there are here and there fertile spots, proportioned in size to the springs which produce them. For some distance the breadth of the Arabian desert is considerably diminished by the Red Sea, which reaches almost as far as the Mediterranean. This enormous bay is formed by the Indian Ocean, and terminates in two smaller gulfs, which enclose a portion of the Arabian desert, and give it the character of a peninsula. Both of these gulfs receive their ancient, as well as their modern names, from towns which stand, or have stood, in the neighbourhood. The western arm was formerly called the *Heroopolitan* gulf, the eastern the *Elanitic* ; at present the former is called the *gulf of Suez*, the latter the *gulf of Akabah.* The *mountains of Idumea* (Mount *Seir*) stretch from the Elanitic gulf to the Dead Sea, intersecting the Arabian plateau from north to south, and dividing it into two unequal parts. The western half (the smaller of the two), including the mountains of Idumea, has been known since the time of the Romans as *Arabia Petræa.* This name is not derived from the rocky nature of the soil, as is commonly, though erroneously, supposed, but from the strong city of *Petra* in the land of the Edomites. Under the last of the Emperors Arabia Petræa was called *Palæstina tertia.* The name was given on correct geographical grounds, the whole district being apparently an integral part of the mountainous region of Palestine (the provinces of Judah and Ephraim were named *Palæstina prima*, and Galilee, with the country beyond Jordan, *Palæstina secunda*). It was also designated *Palæstina salutaris* on account of the healthy nature of the climate in the mountains of Edom. The northern boundary of Arabia Petræa, from the mouth of the Pelusiac arm of the Nile as far as Gaza, is formed by the Mediterranean Sea ; from Gaza to the southern extremity of the Dead Sea, it is bounded by the mountains of Judah, which are already known to us by the name of the mountains of the Amorites (Vol. i., § 40. 4). Towards the south, it runs between the two arms of the Red Sea, and terminates in the promontory of *Ras-Mohammed.* The larger or eastern half of the Arabian desert, to which the Romans gave the name of *Arabia deserta,* commences on the other side of the Idumean mountains. It stretches eastward as far as the Euphrates, northward to Damascus, running by the side of the fertile highlands of the country beyond Jordan (§ 42), and southward to a considerable distance into the heart of Arabia proper (*Arabia felix.*) The last-named portion of the Asiatico-African desert, and also the portion first referred to (the Sahara with the Lybian desert) lie altogether beyond the province of our history, the first stage of which belongs to *Egypt,* the second to *Arabia Petræa,* and the third and fourth *to Palestine.* Palestine has already been described (Vol. i., § 38—43). The only portion of Egypt with which we are concerned is the eastern part of the country, viz., the province of *Goshen,* for which see § 1. 5, and § 37—42. It only remains for us to take a survey of the characteristics of Arabia Petræa. At present, however, we shall content ourselves with the most general features. A more particular description will be given, as the history brings the different localities under our notice.

§ 11. In the heart of the *peninsula,* which is enclosed by the Heroopolitan and Elanitic gulfs, somewhat towards the south, rise *the mountains of Sinai* (Jebel el Tur), from which the whole country has received the name of the *peninsula of Sinai.* Sinai consists of a nearly circular group of mountains from forty to sixty miles in diameter. The average height of the mountains composing this group is six or seven thousand feet above the level of the sea, about 2000 feet above the surrounding valleys and plains. Two of the highest points are almost in the centre of the range, *Sinai* itself (Jebel Musa, 7097 feet high) and *Mount Catherine* (Jebel el Homr, 8168 feet). As soon as the traveller leaves the burning heat of the sandy desert, and enters within the limits of these mountains, he finds a genial Alpine climate, and a cool refreshing breeze. Copious streams of water flow down from the mountains, and fertilize the soil, causing it to produce a most luxuriant herbage. Date-palms, acacias, dense bushes of tamarisks, white thorns, mulberry trees, vigorous spice plants, and green shrubs are found on every hand, wherever the bare rock is not entirely destitute of soil. And where the hand of man has done anything to cultivate the ground, there are apricots and oranges in rich profusion, with other valuable kinds of trees. It is true, there is a striking contrast between the richly wooded valleys and the steep, barren rocks by which they are so closely confined; but so much the more majestic is the aspect of these mighty masses of rugged rock. The mountains are also frequented by great quantities of game and fowl of different descriptions; among others by antelopes and gazelles, partridges, pigeons, and quails. The geological base of this range consists of large masses of primary rock, principally granite, porphyry, and syenite. The promontories are chalk, limestone, and sandstone. There is another large group of mountains on the north-west of the mountains of Sinai, called the *Serbal Mountains*, which rise like an island between the lower coast-line of el-Kaa and the deep valley of Feiran, by which they are bounded on the north. They reach the height of 6342 feet. The *Serbal* itself, a mighty giant of the desert, crowned by five peaks, is surrounded by lower mountains; the whole group deriving its name from the lofty mountain in the centre. This cluster is connected with that of Sinai by the Saddle-mountain, *Jebel-el-Kaweit*. For further details see Vol. iii., § 5—8.

§ 12. In the northern part of this cluster of mountains, there is a waste and sandy tract of table-land, *Debbet-er-Ramleh*, about 3000 feet above the level of the sea. It is nearly semicircular, and runs diagonally across the peninsula (from E.S.E. to

W.N.W.), reaching almost from the one gulf to the other. On the north of this are the *limestone et-Tih mountains*, which rise to the height of 4300 feet, and run like a crescent-shaped wall, parallel to the tract of table-land, from the Elanitic gulf, almost to the gulf of Suez. At this point they turn towards the N.N.W., and follow the line of the coast. The latter portion of the range is called *Jebel-er-Rahah*. This long mountain wall, of about sixty German miles in length, forms a second section of Arabia Petræa. On the northern side of the *et-Tih* mountains, and the eastern side of those of *Jebel-er-Rahah*, there is an extensive tract of table-land called the desert of *et-Tih-Beni-Israel* (*i.e.*, the confusion of the children of Israel). The Arabs still make a distinction between this and the *desert of Jifar*, and confine the latter name to the western and north-western edge of the tract, which lies at a lower level, and extends to Egypt and the Mediterranean Sea. Properly speaking, these two deserts form the (Asiatic) continuation of the Sahara, which is interrupted by the Nile. Barren rocks of lime and sandstone, hills of dazzling chalk and red sand, form almost the only variation in this dreary desert, which is thickly strewed with black flints and gravel. It is only in the recesses of the Wady, that sufficient water is collected in the rainy season to enable a few miserable plants to yield a meal to the passing herds ; and there are a few springs, surrounded by trees, which furnish to the travelling caravans a welcome place of encampment. (For further particulars see Vol. iii., § 23—31). On the north a wide valley, the *Wady Murreh*, separates the desert from the mountainous district of Palestine. Towards the east it slopes off into a broad, deep valley, the so-called *Arabah*, which extends from the southern points of the Dead Sea to the northern end of the Elanitic gulf, a distance of more than a hundred miles. This valley is like a continuation of the valley of the Jordan, the *Ghor* (see Vol. i., § 39. 5), and in the Old Testament they are called by the common name, *Arabah*. It is a broad sandy desert, the surface of which is covered with innumerable sand-heaps and little hills. Here and there you meet with green oases, shrubs, and palms, and even with the ruins of ancient towns. The water-shed of the Arabah is twenty-five miles from the Elanitic gulf. Further to the north the waters flow through the Wady *el-Jîb* into the Dead Sea. The low level of the Dead Sea (Vol. i., § 39. 6) is a sufficient proof that the northern part of the Arabah is below the level of the ocean.

§ 13. On the east of the Arabah rise the steep and rugged *mountains of Idumœa* (or Mount *Seir*, now *es Sherah* or *Jebal)*, which are almost of the same length as the Arabah itself, stretching from the Dead Sea to the Gulf of Akabah, with an average breadth of fifteen or twenty miles. The loftiest peaks are hardly 3000 feet high. They are steep and rugged cliffs of porphyry, which protrude themselves from the chalk formation, and are again surrounded by immense masses of sandstone. Among the shattered fragments of rock, there are valleys covered with trees, shrubs, and flowery meads. The higher ground is sometimes sown with corn. The vines in these valleys are as large, and the grapes as sweet, as in any part of Palestine itself. In some places there are woods, or what pass for woods in these countries, and spice-bearing plants, growing out of clefts in the rock, which furnish a plentiful supply for the sustenance of wild goats and gazelles. But while there are isolated examples of great fertility, the general aspect of the mountains is wild and bare, and the western mountains especially are described as altogether barren and unfruitful (Vol. i., § 73. 1).

On the eastern side, the mountains of Idumæa slope off just as smoothly and gradually, as they rise abruptly on the western. Following the range on which Idumæa is situated, we arrive at the mountainous country of the *Moabites*, the modern *Kerek*, which lies to the north of Idumæa, on the east of the

Dead Sea. The *southern* boundary, by which this district is separated from the mountains of Idumæa, is the Wady *el-Ahsy* (el-Kurahy), which opens at the southern end of the Dead Sea. On the *north* it is bounded by the deep rocky valley, through which the brook *Arnon* flows, which enters the Dead Sea near the centre of the eastern side. The Arnon divides the Kerek from the highlands of *el-Belkah* on the east of the Jordan (Vol. i. § 42, 3). In the nature of its soil the Kerek forms a link between the highlands of Palestine beyond the Arnon, which consist for the most part of table-land, and the mountains of *es-Sherah*, the aspect of which is most rugged and grotesque. But the conformation and geological character of the Kerek are far from being sufficiently known, to enable us to describe its details with accuracy, or to employ all the Old Testament data with any degree of certainty.

FIRST STEP

TOWARDS THE

DEVELOPMENT OF THE NATION.

ISRAEL'S SOJOURN IN EGYPT;

OR

THE PREPARATION OF THE PEOPLE OF THE COVENANT,

A PERIOD OF 430 YEARS.

CONDITION OF THE ISRAELITES
AND
DEVELOPMENT OF THE NATION DURING THE PERIOD SPENT IN EGYPT.

§ 14. (Exodus i.)—The historical records of the Old Testament pass very quickly over the first three centuries and a half of the *period of* 430 *years* (1), to which the sojourn of the children of Israel in Egypt extended. Still there is no ground for attributing to these records *either faultiness or omissions*, provided we do not measure them by such a standard, as is foreign both to the intention of the records and to the circumstances of the case (2). In accordance with both of these, the historian is content to relate the extraordinarily rapid *increase of Jacob's descendants* in general but characteristic terms (ver. 6, 7), and then passes at once to a description of the circumstances, which eventually led to Israel's departure from Egypt. The rapidity with which their numbers increased may be learned from the census, taken shortly after the Exodus, from which we may infer that there were in all about two million souls (3). So long as there was a continuance of the good understanding, established by Joseph between the ruling dynasty in Egypt and the Israelitish settlers,—so long, that is, as the former could ensure the faithfulness and attachment of the latter,—this rapid increase in the number of the Israelites must have been a most welcome thing to the Egyptian rulers; for it enabled them with the greater ease to fulfil the task which the policy of Egypt imposed upon them, of guarding carefully against incursions on the part of the hostile hordes to the East.—But the government of that time was apparently overthrown by force, and a *new dynasty* (4) arose. As this put an end to the relations of confidence and devotion, which had existed from the time of Joseph, between the government and the nomadic settlers in the land of Goshen, the extent to which the latter were increasing could not but suggest the possibility, that opposing interests might one day give rise to political difficulties. On the one hand, for example, it must have appeared a dangerous thing to have so powerful and numerous a body of men, estranged from the ruling government, just in that border province of the kingdom, which was continually threatened by the tribes on the East, who were ready to invade it for the purpose of plunder or conquest. How easily might it happen, that the latter would find in the Israelites, not protectors of Egypt, but confederates in their enterprise. On the other hand, it was to the interest of the government to prevent the settlers from leaving the country, that they might not lose so considerable a body of useful subjects; and it became all the more important to put a timely check upon their wish to emigrate, on account of the increasing desire of the descendants of Jacob to possess the promised land, which they regarded as their proper home. Under these circumstances it seemed most advisable to break the free and independent spirit of the shepherd-tribe, and to set bounds to the excessive rate at which they were increasing, by forcing them to hard labour and *tributary service* (5). But this was so far from accomplishing the end desired, that the dreaded increase went on at a still more threatening rate. This partial failure in their plans only drove the government to adopt severer measures still. The Hebrew *midwives* received secret orders from the king, to put the Hebrew boys to death in some private way, as soon as they were born. But these measures were also unsuccessful, and, therefore, the king of Egypt made known his ruthless policy in the most undisguised manner, by issuing a command to all the Egyptians, to drown the new-born sons of the Israelites in the river Nile (6). It is not known how long this command was strictly enforced, but its extreme inhumanity is sufficient to warrant us in believing that it could not be carried out for any considerable length of time. Moreover, the Egyptians knew well, that whilst it was policy on their part to weaken, it was highly impolitic to exterminate the Israelites.

(1). *The length of their stay* in Egypt is clearly and unequivocally stated in Exodus xii. 40 to have been 430 years: "*Now the sojourning of the children of Israel, who dwelt in Egypt, was four hundred and thirty years.*" In the *Septuagint*, however, (*Codex Vaticanus*) we read: **Ἡ** δὲ παροίκησις τῶν υἱῶν Ἰσραήλ, ἣν παρῴκησαν ἐν γῇ Αἰγύπτῳ καὶ ἐν γῇ Χαναὰν ἔτη τετρακόσια τριάκοντα. In the Alexandrian Codex the word παρῴκησαν is followed by the clause αὐτοὶ καὶ οἱ πατέρες αὐτῶν. We find the same reading in the Samaritan texts and the Targum of Jonathan. Hence, according to these, the 430 years included the 215 years, during which the three patriarchs sojourned in Canaan.

We must first enquire, therefore, which is the *reading of the original text*: whether the words in question have been omitted from the Hebrew, by accident or design, or whether they have been interpolated in the versions in which they occur. To this we reply, that an impartial examination of all the arguments *pro* and *con* yields the most decided and indisputable testimony to the genuineness of the Hebrew text. There are no various readings in the Hebrew MSS. (*vid. Rosenmüller* Comm. ii. p. 222), which might lead us to doubt the authenticity of the received version; and whilst the Hebrew is recommended by its simple, natural, and inartificial construction, the *Septuagint* is rendered just as suspicious by the opposite qualities. At the very first glance these additions look like artificial emendations of the text, which have been made on the supposition that 430 years was too long a period for the stay in Egypt. Starting with this assumption, it was very easy to include the period spent in Canaan, especially as this embraced exactly half of the 430 years. But this rendered it necessary to add the clause ἐν γῇ Χαναάν. Moreover, we see the evidence of a guilty conscience in the unskilful clause αὐτοὶ καὶ οἱ πατέρες αὐτῶν, which is introduced for the purpose of removing the apparent incongruity, of reckoning Abraham, Isaac, and Jacob, among the children of Israel ; for this inaccuracy would no more have given offence to an unprejudiced mind, than the similar one in Gen. xlvi. 8, where Jacob is reckoned as one of the children of Israel. Moreover, the alteration is a very unfortunate one, for it does not entirely answer its purpose, as the principal clause, "the sojourning of the children of Israel was 430 years," still remains. It is not likely that the words were written by the translator himself, since *Theophilus of Antioch*, who always follows the *Septuagint*, frequently speaks of a 430 years' sojourn in Egypt (ad Autolycum iii. 9. 24). But if they were, we know what liberties he took with the text, and how often he has altered it, especially in chronological statements, probably to suit some preconceived system. *Seyffarth's* hypothesis, that the chronological accounts in the Hebrew text originally tallied with those of the *Septuagint*, but that they were altered by the Jewish academy at Tiberias, for the purpose of sustaining their Messianic expectations, is too arbitrary and unfounded to meet with support (*vid.* his *Chronologia sacra* p. 218 sqq.). The agreement between the Samaritan and Chaldee and the *Septuagint* only proves that in their case there was the same reason for shortening the 430 years. The apostle Paul, it is true, also reckons 430 years from the call of Abraham to the giving of the law (Gal. iii. 17), but as his statement is founded upon the *Septuagint*, it cannot be regarded as an independent authority. Paul was writing for Greeks, who were only acquainted with the *Septuagint*, and as the question of chronology did not in the least affect his argument, it would have been as much out of place on his part to correct the *Septuagint*, as it is on the part of his expositors to appeal to the doctrine of inspiration in connexion with this passage. *Josephus* also says (Ant. ii. 15, § 2), that the Israelites left Egypt 430 years after the entrance of Abraham into Canaan ; but we know how little dependence can be placed upon his chronological statements with reference to the earlier times, and in this case they lose all their worth, on account of his having spoken in two other places of 400 years as the duration of the oppression of Israel in Egypt (Ant. ii. 9, § 1, and De bello jud. V. 9, § 4). In addition to the arguments already adduced in favour of the authenticity of the reading in the Hebrew text, we may also mention the circumstance that it is impossible to see what end could be served by an intentional omission of the words in question; whereas, as we shall presently show, it is by no means difficult to ascertain the motives for an artificial emendation of the passage by the introduction of the clause. And if that be the case, the agreement between the Samaritan, the paraphrase, and the *Septuagint* loses all its importance, though they are apparently independent of one another.

By the influence of the authorities just named, the notion, that the 430 years were to be reckoned from Abraham, became a settled tradition both among Jews and Christians, and was adopted even by expositors, who followed the Hebrew text in every other case, and admitted its authenticity in the present instance. The fetters of this tradition were first broken by *J. B. Koppe* (progr. quo Israelitas non ccv. sed ccccxxx. annos in Aegypto commoratos esse efficitur. Göttingen 1777), and he was immediately followed by *J. G. Frank* (novum syst. chronol. fundam. Göttingen 1778). Since then the opposite view has become the prevailing one. It has been supported by *Rosenmüller* (ad. h. l. p. 220 sqq.), *Hofmann* (in the *Studien* u. *Kritiken* 1839, p. 402 sqq.), *Tiele* (Comm. ad. Gen. xv. 13 sqq., and his Chronol. d. A. T. p. 33 sqq.), *Ewald* (Geschichte i. 454 sqq.), *Bunsen* (Aegypten i. 214 sqq.), *Delitzsch* (Genesis Ed. 2. 1. 363 seq.), *L. Reinke* (Beitr. zur Erklärung d. A. Test. Münster 1851), and many others. *M. Baumgarten*, however, has revived the old traditional explanation (theol. comm. i. 474 sqq.)

We will commence by examining the arguments of those who are of opinion that the call of Abraham must be taken as the *terminus a quo*. They are founded upon Gen. xv. 13—16, Ex. vi. 16—20, and Num. xxvi. 59, all of which are said to be irreconcileable with the notion that the stay of the Israelites in Egypt lasted 430 years.—The *first* passage cited is Gen. xv. 13—16. Jehovah announces to Abraham: "*thy seed shall be a stranger in a land that is not theirs, and shall serve them, and they shall afflict them four hundred years.* And also that nation, whom they shall serve, will I judge; and afterward will they come out with great substance. And thou shalt go to thy fathers in peace, thou shalt be buried in a good old age. *But in the fourth generation they shall come hither again.*" The argument founded upon this by *Bengel (ordo temporum*, ed. ii. p. 53 seq.), and *Baumgarten* (i. 190 seq.), rests upon the assumption that the announcement of a 400 years' sojourn in a foreign land refers to a definite chronological period, to be reckoned from the birth of Isaac (viz. from that time to the birth of Jacob sixty years, thence to the migration into Egypt 130 years, and lastly the time spent in Egypt 210 years, in all 400), whereas the 430 years, mentioned in Ex. xii. 40, are supposed to be calculated from the first call of Abraham in Haran (which must in that case have taken place five years before he removed to Canaan). But the commencement of the 400 years of service must be looked for, not in Canaan, but in Egypt. This has been shown in a brief but forcible manner by *Hofmann* (p. 402). "Can it be supposed," he says, "that God was here predicting to Abraham something which had already taken place in part, in his own history? . . . . To Abraham's seed Canaan was not 'a land that was not theirs;' on the contrary, it already belonged to his seed by promise, though not by possession. Moreover, there was nothing resembling service and oppression in Canaan." *Baumgarten* replies to this, with some plausibility it must be confessed, that the last argument tells as much against *Hofmann's* own explanation. The actual servitude was confined to the closing period, the reign of only two Pharaohs. And if the whole of the time from Jacob's going down to Egypt to the accession of the new king (Ex. i. 8) must be included in the period of Israel's servitude and oppression, there is no reason why the same designation should not apply equally well to the history of the last two patriarchs. The reason why it must be so applied is, that the most important part of the announcement is the fact of their living as foreigners (גֵּר יִהְיֶה), and that this mode of life commenced with Abraham, and was to continue with Isaac. But even if this were granted, there would still be two difficulties in the way. In the divine announcement only *one* land is spoken of, in which they were to be strangers, to serve and to be afflicted, as in a land that was not theirs; and we cannot, therefore, think of both Canaan *and* Egypt, especially as the words "afterward shall they *come back*" (יָשׁוּבוּ) place the land in which they were to serve and to be oppressed for 400 years in direct antithesis to the land of Canaan, the land of their fathers. The *departure* from the land of bondage (ver. 14) is a *return home* to their own land. Moreover, it is expressly announced to Abraham, in evident contrast with the foreign life, the servitude, and the oppression of his seed, that he shall die *in peace* and in a *prosperous* old age. From this it follows that the remainder of Abraham's life, at least, cannot be included in the 400 years; and just as little can we include the lives of Isaac and Jacob, which in this respect resembled Abraham's. But if we are thus brought to the conclusion, that the 400 years refer exclusively to the period spent in Egypt, there is certainly a difference between this announcement, and the passage in Ex. xii. 40 which speaks of 430 years. But who would think for a moment of calling this a discrepancy? In Gen. xv. 13 we have a prophetic declaration, in which a round number is quite in place. In Ex. xii. 40, on the contrary, we have a definite chronological and historical statement.—With regard to the *four generations*, mentioned in ver. 16, it would be a most arbitrary thing to assign to these a different starting point from the 400 years in ver. 13, and to restrict them to the stay in Egypt, as *Bengel* and *Baumgarten* are obliged to do. The four generations are evidently identical with the four centuries. *Baumgarten* is perfectly right when he says, in opposition to *Tiele*, that דּוֹר does not mean a century, but a generation, an age; but he is just as decidedly in the wrong, when he supposes it to represent the modern artificial notion of a generation of thirty years. *Hofmann* had already given the correct explanation. "דּוֹר," he says, "was not to the Hebrew an artificially calculated γενέα, of which there were three in a century, but embraced, as Gen. vii. 1 is quite sufficient to prove, *the sum total of the lives of all the men who were living at the same time*; and according to the ordinary length of life at that time, this would give a century as the duration of each generation." The meaning of the word דּוֹר is still more apparent from Ex. i. 6, where we read "and Joseph died, and all his brethren, and *all that generation*," especially if we compare Gen. l. 23, where Joseph is said to have seen his grandchildren's grandchildren, all of whom are reckoned in Ex. i. 6 as *one* generation.

The second passage, which is thought to be irreconcileable with a 430 years' stay in Egypt, is Exodus vi. 16—20. We have there a genealogical table of the tribe of Levi, in which Moses and Aaron are said to belong to the fourth generation (Levi, Kehath, Amram, Aaron). Levi was 137 years old when he died, Kehath 133, Amram 137; and when the Israelites went out of Egypt, Aaron was only 83. If from these numbers we deduct Levi's age when they first went down to Egypt, and the age at which Kehath and Amram begat children, the sum of these numbers will fall very far short of the 430 years mentioned in Ex. xii. 40, and consequently, it is said, we must either give to Ex. xii. 40 a different meaning from that which lies upon the surface, or there will be an irreconcileable discrepancy between the two accounts.—*J. G. Franck* endeavours to bring this genealogy into harmony with the 430 years, by assuming that the sons in this family were not born till their fathers had nearly reached the end of their life, and that Levi begat Kehath seventy-five years after he went down to Egypt (*Astron. Grundrechnung der bibl. Gesch. Gottes. Dessau u. Leipzig* 1783, p. 178). But there is something so forced and unnatural in this explanation, that it is not likely to meet with approbation. Moreover, it is impossible to reconcile either this or *Bengel's* explanation with Num. iii. 27, 28, on which we shall presently speak more at large. But we do not want any such artificial aids in order to escape from the difficulty; for the explanation suggested by *Koppe, Tiele, Hofmann*, and others, that some of the members have been omitted from this genealogical table, is perfectly satisfactory. It is well known that such omissions are very common in the biblical genealogies, and in the present instance their occurrence is attested by indisputable proofs. In Num. xxvi. 29 sqq., we find *six* members comprised within the same space of time, viz., from Joseph to Zelaphehad; in 1 Chr. ii. 3 sqq., there are *seven* persons mentioned between Judah and Bezaliel; and in 1 Chr. vii. 22 sqq., there are as many as *ten* named from Ephraim to Joshua. Then, again, from a comparison which *Hofmann* has instituted between the other genealogies of Levi in Ex. vi. and 1 Chr. vi., it is evident that there are names omitted from the former, which have been obtained from other sources and inserted in the latter. The fact that only four names are given in the pedigree of Moses and Aaron, may be simply and satisfactorily explained, as *Hofmann* has acutely observed, if we suppose that the number was selected with an evident reference to Gen. xv. 16, for the purpose of showing that the prediction was fulfilled. "Sometimes particular members are omitted; at other times several are linked together. The four members, which commonly appear, are intended merely to represent the four generations who dwelt in Egypt. And this is the reason why the ages of Levi, Kehath, Amram, and Moses, are given; and not to enable us to calculate how long the Israelites were in Egypt, which they would never enable us to do."

Lastly, we are referred to *Num.* xxvi. 59, compared with Ex. vi. 20. In the second passage, Moses' mother, *Jochebed*, is called the aunt (דּוֹדָה) of her husband Amram, and this is stated even more plainly and decidedly in Num. xxvi. 59: "The name of Amram's wife was Jochebed, the daughter of Levi, whom (his wife) bare to Levi in Egypt." If now Moses' mother was Amram's aunt and Levi's daughter, it is at once apparent that there is no room for the assumption that any members have been omitted from the genealogical list in the sixth chapter of Exodus. But when we look a little more closely into this argument, which is evidently the most important of all, it is quite clear that the expression, "a daughter of Levi," is not to be taken literally. Jochebed may be called a daughter of Levi, in the same sense in which Christ is called a son of David. Nor is there anything more conclusive in the statement that Jochebed was *Amram's aunt*, for דּוֹד and דּוֹדָה may both be used to express blood-relationship in general; for example, on comparing Jeremiah xxxii. 12 with ver. 7, we find דּוֹד applied to the son of the uncle, and also to the uncle himself. But even if there have been several members omitted, the probability of which we pointed out above, Jochebed may still have been Amram's aunt in the strict sense of the word. At the same time we must admit, that the words "*Jochebed a daughter of Levi, whom (his wife) bare to Levi in in Egypt*" (Num. xxvi. 59), as they stand here, cannot mean anything else than his own daughter. But if this be the meaning, Jochebed must have been at least fifty or sixty years old when she was married, even if the stay in Egypt lasted only 210 years; and that would be certainly a most improbable age. There is sufficient, therefore, to suggest the thought, that there may be a corruption of the text or an error of some kind in Num. xxvi. 59; and we might perhaps be justified in coming to the same conclusion on account of the harsh and peculiar form of the sentence, בַּת־לֵוִי אֲשֶׁר יָלְדָה אֹתָהּ לְלֵוִי בְּמִצְרָיִם, in which there is no subject. The *Septuagint* appears to have read אֹתָם instead of אֹתָהּ: *θυγάτηρ Λευί, ἣ ἔτεκε τούτους τῷ Λευὶ ἐν Αἰγύπτῳ.* The word *τούτους* here can only refer to Aaron, Moses, and Miriam, whose names occur immediately afterwards. We cannot certainly make up our minds to pronounce the reading אתם the correct one, on the authority of the *Septuagint*. Moreover אתם does not, strictly speaking, mean *τούτους*, but *αὐτούς*, and would properly refer to persons already mentioned, not to those about to be named. Still even this deviation on the part of the *Septuagint*, when taken in connection with the absence of any subject, is a proof of the suspicious character of the passage in general. To us the whole clause, commencing with אשר ילדה, has the appearance of a gloss, appended to the preceding words בת־לוי; and the author of the gloss seems to have understood בת־לוי in its literal sense, as denoting an actual daughter of Levi, and then to have endeavoured to soften down the improbability of Moses' mother being a daughter of Levi, by appending a clause, to the effect that the daughter in question was born in Egypt. This gloss, we admit, must have been introduced at a very early period, as it is found in every codex and every version. But, in any case, the professedly chronological statement in Ex. xii. 40, confirmed as it is in a most decided manner by Gen. xv. 13, is more deserving of confidence than the suspicious notice in Num. xxvi. 59.

But, to return to Ex. xii. 40, *Baumgarten* holds fast to the reading of the Hebrew text, but thinks it possible to explain it as the *Septuagint* has done. He says: "There is an analogy in the computation of the forty years occupied in the journey through the desert (Num. xiv. 33, 34). In this passage thirty-eight years were reckoned as forty, because the two years, which had already elapsed, were considered as belonging to the same category of years of punishment, as the other thirty-eight, when once the apostasy of Israel had come to light" (p. 475 sq.). And just in the same manner, he thinks, could the 210 years, spent in Egypt, be reckoned as 430, the 220 years, which had elapsed from the call of Abraham to the migration to Egypt, being placed in the same category of servitude and exile, as the subsequent 210.

In this, however, we cannot agree with him. The difference between thirty-eight and forty is not by any means the same as that between 210 and 430. An inaccuracy of expression in the case of the former would not be very striking, but in that of the latter it would be a most startling thing. However, this is not really how the matter stands. The two years spent in the desert, of which great part had already elapsed, might very well be regarded as years of punishment, inasmuch as *the* apostasy, which came to a head at Kadesh, and was followed by the rejection of the people, had really commenced at Sinai in the first year of their journey, when they worshipped the golden calf (see Vol. i., § 51. 2). Now there is nothing resembling this in the circumstances before us. The free, unfettered pilgrimage of an independent nomad-chief in a land, which God had promised him as his own inheritance, could not be placed, without further explanation, in the same category as the residence of a tribe in a state of oppression and servitude in a foreign land. Moreover, in the former case, the two years were spent in the same place as the thirty-eight; but in the latter the 220 years were passed in a totally different place from the 210. Luther spent thirty-eight out of the sixty-three years of his life at Wittenberg; but no reasonable man would think of saying that he lived at Wittenberg sixty-three years, however true it might be that the first twenty-five years of his life were but the "preliminary stages" of his Wittenberg career. The absurdity of the attempt made by *Buddeus* (hist. eccl. i. 455) and others, to save the traditional explanation by translating the passage: "*Peregrinatio* filiorum Israel, qui commorati sunt in Aegypto, fuit 430 annorum," is too apparent on philological grounds, for it to need any refutation.

Lastly, *Baumgarten* brings against such of the modern expositors, as have given up the old, traditional explanation, the very severe charge of "having no eyes for anything but the mere surface of things." He fancies that he has discovered in the essential unity of the whole period, from the call of Abraham to the exodus from Egypt, a reason why it was absolutely necessary, that a chronological statement should be given in Exodus xii. 40, embracing that period in its entire extent. But as the chronological limits of the interval between the call of Abraham and the migration into Egypt had already been described in the book of Genesis, we are quite unable to discover any such necessity.

Another argument against the old interpretation is founded upon Num. iii. 27, 28, and is sufficient in itself to decide the question. It has been brought forward by *Koppe*, *Rosenmüller*, and *Tiele*, and we will give it in *Tiele's* words. In his Chronology (p. 36) he says: "According to Num. iii. 27, 28, the Kehathites were divided into four branches, the Amramites, the Izcharites, the Hebronites, and the Uzzielites, containing together 8600 men and boys, the women and girls not being counted. Of these about a fourth part, or 2150 men and boys, would belong to the Amramites. Now Moses himself had only two sons, as we learn from Exodus xviii. 3, 4. Hence if Amram, the son of Kehath and the founder of the Amramites, was the same person as Amram the father of Moses, Moses must have had 2147 brothers and nephews. But as such a supposition is quite impossible, it must be granted that this is sufficient to prove, that Amram the son of Kehath was not the father of Moses, but that a series of names, whose number cannot be determined, have been omitted between the first Amram and his later descendant and namesake." To this *Baumgarten* replies (i. 2, p. 268 seq.): "this would be trifling with the whole science of statistics, but it is founded upon too hasty a calculation, viz. upon the supposition, that the rate of increase proceeded quite as slowly in the three other branches, as in that of Amram himself, which would be in any case a very extraordinary phenomenon." But this does not by any means remove the difficulty. Are we to believe, then, that Kehath's descendants through Amram consisted of no more than six males, at the time of the census recorded in Num. iii. (viz. Moses and his two sons, and Aaron and his two sons, Eleazar and Ithamar), whilst his descendants through the other three sons consisted, *at the very same period*, of 8656 males (*i.e.* 2885 each). This certainly is a *large* demand upon our faith. Still, as we cannot positively say that it is impossible, we submit, and believe. But we are further required to believe (according to Num. iii. 27) that at this census the *six* Amramites—(what am I saying? there could not have been *six* of them; there could really only have been *two* included in the census, viz. the two sons of Moses; for Aaron and his sons were priests, to whom the Levites were to be assigned as a present, and as it was for this very purpose that the census was taken, they would certainly not be included in it any more than Moses himself);—hence then we are required to believe that the *two* remaining Amramites formed a distinct "*family*," a *Mishpachah* (§ 16), with precisely the same privileges and duties, as the 2885 Izcharites, the 2885 Hebronites, and the 2885 Uzzielites (Num. iii. 27 sqq.)! We must candidly confess, that our faith will not reach so far as this.

Whilst *Bengel*, *Baumgarten*, and others pronounce 430 years much too long a period, according to the standard of their biblico-theological system, for the stay of the Israelites in Egypt, *Bunsen* measures it by the standard of his Egypto-chronological system, and decides that it is much too short. And his conviction, that the statement is not historical, is strengthened by the fact, that 430 years is just double the 215 years of the patriarchs. These 215 patriarchal years he considers historical, because they form part of the tradition. "For the period of the stay in Egypt no historical reckoning was handed down, any more than the history itself. Hence the patriarchal number was doubled, and the number thus obtained was applied to a period of much longer duration, and treated as historical, though not founded upon genealogical tables." *Lepsius*, on the other hand, arrives at the very opposite conclusion, and thinks that he can find in Ex. vi. 16 sqq., a proof of his Egyptologico-chronological statement, that the Israelites did not remain in Egypt more than about ninety years!!! (*vid.* § 43. 1).

(2). *De Wette* complains of the "*immense gap*" between Genesis and Exodus, and expresses his opinion that it is "useless to attempt to restore the history and establish any connexion;" (Beiträge zur Einleitung in d. A. T. ii. 169). On this supposed gap *Vatke* rests his hypothesis, that Mosaism was a later product of the prophetic period, and says that even according to the account contained in the Pentateuch, there was evidently but little foundation for the Mosaic constitution to rest upon; (Religion d. A. T. 1. 204). *Bruno Bauer* (in his Rel. d. a. Test. i. 105 sqq.) says that the historian leaps over the lengthened period without the slightest suspicion of its importance; that even to the present day, commentators have imitated him in taking this leap in an equally unscrupulous manner; and that although there has been at length a revival of the critical consciousness in *De Wette*, apologists have not been able to offer any reply to his arguments, since hitherto they have not manifested the least idea of the importance of the gap itself. Yet the remarks of *Hävernick* (Einl. i. 2, p. 173), and especially of *Ranke* (Unterss. ii. p. 2), are not so irrelevant after all. The latter observes, "the work would be faulty, if it had been the intention of the writer to give a complete history of all the events which happened to the Israelites. But as the express design of the work embraced merely the relation of Israel to Jehovah, he was content to pass over the whole interval, during which the chosen people were growing into a great people according to the prophecies in the book of Genesis, and simply state that those prophecies were fulfilled. This was all that the centuries in question contributed to the development of the theocratic plan, and in this respect they stood far behind the few days, in which Jehovah magnified himself in his people before the eyes of the Egyptians." We may also quote the general remarks of *Bertheau* (zur Gesch. d. Isr. p. 202) as both striking in themselves and applicable here. "There is no historical work," he says, "in which the selection and arrangement of the events narrated are so exclusively and unmistakeably regulated by *one* idea as in the historical books of the Old Testament. Everything is looked at from one point of view; prosperity and misfortune, slavery and redemption, joy and sorrow, are all regarded as operations of God on behalf of his people. *There is nothing mentioned, which does not admit of being easily and intelligibly described from this point of view.* This will explain the fact that nothing is said of the lengthened period, during which the Israelites were in Egypt, and so little of the period of the Judges. The historical writings of the Hebrews are as different as they possibly can be from chronicles and annals, or a mere recital of naked facts." Even *Lengerke* expresses himself in the same considerate manner (i. 368): "A description of this period formed no part of the plan proposed by the authors of the Pentateuch. The prediction in Gen. xv. 3 (? 13) contained all that was necessary. Whatever did not serve to exhibit the fulfilment of the promises of God is either treated very briefly, at least by the original work, or else passed over in perfect silence. The intention is merely to write a history, having a particular reference to the possession of Palestine. And even of the period of the captivity in Eastern Asia, which occurred in an age of letters, the reminiscences are very few."

If we look into the question a little more closely, we find that in reality everything has been given, which from the nature of the case *could* be given, or which from the tendency and design of the record *ought* to be given ; and it soon becomes apparent, that it is unreasonable to require anything further, or at all events to speak of it as a necessary thing. (1). One of the principal facts of historical importance, connected with this period, was the multiplication of Jacob's descendants. It is evident that this was an important subject to introduce into the record, since it was both the result of the foregoing history and the fulfilment of its predictions, and also the substratum for the history of the time to come. And have we not all that is required in the account contained in Exodus i., which, however summary it may be, gives a lively and graphic description of the rate at which this increase took place? There is no one, surely, who would demand complete genealogical evidence of this increase! —(2). The history, which immediately follows, contains an account of the exodus from Egypt, in which Jehovah first manifested himself in so glorious a manner as the *deliverer* of his people; and it was quite as indispensable that this should be preceded by a historical description of the change which occurred in the policy of the Pharaohs, when the favoured foreigners became an object of hatred, mistrust, and ever-increasing oppression. And in our opinion, *this* demand has been amply met, so far, at least, as the intention and standpoint of the author were concerned.—(3). Another object of importance in the history of this period would be a sketch of the lives of prominent individuals. But it is a question, whether there were any persons of peculiar distinction, and if there were, whether the events of their lives were handed down by tradition in the same vivid manner as those of Abraham, Isaac, and Jacob; and even if this were the case, whether they were of such a nature, that the author could regard them as bearing sufficiently upon the design of his work to be worth preserving. The last question may be answered decidedly in the negative; the second may probably be so answered, and possibly the first also. The biographical sketches, which have been handed down to us in the patriarchal history of the book of Genesis, were regarded by the author as interesting and important; merely because, and so far as, they were proofs of the special care and guidance of God. If, then, this special guidance of God was not apparent during the period in question, because it was not required; however interesting the lives of particular persons might be from other points of view, for our author's purpose they would not be of any importance. But on the whole it is very probable, that there were no memorials of any particular note handed down by tradition, and perhaps there were no persons of any particular note during that period; for the peculiar circumstances which gave so much importance to the persons of the patriarchs, and impressed their history upon tradition in so indelible a manner, were altogether wanting during the period spent in Egypt.—(4). This period evidently derived great importance from the fact, that Israel was then brought into contact with a state, which had reached the highest stage of development both in a religious and political point of view; and this contact could not fail to exert a considerable influence, either of a beneficial or an injurious character, upon the early history of a people, which was just then in a condition to receive and to require cultivation. We have already said (§ 1. 7), that one of the principal reasons why Israel was led by God into Egypt, must in our opinion have been, that the Israelites might there undergo such *human* preparation as would fit them to receive a theocratic constitution. Should we not then be justified in expecting that the author would mention this, and give some information respecting it? Most certainly, if his manner of writing history had been the same as that of the 18th and 19th centuries. A historian of our age would no doubt feel it to be his duty, and a necessary part of his work, to enter into the peculiar nature of Egyptian culture, its science and religion, its industry and politics, and to search for the traces, unfortunately too few, of the influence exerted by these upon the culture and development of Israel; but this formed no part of the plan of the Israelitish historian, who had no eye for anything but the movements, which took place under the immediate guidance *of God.*—And (5), lastly, with regard to the condition and progress of Israel in matters of religion and worship, and in the arrangements of domestic and civil life, we must not overlook the fact that it is never the custom of Israelitish historians to enter into any minute description of such points as these, or to notice their historical development; so that we must gather our information respecting them from such occasional data as we possess, just as we are obliged to do in the case of the patriarchs themselves (§ 5 sqq.). On the other hand we must equally bear in mind the fact, that to an Israelite the theocratic legislation at Sinai appeared so much like a *new creation* on the part of *Jehovah*, that he lost sight altogether of the other, viz., the natural side of that legislation, that is to say, of its connexion with any manners, customs, and circumstances, which had existed before. And however little *we* may regard the giving of the law at Sinai as a *Deus ex machina*, however *we* may be disposed to recognise the important bearing of previous circumstances upon that legislation, we can easily understand how an Israelitish historian might overlook that importance, and undervalue the human basis, on account of the high estimate which he formed of the part performed by God in the giving of the law.

(3). From the census taken at Sinai (Num. i.) it appeared, that the whole number of men, "from twenty years old and upward, all that were able to go forth to war in Israel," was 603,550. If to these we add 400,000 male children under twenty years of age, and suppose the females to have been about as numerous as the males, we find that the entire mass of the people of Israel amounted to more than *two million souls*. But it is a gross mistake to suppose that the two millions were all the direct descendants of Jacob. When Jacob and his sons went down to Egypt, they must certainly have taken with them all their men-servants and maid-servants, as well as all their cattle, for these formed a portion of their wealth. We have no information as to the exact number of the latter. But we know that Abraham had 318 servants fit for war and trained to arms; his nomadic household, therefore, must have contained more than a thousand souls. Jacob, again, who inherited all these, brought with him from Syria so many men-servants and maid-servants, and so much cattle, that, when he was afraid of an attack from Esau, he divided them into two armies. With such data as these, then, we are justified in assuming that the number of those who went down with Jacob to Egypt was not limited to his sixty-six children and grandchildren, but consisted of several thousand men-servants and maid-servants. But according to Gen. xvii. 12, 13, these had been all received by circumcision into the religious community of the children of Israel, and thus the distinction between master and servant, which is never very marked among nomads, must have been still further softened down. In Egypt, where the striking contrast between Israelites and Egyptians was necessarily a great impediment in the way of intermarriages, the descendants of Jacob will no doubt have married the descendants of his servants. And under such circumstances the distinction must gradually have worn away. Hence we regard the two million souls, who left Egypt after the lapse of 430 years, as the posterity of the whole of the people who went down into Egypt with Jacob. But even then, this increase to two millions would be unparalleled in history. We must look upon this fact therefore in the light of divine providence, and regard it as a special blessing from God, the fulfilment of the promise given to Abraham, Isaac, and Jacob. In addition to this we may also quote both ancient and modern witnesses, who all agree, that the productiveness of both men and animals is far greater in Egypt than elsewhere. *Aristotle*, for example, says (hist. animal. 7. 4): *Πολλάκις καὶ πολλαχοῦ (τίκτουσι γυναῖκες) οἷον περὶ Αἴγυπτον, τίκτουσι δὲ καὶ τρία καὶ τέτταρα, πλεῖστα δὲ τίκτεται πέντε τὸν ἄριθμον, ἤδη γὰρ ὤπται καὶ τοῦτο καὶ ἐπὶ πλειόνων.* *Columella* writes to the same effect (de re rust. 3, 8): Aegyptiis et Afris gemini partus familiares et paene solemnes sunt; and *Pliny* (hist. nat. 7, 3): Et in Aegypto septenos uno utero simul gigni, auctor est Trogus. For more modern accounts consult *Rosenmüller's* altes und neues Morgenland i. p. 252. From this we may see that, even if we deduct something from the accounts as being greatly exaggerated, Egypt must in this respect have been peculiarly fitted for effecting *the* purpose, which it was intended to accomplish in connexion with the house of Israel.

(4). We are of opinion that the statement in chap. i. 8: "*there arose a* NEW KING *in Egypt who knew not Joseph,*" indicates not merely a change of government within the same dynasty, but the suppression of a former dynasty. It was so understood by *Josephus* (ant. ii. 9. 1.: *τῆς βασιλείας εἰς ἄλλον οἶκον μετεληλυθυίας*); and the following reasons lead us to the same conclusion. (1). The word וַיָּקָם requires it. Let any one take a concordance in his hand, and he will find that קוּם and הֵקִים, when used in such a connexion, always denote an entirely fresh commencement, and *never* a regular advance of the same description, or a renewal of something which existed before. (2). This explanation is supported by the expression, לֹא יָדַע אֶת־יוֹסֵף "he knew not Joseph." For these words must mean, either that the new king actually did not know, or that he would not know anything of Joseph's services on behalf of Egypt. If the latter be the meaning, we must necessarily assume that some kind of hostility existed between the new king, who now arose, and his predecessors, to whom Joseph had rendered such services; and this would be most simply explained on the assumption, that there had been a forcible change of dynasty. In the former case, we should either have to seek an explanation of the ignorance of the new king with regard to Joseph's history, in the fact that the Egyptians had entirely forgotten it and therefore the new king had never heard of it at all; or else to assume that there was some other cause, which prevented the new king from becoming acquainted with what Joseph had done. The former is absolutely inconceivable, when we consider the diligence and zeal, which the Egyptians are well known to have displayed in the preservation of their history. And we cannot think of any other cause, unless the new king had moved in a totally different sphere from his immediate predecessors; which brings us at once to the assumption, that he was the founder of a new dynasty. Some light is thrown upon the meaning of the word ידע "to know," in such a connexion, by Deut. xxviii. 36. The lawgiver there announces to the people, that the punishment of their apostasy from Jehovah will be, that they will be brought into slavery, "unto a nation, which neither thou nor thy fathers *have known*." From this passage we clearly see, that the word ידע in such a connexion does not denote a mere historical acquaintance with any object, but an acquaintance founded upon friendly intercourse with each other. The nation, to whom Israel was to be given up as a prey, would be an entirely foreign nation, which would have no regard whatever for the Israelites. And this was the case here; the new king, who rose up in Egypt, had no regard for Israel, and took no interest in its welfare.—(3). The connexion of this passage with ver. 6, 7, is to our mind *completely* decisive: "and Joseph died, and all his brethren and all that generation, and the children of Israel were fruitful, and increased abundantly, and multiplied and waxed exceedingly mighty, and the land was filled with them; and there arose up a new king, &c." In this passage all the kings, who reigned from the time of Joseph to the period in question, are evidently placed together under one point of view, and in a common relation to the *new* king. The new king must therefore have been *new*, in a totally different sense from that in which every one of the successors of the earlier Pharaoh had been a *new king*. In the writer's view they all formed *one Melech*, in contrast with *the* king, who now came to the throne; *i.e.*, they were *one* dynasty by the side of the founder of a *new* dynasty. In support of this, also, we may appeal to Deut. xxviii. 36: "Jehovah will bring thee, and *thy king which thou shalt set over thee*," into subjection to a foreign nation. The general and particular use of the word *Melech* are here fused together. For the meaning of the legislator was evidently not that *the very* person, whom the people should first set over the kingdom, would be led into captivity, but that the government, which the people would establish in connection with the theocratic constitution, should go into captivity in the person of one of its administrators.—Although *Hengstenberg* maintains, in his Egypt and the Books of Moses (p. 252 transl.), that "the reason why the king is called new is given in the phrase, 'who knew not Joseph;'" every unbiassed reader must at once perceive, that the very reverse is the truth, namely that he knew not Joseph just because he was a *new* king.

For the history of the Israelites, it is of no importance whatever in what sense the king, who began to oppress them, was a new king. The question is of more importance, for the determination of contemporaneous events in connexion with the history of Egypt. And if our explanation be correct, we have a most important *datum* in Ex. i. 8, which may serve us as an Ariadne-thread in the confused labyrinth of Egyptian history and chronology. But we shall return to this question again. (*Vid.* § 45. 4).

(5). The TRIBUTARY SERVICE, which the Israelites were forced to render, consisted chiefly in *brick-making* and *field-labour*. By the latter we are undoubtedly to understand the severe labour of watering the land in the more elevated districts (see § 15. 2); and from the former we learn that the Israelites were employed both in the erection of the colossal monuments, and in the building of cities and fortresses (Ex. i. 11: Pithom and Raemses, *vid.* § 41. 2). The preparation of the incalculable number of bricks, which were required, must, no doubt, have taken up the greatest amount of time, and demanded the greatest exertion, and therefore this is mentioned *instar omnium*. As the Egyptians prided themselves, according to *Herodotus* (i. 108) and *Diodorus* (i. 56), on the fact that not a single native was employed in the erection of their monuments, but that they were built entirely by captives and slaves, *Josephus* is probably right in associating the tributary service of the Israelites with the construction of the pyramids (Ant. ii. 9. 1).—On the manufacture of bricks in Egypt see § 22. 2. It is a memorable fact, that to all appearance a contemporaneous testimony to this tributary service of the Israelites is still in existence in a *picture found in the tomb of Rochscerê at Thebes. Rosellini*, by whom it was first discovered, has given a copy and description of it in his great Egyptological work, under the heading: "Explanation of a picture representing the Hebrews making bricks." (*Vid. Hengstenberg* Egypt and the books of Moses p. 80 transl.). According to *Rosellini's* description, which we copy from *Hengstenberg's* work: "Some of the labourers are employed in transporting the clay in vessels, some in intermingling it with the straw; others are taking the bricks out of the form and placing them in rows; still others, with a piece of wood upon their backs and ropes on each side, carry away the bricks already burned or dried. Their dissimilarity to the Egyptians appears at the first view; their complexion, physiognomy, and beard, are proofs that we are not mistaken in supposing them to be Hebrews. They wear at the hips the apron, which is common among the Egyptians, and there is also represented as in use among them a kind of short trowsers, after the fashion of the *Mikbesim*. Among the Hebrews, four Egyptians, very distinguishable by their mien, figure, and colour, are seen; two of them, one sitting and the other standing, carry a stick in their hand ready to fall upon two other Egyptians, who are here represented like the Hebrews, one of them carrying on his shoulder a vessel of clay, and the other returning from carrying brick, bringing his empty vessel for a new load. The tomb belonged to a high court-officer of the king, *Rochscerê*, and was made in the time of Thothmes IV., the fifth king of the eighteenth dynasty. The question, "how came this picture in the tomb of Rochscerê?" *Rosellini* answers as follows: he was the overseer of the public buildings, and had, consequently, the charge of all the works undertaken by the king. To the question, "how came the representation of the labours of the Israelites at Thebes?" it is answered: we need not suppose that the labours were performed in the very place where they are represented, for Rochscerê was overseer of the royal buildings throughout the land, and what was done in the circuit of his operations could, wherever performed, be represented in his tomb at Thebes. It is also not impossible that the Hebrews went even to Thebes. In Ex. v. 12 it is said, "that they were scattered abroad throughout all the land of Egypt to gather straw" (p. 80, 81 transl.).—*Wilkinson* has again carefully examined this painting on the spot, and confirmed *Rosellini's* account. It is true that he disputes the reference to the Israelites, but on grounds which *Hengstenberg* justly pronounces inconclusive. As the eighteenth dynasty undoubtedly ruled over *the whole* of Egypt, we may very well imagine that the Israelites were sent away as far as to Thebes to work, for it was the interest of their oppressors to distribute them as widely as possible through the land, and to the present day the Fellahs are brought in droves from the most distant parts of Egypt, whenever any great work is going on. This at once removes *Wilkinson's* principal objection, that according to the inscription the bricks were intended for some building in Thebes. *Wilkinson* also relies upon the fact that the majority of the workmen are without any beard. But this may be explained on the assumption, which is quite admissible, that most of the Israelites had adopted this custom either voluntarily, or on compulsion. And the decidedly Jewish cast of countenance, which even *Wilkinson* cannot deny, is a most powerful argument in favour of *Rosellini's* views.

We have already pointed out the important bearing of the Egyptian oppression and compulsory service upon the sacred history of the Israelites (§ 1. 7). The importance of this is the more obvious, since it is unmistakeably implied in the biblical record. In proof of this we refer, not merely to the fact that the record lays so much stress upon the character of a *redeemed people*, which Israel acquired in consequence of their oppression, but also to the prominence given to it in the announcement made to Abraham (in Gen. xv. 13).

(6). *Josephus* (Ant. ii. 9. 2) attributes the murderous edicts of the king to a prediction made known to him by one of his scribes, that a Hebrew boy would inflict great injury upon the Egyptians. There is no notice of anything of the kind in our record. Moreover we do not believe that Josephus found this in any ancient tradition. It is most likely an invention of his own, intended to place the hero of the Hebrew nation upon a level with *Cyrus* and others, for the benefit of Gentile readers.—*Josephus* speaks of the *midwives* as Egyptian women, evidently in direct contradiction to the Scripture record, which describes them as *Hebrew* midwives. Moreover, it is said that they feared God, and that God made them houses, and this would hardly be said of heathen women.—The midwives defended themselves before Pharaoh, on the ground that the Hebrew women were generally delivered without requiring their assistance, and we are not justified in questioning the truth of their assertion. It is well known that in warm climates the births are generally quicker and easier; and we can very well imagine that the different mode of life adopted by the Hebrew women may have given them an advantage in this respect even over the wives of the Egyptians. Still it is expressly stated in ver. 17, that "the midwives feared God and did not as the king of Egypt commanded them," but saved many children alive, whom they ought according to the king's orders to have killed. Hence their answer looks like a subterfuge, which on the strict ground of morality must be condemned. They were *not* bound to obey the king, when he required that which was ungodly, but they were bound to speak the truth by giving a direct refusal (as in Acts iv. 20, 21). But on this standpoint they did not and could not stand, for such a standpoint had never yet been reached. Nevertheless their *fear of God* was genuine, and as such it was followed by the approbation and blessing of God. Still what they did from fear of God is not on that account to be confounded with what they did from fear of Pharaoh.—The biblical record has preserved the names of two of the midwives, *Shifrah* and *Puah*. It is evident from the number of the people and the frequency of the births, that there must have been others. Whether these two were superintendents of the whole class, or whether there was some other reason for their names being handed down, it is impossible to determine.

§ 15. Jacob and his descendants came into Egypt as *nomads*. So long as they dwelt in Palestine, where they lived as pilgrims and strangers, they were compelled to adopt this mode of life by the circumstances in which they were placed. But even there, whenever it was practicable, they combined agriculture with the rearing of cattle. When Isaac dwelt in the land of the Philistines, he sowed corn there, and reaped the same year a hundredfold (Gen. xxvi. 12). And even if this is to be regarded as an exceptional case, it proves that the patriarchs were not such nomads by nature, that a settled mode of life was intolerable to them, or that they would rather suffer hunger and destitution, than take the trouble to cultivate the ground. It was to be expected, therefore, that when they came down to Egypt, where the circumstances were entirely different, they would soon exchange their wandering habits for a settled mode of life, and add to the rearing of cattle the *cultivation of the soil.* The land of Goshen, which embraced the garden-ground of the Nile on the one hand and the pasture-land of the desert on the other, provided the means and offered an inducement to both of these occupations. The intention of Joseph from the very first was, undoubtedly, to pave the way for such an improvement in his brethren's mode of life. He obtained not only the king's consent to their leading a nomad life with their flocks in the tracts of pasture-land on the east of the land, but also a grant of certain fixed hereditary possessions (אֲחֻזָּה) in the best portion of the country (בְּמֵיטַב Gen. xlvii. 11, 27). The name *Metab* is in itself a proof that the district assigned them was not merely pasturage, but contained also some of the fertile soil, which is watered by the Nile and its branches; and this supposition is confirmed in many passages by express statements to that effect (§ 1. 5). The much more remunerative character of agriculture must have been sufficient to lead the Israelites, if not to prefer agriculture, at least to associate it with the rearing of cattle. For there is no country in which agriculture is more remunerative than in Egypt. No doubt it requires much preliminary labour and many contrivances, which are not needed elsewhere. But as the land was given to the Israelites as an hereditary possession, and they had therefore a guarantee that whatever trouble they might take in cultivating the land would be for the benefit of their children and children's children, the difficulties did not present an insuperable obstacle. There was also another strong impulse to the adoption of agricultural pursuits, in the extraordinarily rapid multiplication of the people, which rendered it necessary that they should search for productive land in every direction. And lastly, the disgust, excited by nomad-shepherds in the minds of the Egyptians, must have contributed to wean the Israelites from their wandering mode of life. These expectations completely tally with the actual condition of the Israelites, as we find it incidentally referred to in different passages of the Pentateuch (1). We meet with no intimation of life in tents, which is characteristic of nomads. The Israelites live in houses and cities, and even in the royal cities (Ex. xii.). They cultivate fishing and gardening (Num. xi. 5), and water the soil in an artificial manner for the sake of the crops (Deut. xi. 10) (2). Even the tributary service, to which they were forced, presupposes such a change in their mode of life as we have described. It would hardly have been possible to compel a nomad-race to perform this labour, at least so generally as Ex. i. 13, 14, and chap. v. describe; for the words of *Maillet* (quoted by *Heeren*, Ideen über Aegypten, p. 148) with reference to the nomads of eastern Egypt in the present day, were undoubtedly quite as applicable then: "they only need, in fact, to go a day's journey into the desert to ensure themselves against any kind of retaliation." Lastly, this is attested by the legislation of Moses, which is framed exclusively for an agricultural mode of life, and instead of containing the slightest indication of having been intended to bring about a transition from wandering habits to agricultural pursuits, presupposes that the change had already completely taken place. The fact that the Israelites returned to a nomad life after they had left Egypt, and continued it during the forty years which were spent in the desert, of course proves nothing. This was an affliction, the removal of which was longed for and anticipated as a mark of the favour of God. The great mass of the nation had become an agricultural people long before the exodus from Egypt; and having been accustomed to the enjoyments and fruits of a settled agricultural life, they were doubly sensible of the privations which their life in the desert necessarily involved (Num. xi. 5). Still there was one portion of the nation, which seems to have retained its nomad habits even till the time of the departure from Egypt, viz., the tribes of Reuben and Gad, and part of the tribe of Manasseh. At any rate, there was a striking contrast between these two tribes and a half and the rest of the tribes, when we consider the number of cattle possessed by the former (Num. xxxii. 1—4). Such wealth in cattle leads to the conclusion that the rearing of cattle was the only industrial occupation known among them, and this is inconceivable except in connexion with a nomadic mode of life. We feel justified in assuming, therefore, that these tribes had dwelt along the eastern border of the land of Goshen, and that their habits were to be attributed not to any particular preference for a wandering life, or any natural disinclination to settled habits, but simply to the peculiarity of the district assigned them, which was not fitted for cultivation.

The adoption of Egyptian agriculture was necessarily followed by a participation in Egyptian civilization. The peculiar nature of the agriculture of Egypt encouraged this, requiring, as it did, machinery and contrivances of various kinds, which again gave an impulse to arts and manufactures. But their settled life contributed still more to bring about this result. Fixed habitations are always promoters of industry ; they foster both a love of comfort and a want of the means of enjoying it. Many things, which a nomad regards as luxuries, become matters of daily and indispensable necessity. But the greatest influence of all must have been exerted by the fact that the Israelites lived in the same towns, and sometimes even in the same houses, as the Egyptians. In this respect also we find our expectations confirmed by the data of history. For example, we learn from 1 Chr. iv. 14, 21, 23, that in some of the families of the tribe of Judah there were carpenters, byssus-weavers, and potters on a very large scale. And since these are only incidentally alluded to, we may assume that other trades and arts were carried out to the same extent. From the work which the people performed in the deserts, we may estimate the various departments of industry in which they had been trained, and the perfection they must have reached. What a variety of arts and handicrafts, and what eminence in both of these, does the mere erection of the tabernacle presuppose ! The finest and most beautifully woven cloths were used, and the most accurate knowledge and skill, in the working of precious as well as common metals, in the grinding and engraving of precious stones, and in many other pursuits, must have been indispensably requisite.—So much, at least, we may clearly and certainly discover, that the time spent by Israel in Egypt, the land of highest culture, had not been lost. They had acquired considerable knowledge, they had been initiated into the advantages of civilization, and had learned how to apply the culture they received. Their *natural* development had been advanced to an incomparably higher stage; the natural foundation had been laid there for a fresh and more glorious revelation from God, and the natural pre-requisites had been attained for a new and nobler form of covenant with God. The announcement made to Abraham

(Gen. xv. 14), "they shall come out *with great substance,*" was thus fulfilled in a much higher sense, than by their coming out of Egypt with vessels of gold and silver (Ex. xii. 35, 36).

(1). In connexion with what we have said above consult the complete and searching investigations of *Hengstenberg* (Beitr. ii. 432—439), and the remarks of *v. Lengerke* (Kenaan i. 369 seq.), who arrives at the same conclusions.—*Heeren* has clearly pointed out, in his *Ideen* (hist. Werke xiv. 161), how thoroughly Egypt was adapted by nature to elevate the lower habits of a nomad-life into the superior habits fostered by agricultural pursuits. He says:—"The objects, which the founders of the Egyptian state naturally kept in view, were to promote the cultivation of the soil and to accustom the nomads to settled places of abode. In doing this they had the great advantage, that nature had already performed more for them, than in any other part of the world. The transition from a nomad-life to agriculture, however difficult of explanation it may generally be, was at any rate nowhere easier than in Egypt, where field-labour required scarcely any exertion, and nearly all that had to be done was to scatter the seed and reap the harvest." *Robinson* calls attention to the fact, that even now the nomads, who settle in Egypt, are almost involuntarily changed into farmers (Palestine i. 77). It is a very remarkable fact that there is not the slightest allusion to camels in any part of the history of Israel in Egypt and the desert, whereas according to Genesis they formed part of the cattle possessed by the patriarchs in Palestine. (See *Ritter* Erdkunde xiv. 739, and xiii. 701. 704).

(2). However easy the cultivation of the soil may be in the lower districts of the Nile-country, where the river overflows the land, and both waters and manures it without any interference on the part of man; in the higher ground there are peculiar difficulties to be overcome. The water must be raised by artificial means, before the land can be irrigated. That the Israelites were accustomed to make use of these means is apparent from Deut. xi. 10: "for the land, whither thou goest in to possess it, is not as the land of Egypt, from whence ye came out, where thou sowedst thy seed, and wateredst it with thy feet, as a garden of herbs." *Philo* gives a more minute description of the process here referred to (in his *de confusione linguarum* T. i. p. 410 ed.

Mangey): "the same may be said of the *pumping-wheel* (ἕλιξ). There are several steps (βαθμοί) in it, by treading on which the wheel is turned, and the water raised for the irrigation of the land. But in order that the man may not fall, he holds by his hands to some fixed object connected with the machinery, so that the whole body is suspended. Thus, instead of the hands he uses the feet, and instead of the feet the hands; for he stands with the hands, with which *we* are accustomed to work, and works with the feet, with which *we* are acustomed to stand." According to *Diodorus Siculus* (l. i. c. 34) this machine (which was called κοχλια, *i.e.*, a snail with a twisted shell, on account of its shape) was invented by Archimedes; but of course there is no ground for such an assertion, as Archimedes was the mythical centre of all mechanical inventions. The miners in Spain made use of similar machines for pumping water out of the pits. Of these *Diodorus* gives a detailed description in Book v. chap. 37: "When water flows in, it is pumped out with the so-called *Egyptian Kochlia.* With this they draw it out in a continuous stream till the pit is dry. By means of this extremely scientific contrivance an immense mass of water is pumped up with very little exertion, and all the water that may have come into the mine is easily raised from the bottom to the top." Pumping-wheels are still used in Egypt to water the higher ground, though they are constructed somewhat differently from those described by *Philo* and *Diodorus* (see *Niebuhr* Reise-beschr. 1. 148, and Abbild. Taf. 15). *Robinson* says (i. 541): "The water-wheel, *Sâkieh*, is usually turned by an ox, and raises the water by means of jars fastened to a circular or endless rope, which always hangs over the wheel." *Hengstenberg* (Egypt and the Books of Moses, p. 221) hesitates to apply the words of Deut. xi. 10 to the watering machine, because there is no representation of such a machine in any of the sculptures, and therefore it is most probably of later origin. And as there are representations, on the other hand, upon the monuments of persons carrying water, he thinks it more advisable to explain the passage as referring to this occupation, seeing that in the *carrying* of water "the *feet* have the most to do and to bear." We must refuse our support to this interpretation, for it would hardly occur to any one to describe such a method of watering, as watering *with the foot.* The omission of the machine from the monuments may be accidental.

(3). In proportion as the Israelites laid aside their wandering habits, and adopted the civilized customs of the Egyptians, the latter ceased to regard them with that abhorrence which they had felt towards them as nomads. Thus it came to pass that the Israelites were allowed to live in Egyptian towns, and even in the same houses with the native Egyptians (Ex. iii. 22). As the Israelites possessed houses of their own (Ex. xii. 4—7), it may sometimes have happened that Egyptians lodged in their houses. But *Hengstenberg* seems to lay too much stress upon the expression in Exodus iii. 22: "Every woman shall ask of her neighbour, and of *her that sojourneth in her house* (מְגָרַת בֵּיתָהּ), vessels of silver and vessels of gold," when he concludes from this passage, that Egyptians of great wealth and eminence lodged with the Israelites (p. 434). Persons who had a superfluity of gold and silver ornaments were most likely to have houses of their own. "*Her* house" need not be understood as meaning a house of which she was the owner; the house may have belonged to another, whilst she was the tenant at the time. —Closely as the Israelites approximated to the Egyptians, and greatly as their mode of life was changed in consequence, the difference of religion and of nationality always raised a sufficient barrier between them to prevent intermarriages. Yet there are cases on record in which this barrier was broken through, and that in a most striking manner. Thus, *e.g.*, according to 1 Chr. iv. 18, a daughter of Pharaoh, named *Bithjah*, was married to a man of the tribe of Judah, named *Mered*. But her name *Bithjah*, which is not only not Egyptian, but is a Hebrew word formed from the name of the God of Israel, must have been received at the time of her marriage, and is a sufficient proof that this unusual step was attended by the relinquishment of her Egyptian nationality and religion. She may possibly have been an Egyptian *Ruth*, with faith as strong as that which dictated the words, "Thy God shall be my God, and thy people my people."

§ 16. The Israelites entered Egypt as a single family, whose unity was represented by the one common father. As their numbers increased, it was both natural and necessary that the entire body of the people should be arranged in classes. From the independent manner in which the Pharaohs allowed the Israelitish community to develope itself, there was no necessity for this classification to be made according to the artificial division into castes required by the principle of the Egyptian state; on the contrary, full liberty was granted for the adoption of a strictly Hebrew classification. This was a purely natural one, founded upon the idea of a family. It was merely an expansion of the family ties which existed already. The connexion was closer or more distant, just according to the nearness or distance of the relationship. From the patriarchal unity there first proceeded a plurality of tribes (מַטּוֹת or שְׁבָטִים, also בָּתֵּי אָבוֹת), of which the sons of Jacob were the founders. The increase proceeded with such regularity and rapidity, that in the next generation the tribes began to divide themselves into different *clans* (*Geschlechter*, מִשְׁפָּחוֹת). As a general rule the grandsons of Jacob are to be regarded as the founders of these *Mishpachoth;* but in reality new *Mishpachoth* continued to be formed for several generations. This is evident from Num. xxvi. The number of *Mishpachoth* at that time was about sixty, and their numerical strength varied from four to sixteen thousand men who were capable of bearing arms. Such numbers as these would lead us to expect the principle of natural classification to be carried out beyond the *Mishpachoth.* And this was really the case. The *Mishpachoth* were divided into *families* or *houses* (בָּתִּים). This was the smallest division of the tribe, for the next in order were the גְּבָרִים, *i.e., individual men*, with their wives and children. The fourfold division is most clearly and fully exhibited in Josh. vii. 14, 17, 18. It is true that reference is there made to the state of things which existed in the time of Joshua; but we are perfectly justified in assuming that the same arrangement existed both in the Mosaic and the pre-Mosaic times, for there were the same elements for the division of the tribes in the days of Moses, though they may not be so clearly described ; and there is not the slightest intimation anywhere, that Moses made any alterations, or introduced any fresh organisation in this respect. On the contrary, the existence of a complete and final classification of the tribes is always presupposed. At the head of the tribes, and sections of the tribes, there were *princes* and heads, who occupied their position by right of primogeniture. They represented the unity of the tribe, or of the section, and in that capacity had undoubtedly corresponding magisterial rights and duties. The common name for these chiefs of every grade was רָאשֵׁי בֵּית-אָבוֹת heads of fathers' houses (generally written elliptically רָאשֵׁי-אָבוֹת). Those of them who stood at the head of a whole tribe were called princes, נְשִׂיאִים (נְשִׂיאֵי הָעֵדָה, נְשִׂיאֵי מַטּוֹת); see Num. i. 4, 16. So far as the command of the tribes was in their hands, Israel was under a federal, aristocratic .government. The *elders* (זְקֵנִים) are mentioned in connexion with the heads of the tribes, and are much more frequently referred to than the latter. There is not the slightest appearance anywhere of their being identical with the heads of the tribes, of either the higher or lower grades; on the contrary they are expressly referred to as distinct from these (Deut. xxix. 9). Their name may have lost its strictly literal signification, but it always indicated that they were the *élite*, of those who were distinguished for their age, their experience, and the general esteem in which they were held. Hence, in addition to the hereditary nobility of the heads of tribes, we find in these men a personal nobility, or nobility of merit belonging to the people. And whilst the former were nobles by birth, the latter were elevated to their rank and official standing on account of their wisdom, prudence, and experience, and were no doubt appointed by a free popular election. They always appear as the representatives of the people (Ex. iii. 16, 18, iv. 29, xii. 21,

xvii. 5, 6, xviii. 12, xix. 7, xxiv. 1, 9, 14, &c.). Whenever any communication had to be made to the people generally, or it was necessary that they should be represented, the *elders* were always convened. Hence they formed, to a certain extent, a democratic element in the otherwise aristocratic constitution. For want of farther information, it is impossible to give an accurate description of the nature of their office. In addition to their duties as representatives of the people, they seem to have possessed a peculiar kind of judicial authority. They were very numerous, for Moses appointed seventy of them as a council, to assist him in the general superintendence of the nation (Num. xi. 16). Probably every *familia* in the more general sense (as the smallest subdivision of the tribes), or at least every *gens (Mishpachah)* had its own council of elders, who were chosen from the wisest and most esteemed of the fathers of a family (גְּבָרִים).—Under the influence of Egyptian customs a new office was created, viz. that of *Scribe* (שֹׁטְרִים; LXX. γραμματεῖς; *Luther*, *Amtleute*). There was no country of the ancient world in which so much writing was done as in Egypt. For every trifling occurrence of public and private life, pen and ink, pencil or chisel, were close at hand, and everything, however unimportant, was written down. As soon as the Israelites began to adopt the civilized customs of Egypt, they felt the want of written documents, and men were quickly discovered to meet the want. These men acquired an official character, which gave authority to what they wrote. It is probable that one of their duties was to draw up the genealogical tables. When the Egyptian oppression commenced, and the people were required to render tributary service, the Israelitish *Shoterim* were commissioned by the government to distribute the labour, and were held responsible for its performance (Ex. v. 10, 14).

(1). According to Josh vii. 14, 17, 18, the whole body of the people were divided into *tribes*, the tribes into *Mishpachoth*, the Mishpachoth into *Bottim*, and the Bottim into *Gebarim*. For the reason already assigned, we consider ourselves justified in assuming that this classification existed in the Mosaic and pre-Mosaic times, though the last two subdivisions are not mentioned in the Pentateuch, where the people are always numbered and classed according to tribes and Mishpachoth. In this opinion we differ from nearly every modern commentator, the general opinion being that the בֵּית אָבוֹת (father's houses) correspond to the בָּתִּים (houses) of the book of Joshua, whereas we regard the former as a designation of the leading tribes.

In order to get at the idea of *Beth-aboth*, we start from the meaning of the word *Aboth*. Two explanations of this are possible. It may either denote the fathers, who were still alive,—those who had become fathers by begetting children, in contradistinction to the unmarried men,—or it may refer to the forefathers *(Majores)*, as distinguished from the existing generation. It appears to us, that there can be no great difficulty in deciding which of these two are meant. There are innumerable passages in the Pentateuch, as well as in the other books of the Old Testament, in which the term Aboth occurs with the meaning *Majores;* and, so far as we know, there is not a single instance in which it is used, without further explanation, with the meaning *husbands*, or *fathers of a family*. The usage of the language had so thoroughly associated the meaning *Majores* with the plural of the word אָב, that it was necessary to select another word, if it was to be employed with a different signification; and thus we find the word גְּבָרִים substituted in the book of Joshua.

If, then, the term Aboth, whenever it occurs, and therefore in the compound word Beth-aboth, denotes, not the fathers then living, but their ancestors and forefathers; it certainly follows, that a Beth-aboth must be an association comprising all the families and individuals, descended from the Aboth referred to at any particular time. But the question then arises, how far back the term Aboth extends, for this must be determined before we can tell whether a Beth-aboth was one of the earlier or later divisions, in other words, whether it was a *familia* (a בַּיִת in the sense of Josh. vii. 14) a *gens* (= Mishpachah), or lastly a *tribus* (מַטֶּה). If we enquire into the general usage, we learn that as a rule the Aboth denoted the earliest ancestors of the people; and therefore a Beth-aboth was most probably one of the earliest of the divisions of the people, viz. *a tribe.* This conjecture of ours is raised into a certainty, when we examine the following passages:

1. Num. i. 4, 16. Here the same persons are mentioned singly in ver. 4, as "every one *head of the house of his fathers*," and are classed together in v. 16, as "princes of the *tribes of their fathers*," from which it necessarily follows, that "the *house of the fathers*" and "*the tribe of the fathers*" were one and the same, *i.e.* that a *Beth-aboth* was one of the tribes.

2. Num. i. 20, 22, 24, 26, 28, &c. These passages are just as conclusive as the former. The census of the twelve tribes is here described, and the same formula is repeated in the case of every tribe; viz. "of the children of (Judah, &c.) their generations were according to their families, *according to the house of their fathers*, according to the number of their names" so many. The evidence in favour of our interpretation is to be found here in the constant recurrence, without any exception, of the singular *Beth-aboth*, *house* of the fathers (never *Botte-aboth*), whereas the *Mishpachoth*, families, are always in the plural. If the *Beth-aboth* were a subdivision of a Mishpachah, it would necessarily be always used in the plural also. We see, therefore, that the plurality of the *Mishpachoth* passed into the unity of a *Beth-aboth*, and hence that the *Mishpachoth* must have been subdivisions of a *Beth-aboth*, in other words, that a *Beth-aboth*, a house of the fathers, must have been a *tribe*. It is true that *Gesenius* and many other expositors answer the argument, founded upon the use of the singular, by saying without explanation that *Beth-aboth* is a plural, (equivalent to בָּתֵּי אָב): "quae pluralis formandi ratio in nominibus compositis apud Syros usitatior est." But it is not proved, and cannot be proved, that this *formandi ratio* was a *usitata* in Hebrew; least of all can בית-אבות be adduced to establish it, since this always makes good sense, when interpreted as a singular in its simplest and most natural meaning. Moreover Beth-ab has been proved to mean something entirely different from Beth-aboth.

3. Num. iii. 15, sqq. throws peculiar light upon this question. In ver. 15 we read, "number the children of Levi,

according to the house of their fathers, according to their *Mishpachoth.*" This is done with the following result: (1), "*according to the house of their fathers,*" the children of Levi are Gershon, Kehath, Merari" (ver. 17); these three therefore form the Beth-aboth of the children of Levi; (2) *according to their Mishpachoth,* the names of the sons of Gershon, Kohath, and Merari are given, as the founders of the *Mishpachoth* of the tribe of Levi (ver. 18, 20). The enumeration concludes with the words "these are the *Mishpachoth* of Levi according to the house of their fathers." Thus the *Beth-aboth* of the children of Levi included the whole of the tribe of Levi, and the *Aboth* were Gershon, Kehath and Merari. The *Beth-aboth* of the children of Levi was divided into three sections, each of which was called a *Beth-ab*, and every *Beth-ab* was subdivided into a certain number of *Mishpachoth.* This is indisputably proved by the following passages: ver. 24: "and the prince of the *Beth-ab* of the Gershonites shall be Eliasaph;" ver. 30: "and the prince of the *Beth-ab* of the families of the Kehathites shall be Elizaphan;" ver. 35: "and the prince of the *Beth-ab* of the families of Merari shall be Zuriel;" ver. 32: "and the prince of the princes of Levi shall be Eleazar."—Here then we have an authoritative explanation of the difference between *Beth-ab* and *Beth-aboth.* The expression *Aboth,* as indicative of the point from which the division of the tribes started, carries us back to the sons of the twelve patriarchs; in other words, only such of the descendants of Jacob, as were the founders of the nation in the land of Egypt, and are expressly mentioned as such in Gen. xlvi., were *Aboth* (fathers) *κατ' ἐξοχήν.*—The following was the classification of the tribe of Levi: the tribe, or *Beth-aboth,* was divided into as many houses *(Beth-ab)* as the *patriarch* (Levi) had *sons;* and every *Beth-ab* was then subdivided into single *Mishpachoth* according to the number of the patriarch's *grandsons.*

In the case of the other tribes, indeed, the classification was not so completely carried out, or at any rate was not so perfectly maintained. For instance they had no *Beth-ab* between the *Beth-aboth* and the *Mishpachoth.* At least, in the two numberings described in Num. i. 20 sqq. and Num. xxvi. the people are merely classified under these two heads. In the second census (Num. xxvi.) the different *Mishpachoth* are mentioned by name. By far the greater number of these derive their name and their origin from the sons of the twelve (or rather, since the adoption of Joseph's sons, thirteen) patriarchs, very few of them from their grandsons or great grandsons. The latter are always co-ordinate with the rest, not subordinate to them. Hence there was no room for the name *Beth-ab.* The tribe of Levi formed the only exception in this respect. The intermediate class, *Beth-ab*, which was omitted in all the other tribes, was restored in the case of this tribe (probably by Moses, Num. iii.), and, as this chapter most clearly shows, it was done for the purpose of securing regularity in the order of encampment, and a better distribution of their duties in the sanctuary.

4. We have thus discovered from Num. iii., that the name *Aboth* (in its highest sense) only reached as far back as the grandsons of Jacob, *i.e.* to those who went down with Jacob to Egypt, and there became the founders of the nation. A *Beth-ab* was a division of the people, springing from one *individual* among these *Aboth*; a *Beth-aboth* was a division of the people, in the formation of which *several Aboth* were concerned. Thus a *Beth-aboth* included several *Beth-abs.* In this manner *Beth-aboth* became fixed as the name of a *tribe.* But as the sons of Jacob and Jacob himself were *Aboth*, and not merely his grandsons (see Gen. xlvi.), *Beth-aboth* may have been employed in a wider sense, to denote the house of the (12) *sons of Jacob*, *i.e.* all the descendants of Jacob, and may thus have been equivalent to the congregation. It occurs in this sense in Ex. vi. 14. A genealogical section is there introduced by the heading: "these be *the heads of Beth-Abotham.*" It then proceeds: "the sons of Reuben are Hanoch, Pallu, Hezron, and Carmi. These be the *Mishpachoth* of Reuben." The children of Simeon and Levi are then named in the same way. The genealogy ends with Levi, as the author was merely writing about Levi, and there was therefore no reason for carrying it farther. The heads (*i.e.* the founders, originators) of the *Beth-aboth* were Reuben, Simeon, Levi. Hence the *Beth-aboth*, here referred to, was formed by a combination of the sons of Jacob.—At all events this passage most decidedly proves, that a *Beth-aboth* was not a section of a *Mishpachah.*

5. If then, as Ex. vi. 14 shows, the expression *Beth-aboth* may be used to designate a combination of all the tribes, it

follows that *Beth-ab* (the house of one of the fathers referred to above) may also be used for a *tribe*. And, undoubtedly, it is so used in Num. xvii. 2: "take of the children of Israel *twelve* rods, *one rod for each Beth-ab*, of all their princes according to *Beth-abotham, twelve rods."—Beth-abotham* is probably used here, as in Ex. vi. 14, to denote the twelve-membered unity of the whole people; and *there can be no possible doubt* that *Beth-ab* is to be regarded as a designation of each one of the twelve tribes.

From the passage referred to it is evident, that although the meaning of the words *Beth-ab* and *Beth-aboth* is not sharply defined or invariably the same, they never can be explained as denoting subdivisions of a *Mishpachah*, that, on the contrary, the *Mishpachah* must be a subdivision of the *Beth-ab* and *Beth-aboth*. This is so certain and so plain, that it is almost inexplicable, how so many excellent commentators can have overlooked their proper relation. It does admit of explanation, however, seeing that there are many passages, which *appear* to favour the opposite view. The first thing, which strikes us as at variance with our conclusion, is the fact that very frequently a number of heads of *Beth-aboth* (or still more frequently by ellipsis heads of *Aboth*) are mentioned, and *that* evidently within the limits of a single tribe, so that it seems necessary to render the *Beth-aboth* as a plural, indicating the sub-divisions of the tribes and *Mishpachoth*. When, for example, the *Mishpachah* of the Belaites is spoken of in 1 Chr. vii. 7, as containing five heads of *Beth-aboth*, and in 1 Chr. vii. 40 a large number of descendants of Asher are called heads of *Beth-aboth;* when again the *Mishpachah* of the Gileadites is referred to in Num. xxxvi. 1 as containing a plurality of heads of *Aboth*, and the same occurs in many other passages; it appears that we are justified in assuming, or rather actually compelled to assume, that the term *Beth-aboth* is used to describe a number of minor divisions, subordinate to the *Mishpachah*. Yet the whole difficulty vanishes before the simple observation, that tribe-leaders *(Rashe-Beth-aboth)* were not necessarily heads *of* the tribe, but might also be heads *in* the tribe, that is, not those who presided over the whole tribe, but over certain of its sub-divisions. The *Rashe-Beth-aboth*, or, in the abbreviated form, *Rashe-aboth*, were all those who were by birth the leaders of the people within the limits of a *Beth-aboth*, whether they stood at the head of an entire tribe, of a gens, or of a family in the less restricted sense. This is so clear and indisputable, that we scarcely think it necessary to bring forward analogous cases in proof of it. Let it suffice, therefore, to point out the expression, princes of the congregation, which so frequently occurs in the Pentateuch, and by which we are to understand not princes over the entire community, but princes over particular sections of the community.

In the foregoing remarks we have shown, that there are a number of passages, in which the meaning is so clear that we are necessarily forced to the conclusion, that the term *Beth-aboth* is the name of a whole tribe, if not of the entire community. In all the other passages, in which the expression occurs, it may easily be so explained as to admit of this meaning. The most likely passage to create a difficulty is Ex. xii. 3: "speak ye unto all the congregation of Israel, saying, in the tenth day of this month, they shall take to them every man a lamb, *according to the house of their fathers*, a lamb for a house." But I do not see why *l'beth-aboth* should not be rendered κατὰ φυλήν. At any rate so much is certain, that the passage does *not* compel us to adopt our opponents' explanation of *Beth-aboth*.

There is only one passage in which we have been unable to see our way clear, to the removal of every difficulty. We refer to 1 Chr. xxiii. 11. In this passage it is said of the two grandsons of the Levite Gershon: "they had not many sons, therefore they were לְבֵית אָב, in *one* reckoning." The passage is apparently all the more important, as treating of the period, in which, according to the views of our opponents, the "*fathers' houses*" began to be formed. But if the *Beth-ab* in this passage is to be regarded as a fixed genealogical term, in the sense of a sub-division of the *Mishpachah*, there is evidently an undisguised and irreconcileable discrepancy between the statement here made and Num. iii. 24, where, as we saw above, Gershon himself is the founder of a *Beth-ab*, and the *Mishpachoth* subordinate to it are founded by his sons, whilst *here* the grandson of Gershon lays the foundation of a *Beth-ab*, as a minor section of a *Mishpachah*. With such a discrepancy before us, we should decidedly feel bound to give the preference to the authentic, and at all events more trustworthy account in the Pentateuch, and to leave the statement in the Chronicles alone.

Great stress is laid by our opponents upon the fact that we only meet with the words *Beth-ab* and *Beth-aboth* (never *Botte Ab* or *Botte Aboth)*, as justifying, *if not* necessitating the conclusion, that *the latter* is the plural of *the former (Beth Aboth* for *Botte Ab).* But so long as not a single example can be adduced from the whole of the early Hebrew thesaurus of a plural so formed in the case of a compound word, whereas in *every case* the *nomen regens*, as the more important of the two, naturally takes the plural form, I adhere to my opinion that *Beth-aboth* can only mean "house of the fathers," not "houses of the father," especially as the former meaning, as I have shown, is admissible in every passage in which the word occurs. It has been already apparent from Num. i. 16, compared with ver. 4, that the plural *aboth* is not a dependent word, governed by the *nomen regens.* In ver. 4 the plural *Nesi'e Mattoth Abotham* is substituted for the singular *Rosh-l'beth Abotham.* If, then, *Beth-aboth* were used in ver. 4 for *Botte Ab*, we should necessarily find *Mattoth-ab* in ver. 16. But it is just this passage, which apparently proves, that the plural forms *Botte-ab*, and *Botte-aboth* were intentionally avoided, and that, wherever the context required a plural, some other form was selected in preference to *Botte.* It is impossible to decide with certainty, what gave rise to this wish to avoid using the forms *Botte-ab* and *Botte-aboth*,—it probably arose from the fact that *familiae* was regarded as the fixed meaning of *Bottim* (as Josh. vii. clearly shows).

§ 17. All divine *revelation*, both direct and indirect, by *prophetic* discourse and visions, as well as by the words, and acts, and appearances of *God* himself, had ceased since the days of Jacob. At least we cannot find the slightest trace of its continuance. It was not till the end of their stay in Egypt, that the Israelites began to receive it again, as a preparation for their entrance upon a fresh and more advanced stage in their history (1). Even the birth of Moses, the hero of God, and the greatest of all the heroes of the Old Testament, was not attended by any such divine manifestation, as we should expect from other analogous cases.[1] The reason of this interruption of divine revelation for 400

[1] We take the liberty, in opposition to the mythical theology, of calling attention to this omission as a fresh argument against the mythical theory.

years, appears to us to have been that the peculiar end to be answered by the sojourn in Egypt, was one which could be attained by purely natural means. When once the grace, which worked in Abraham, Isaac, and Jacob, had overcome the natural curse of barrenness which rested upon this family, the growth of the nation could be effected by the simple process of nature, which merely required the general superintendence of divine providence for its successful results. And so far as the training of Israel as a cultivated nation was concerned, Egypt was to be its tutor. In this no special assistance from God was required. It is true that the civilization of Egypt, in which Israel was to participate, was thoroughly impregnated with the worship of nature, which Israel was to avoid; but it was not impossible to take the one without the other. In the religious consciousness which they had inherited from the fathers, in the recollection of the revelations and promises which they had received, and in the consequent hope of a coming day, when their independence as a nation would be secured, the Israelites were furnished with safe and powerful re-agents, by which to test and separate all that was ungodly in the customs of Egypt.—We have no direct information with regard to the *worship* of the Israelites during their stay in Egypt, but there are incidental allusions from which many things may be inferred. We may lay it down as *a priori* certain that they were not entirely without forms of worship; for where was there ever a nation of antiquity which did not stand in an acknowledged relation to the Deity, and did not express that relation in some mode of worship? The only question that can arise is whether, and to what extent, the Israelites adhered to the mode they had inherited from their fathers, or adopted the ceremonies of Egypt. From the vivid recollections of the history of the fathers, which were universally preserved in the consciousness of the people, as we may infer from the elaborate description of that history contained in the book of Genesis, we should be led to imagine that they remained true to the forms of worship inherited from the patriarchs. But the comparative poverty of the patriarchal forms, when compared with the gorgeousness and variety of the ceremonies of Egypt, with which they came into such close contact, would also lead us to expect that the latter exerted a constantly increasing influence upon the former. There are two ways in which the Israelitish forms of worship may have been enriched by elements of Egyptian origin. No harm could result, so long as they adopted only forms and symbols in harmony with the religious views which they had inherited from their fathers, *i.e.*, such as were adapted to give a more fitting expression to those views, to display them in richer and more various ways, without destroying or in any way detracting from their peculiar and distinctive (theistic) character. This, in fact, was one of the services to be performed by Egypt for the chosen people of God. The history of the giving of the law proves that their worship must have been so enriched, and that in no slight degree. How many religious customs, symbols, and institutions are there referred to as familiarly known, the relation of which to the ceremonies of the Egyptian worship cannot be disputed (*e.g.* the Urim and Thummim). With an impartiality, which presupposed that these forms and symbols were already current among the people, the lawgiver did not stop to give any detailed description of them, whilst others, of which this could not be assumed, were described by him in the most minute manner, one might almost say with trivial carefulness. All that the law had to do, in such cases as these, was when necessary to improve, legalise, and regulate what had been already adopted, and to assign to each its proper place in the whole system of religious symbolism, of which it was to form a part. But there was *another* way in which the worship might have been enriched, and which would not have been so harmless. The Israelites might have adopted religious forms and symbols together with their heathen signification, or, what is the same thing, have adopted such forms as were *a priori* unfitted to serve as vehicles for theistic ideas and views, on account of their having been created for and adapted to purely heathen notions. In such a case, even if the discordant theistic idea had been forced into association with the form, the latter would naturally and inevitably have turned it into a heathen idea (an example of this was the worship of God under the image of a calf, Ex. xxxii.) The worship of nature possessed a magic power, and presented irresistible attractions to the minds of men in the ancient world. Against these, it is true, such of the Israelites as were spiritually minded were protected by the religious inheritance, bequeathed by the fathers, and by their own promises and hopes; but they were just as seductive to carnally minded Israelites as to any other people. Hence from the power possessed by the worship of nature in those days, it was to be feared from the very first, that the lawful adoption of Egyptian forms and symbols would not be attended by so strict a process of sifting and refining, as would be requisite to prevent their being guilty of mixing up different religions in a false and ungodly manner. How much reason there was for such an apprehension is proved by their history, to a far greater extent, perhaps, than we should expect. Ezekiel (chap. xx. 5—8, cf. xxiii. 3) complains that Israel defiled itself with the idols of Egypt in the days of its youth. So also does Joshua in chap. xxiv. 14. And in the making of *the golden calf* in the desert (Ex. xxxii.), we have an example and a proof of the extent to which this false syncretism had taken root and spread among the people. Again the constantly recurring prohibitions of nature-worship, and of the ceremonies associated with it, on which it was thought necessary to lay such frequent stress in the law, presuppose existing indications of a strong tendency to such worship. Thus from Leviticus xvii. 7, we perceive that the Egyptian goat-worship,[1] especially, had found great favour with the people. We cannot suppose that the people intended by this an express denial by the God of their fathers, or were conscious that it involved an apostasy from their fathers' religion. But the precepts of the law and the discipline of history were required to open their eyes to the dangers of that abyss, into which they were ready to plunge. When we enquire for proofs of the actual employment of forms of worship, which had already been known and adopted by the fathers, our attention is especially directed to circumcision, sacrifice, and the Sabbath. With regard to *circumcision* it is evident from Exodus iv. 24—26 (*vid.* § 21. 3), that this token of the covenant never lost its validity or fell into disuse, and in Josh. v. 5 it is expressly said, that all the people who came out of Egypt had been circumcised. We might think ourselves justified in inferring from Ex. viii. 25—28, that the offering of *sacrifice* had been entirely discontinued during their stay in Egypt, from a regard to the Egyptians, to whom the Israelitish mode of sacrifice was an abomination. But there is a reference in this passage to a particularly solemn festival, in which the whole community was to take part, and which would therefore necessarily attract universal attention. Hence it could not but appear unadvisable to celebrate such a festival within the limits of the Egyptian territory (§ 29. 3). But it does not follow from this, that it was impossible to offer sacrifices within the walls of private houses, without attracting attention or assuming the character of a demonstration, and therefore without any hindrance or fear of disturbance. At any rate this passage proves, that the necessity for sacrificial worship had not lost its hold on the religious consciousness of the people, and also that that mode of sacrifice, which had been inherited from the fathers, and was an abomination to the Egyptians, was still

[1] The English version is: "They shall offer no more their sacrifices unto *devils.*" But the word used here is the ordinary Hebrew word for *a buck*, or *he-goat.—(Tr.)*

in force, so that in this respect at least, the Israelites had faithfully preserved the religious peculiarities, which distinguished them from the Egyptians. We find no trace of any special *order of priests.* The existence of such an order cannot be inferred from Ex. xix. 22 any more than from 1 Sam. ii. 27 ; for at Sinai the elders evidently officiated as priests (ver. 7), and the second passage says nothing about the tribe of Levi having held the priesthood in Egypt. If sacrifices were offered, there can be no doubt that the fathers or heads of the families officiated, as in the time of the patriarchs, unless the sacrifice was offered for the whole nation, when the representatives of the nation, *i.e.*, the elders, would officiate. With regard to the *Sabbath*, not only is the mode of its celebration doubtful, but there is reason to question its existence even during the patriarchal age (§ 7. 2), and neither Ex. xvi. 22 seq. nor Ex. xx. 8 furnishes any certain information with reference to the practice in Egypt. We may safely assume, however, that the Egyptian taskmasters (Ex. v. 13, 14) would pay no attention to any Sabbatical institution that might be in existence.

(1). It has been argued from 1 Sam. ii. 27, that there was not an interruption of divine revelation during the stay in Egypt. But the argument is unsound. The meaning of the words : " I plainly appeared unto the house of thy father, when they were in Egypt in Pharaoh's house," &c., is fully exhausted, if we suppose them to refer to the last year of the sojourn of the Israelites there.—At the same time there is a strong proof, that the religious consciousness was kept alive in the hearts of the people, in the fact that in so many of the proper names which were given during that period (Num. iii.), the name of God is found as one of the component parts.

§ 18 (1 Chr. vii. 20—24).—There is no account in the *Pentateuch* of any particular events, which may have happened to individual tribes during the first centuries of these 430 years.

But the passage, cited above, contains some data of a most remarkable kind, from which, if our explanation be the correct one, we learn that some of the Israelites began to think of returning to Palestine at a very early period, and attempted to carry out their intentions by their own power. One portion of the tribe of *Ephraim* returned and settled in the southern highlands of Palestine, even during the lifetime of Ephraim himself. From these settlements they made predatory incursions into the plain of *Philistia,* in which, however, they suffered such severe losses that the whole of their father's house was thrown into the deepest sorrow. This repulse probably weakened them so much that the quixotic undertaking had to be relinquished.—An enterprise of a similar character is referred to in 1 Chr. iv. 22, where some of the descendants of Judah are said to have ruled over *Moab.* The writer of the Chronicles refers to the דְבָרִים עַתִּיקִים, that is, to the ancient accounts belonging to a very remote period. On the relation of the Israelites to the Hyksos-dynasty see § 34 sqq.

(1). In 1 Chr. vii. 21 there are almost as many enigmas as words. The preceding verse contains a genealogy of Ephraim carried down to the seventh generation : "The sons of Ephraim are *Shuthelah,* and his son Bered, and his son Tahath, and his son Eladah, and his son Zabad, and his son Shuthelah, and *Ezer* and *Elead.*" Then follows in ver. 21 : "*And the men of Gath, who were born in the land, slew them, for they had gone down to take their cattle*; (ver. 22) and their father Ephraim mourned many days, and his brethren came to comfort him. (Ver. 23) And he went in to his wife, and she conceived and bare a son, and called him B'riah, for it went evil with his house. (Ver. 24) And his daughter Sherah built lower and upper Beth-horon and Uzzen-sherah."

The first thing that is doubtful is the period here referred to. *Ewald* (i. 490) places it before the migration into Egypt. As *Ewald* thinks he has a right to construct history at his pleasure with oracular authority, it does not of course trouble him in the least, that, according to the book of Genesis, Ephraim was born in Egypt. *Lengerke* (i. 355) and *Bertheau* (Chronik. p. 83) on the other hand assign it to the period immediately subsequent to Moses, and arbitrarily identify the Beriah in chap. vii. 23 with the Benjamite Beriah in chap. viii. 13. Moreover, in reply to the question: "How are we to dispose of the father Ephraim, who mourns for the loss of his sons?" *Bertheau* says, we shall be obliged to regard Ephraim as meaning the tribe, which mourned for the calamity that had happened to two of its sons, *i.e.*, to two divisions of the tribe." Good, we reply, but what are we to understand, then, by the Ephraim, who *after* this calamity goes in to his wife and begets a son named Beriah? Does this mean the whole of the tribe? As we cannot possibly think of any other Ephraim of later date, the account in the Chronicles brings us at the latest to the commencement of the second century of the sojourn in Egypt. But this does not seem to tally with what precedes, provided, that is, we look upon Shuthelah, Ezer and Elead (in ver. 21) as descendants of Ephraim in the seventh degree. Undoubtedly the suffix in וַהֲרָגוּם (and they slew *them)* may refer to the last names only. But it is certainly a mistake to string the *three* last names together and look upon them as sons of Zabad, for in that case we should expect to find "his sons" instead of "his son." The more correct arrangement is that adopted by *Bertheau* (p. 82), who classes the two last-named (Ezer and Elead) as sons of Ephraim himself, who continue the series commenced with Shuthelah in ver. 20.

Again it is doubtful whether the Ephraimites or the Gathites are to be regarded as the subject of יָרְדוּ (they had gone down) and what was the scene of this event. It has generally been supposed by earlier expositors, that the Ephraimites made a predatory attack upon the Gathites, entering Philistia from Egypt. *Calovius* (Bibl. illustr. ad. h. l.) gives the following unsatisfactory explanation of the event: "De Ephraimitis res ita habet: mora impatientes et gloria primogeniturae a Jacobo concessae tumentes tentarunt magnis consiliis eductionem ex Aegypto, adeoque progressi sunt, collecto exercitu, vivente adhuc patre Ephraimo, ex Aegypto affines terrae Canaan. Quo nomine accusat eos Assaph (Ps. lxxviii. 9), quod non exspectato justo tempore terram promissam invadere ausi fuerint fiducia copiarum et peritia sua in re bellica, additque, quod justo Dei judicio temeritatis suae poenas dederint, terga verterint, inque fuga misere perierint." But apart from the fact that Ps. lxxviii. 9 contains nothing at all of what *Calovius* has discovered there, this exposition is rendered impossible by the word ירד, which cannot refer to an expedition from Egypt to the more elevated land of Philistia. If we suppose the Ephraimites to have been in the land of Goshen at that time, we must necessarily regard the Gathites as the aggressors. Or if, on the other hand, we refer the words " they came down" to the Ephraimites, we must assume that they were no longer living in the land of Goshen, but had already established themselves in the highlands of Palestine.

Between these two interpretations we have to make our choice. *Bertheau* and *Lengerke* decide in favour of the latter, though we have already shown that the explanation given by *Lengerke* is inadmissible. *Saalschütz* (Mos. Recht, Berlin 1848, p. 651, seq.) also adopts it, and his interpretation is original and well worthy of consideration. His views are to some extent the same as those advocated by *Calovius*, but he describes and accounts for the expedition in a very different manner. "From chap. vii. 24, we perceive, he says, that a great-grand-daughter of Joseph built both upper and lower Beth-horon in the land of Canaan. If the building of these towns took place during the sojourn of the Israelites in Egypt, as some suppose, and as the context of the passage indisputably implies, seeing that it speaks of Ephraim as still alive, we have a positive proof that a portion of the Hebrews drove their flocks back to Palestine, and that they even went so far as to establish themselves in the land and build cities there." From chap. vii. 21, it follows as he thinks, "that the Ephraimites had settlements in Palestine, before the death of Ephraim; and if these settlements were in the district in which Beth-horon was built, either at that time or a little later, the map will furnish us with the best exposition of this passage in the Chronicles, for the situation of Beth-horon is pretty well known to us, being identical, as *Robinson* thinks, with *Beit-Ur*, which is about five hours' journey to the north-west of Jerusalem, that is, in the mountainous district at a short distance from Gath."

The other view, which makes the Gathites the aggressors, has been advocated by *Lightfoot* (Opp. i. 23, Rotterdam, 1686),

C. B. Michaelis (Annotationes in hagiogr. iii. 370), and many others. As the words "born in the land" must necessarily be understood as applying to the land, into which the incursion was made, the only explanation, which can possibly be given by those who adopt this view, is that the Gathites, by whom the attack was made, had formerly dwelt in the land of Goshen, and that having been forced out by the spread of the Israelites, they retaliated by making this attack upon their oppressors. We admit that the words of the text allow of such an interpretation, but in several respects it appears to us a forced one. First of all, it seems more natural to render the passage thus: "The Gathites slew the Ephraimites, for (כִּי) they had gone down to steal the cattle of the Gathites." Again it appears to us to be much more natural, *i.e.* more in accordance with the context and with history, to understand the words "born in the land" as referring to the land of Philistia. And lastly, there is the unmistakeable testimony of ver. 24, if we are correct in our supposition that the erection of Beth-horon occurred before the time of Moses. For these reasons, then, we are inclined to give the preference to the interpretation of *Saalschütz*.

BIRTH AND EDUCATION OF MOSES.

§ 19 (Exodus ii. 1—22, vi. 16—25).—Just at the time when the oppression was most severe, and when the command to drown the new-born boys of the Israelites was most stringently enforced, a son was born to an Israelite named *Amram*, of the tribe of Levi and the family of Kehath, by his wife Jochebed (1). The child was remarkable for its beauty; and therefore the mother was all the more concerned to save it, if possible, from the threatened destruction. She succeeded in concealing it for three months, but she could not hope to hide it any longer from the keen eyes of the Egyptian executioners. Maternal love, however, is always inventive. Jochebed knew that Pharaoh's daughter was accustomed to bathe at a certain spot in the Nile. This knowledge helped her to form her plan. She reckoned on the tenderness of a woman's heart. She placed the child in an ark constructed of papyrus stalks and securely pitched, and laid it among the reeds in the well-known spot by the side of the Nile, and left her eldest daughter *Miriam* to watch its further fate. The plan was successful. The king's daughter noticed the ark, and had it brought to her; and the sight of the beautiful weeping infant did not fail to produce the desired impression upon her heart. She soon conjectured that it must be one of the Israelitish boys; and as if by accident, *Miriam* came forward. She offered to fetch a Hebrew nurse. Of course she fetched the child's own mother, and Pharaoh's daughter gave her the child with the words: "take this child away, and nurse it for me, and I will give thee thy wages" (2). We look forward with anxiety to the future course of the child that has been so wonderfully rescued, feeling sure that he is destined for some remarkable mission. Nor can we doubt that some such surmise or hope must have been entertained by his parents, and that this increased their anxiety to give such a direction to his mind, as would be most likely to lead to the fulfilment of their own hopes. It is true that the child would only remain a few years in his parents' house, seeing that Pharaoh's daughter intended to bring him up as her adopted son; but even at a subsequent period it could not appear strange if the boy frequently visited his nurse's home. The people, too, to whom he belonged by birth must certainly have gazed upon him with looks full of expectation and hope; or, at any rate, they must have regarded the extraordinary events of his early life, as proofs of an overruling providence and divine call.—After he was weaned, Jochebed brought back the boy to his foster-mother, who gave him the name *Mo-udshe* (*i.e.*, ex aqua servatus, LXX. *Μωϋσῆς*, Hebraized מֹשֶׁה) (3), and had him educated *in all the wisdom of the Egyptians* (4). In this position a splendid career awaited him. The highest honours were within the reach of the adopted son of Pharaoh's daughter. But he felt within him a different call. He had imbibed affection for his people with his mother's milk, and the sufferings of his brethren went to his heart. He believed that he was called to be their deliverer and avenger. Whilst brooding over such thoughts as these, he happened one day to see an Egyptian ill-treating an Israelite. At once he was carried away by his zeal for his people, and, having slain the Egyptian, he buried him in the sand. There was no witness of what he had done except the injured Israelite; but the news soon spread among the rest, and it was probably the Israelite himself who circulated the report. Such a deed was like a general summons to them to rise against their oppressors, and Moses imagined that he had thereby obtained a certain amount of authority over his brethren. A short time afterwards he saw two Israelites quarrelling, and wished to act as arbitrator, but he was rudely thrust aside by the one whom he pronounced in the wrong. "Who," said he, "made thee a prince and a judge over us? Intendest thou to kill me as thou killedst the Egyptian?" On account of this, the report of what Moses had done began to spread among the Egyptians as well. The king heard of it, and determined to put him to death. Under these circumstances—pursued by the king, and forsaken by the people—Moses saw the necessity for *flight* (5). He sought refuge, and found it, in the land of the *Midianites* (6). A prince and priest of this people, named *Reguel* (7), received him into his house on account of the protection he had afforded to his daughters against the rudeness of the shepherds, gave him his daughter *Zipporah* as a wife, and entrusted his flocks to his care. The flight of Moses from Egypt introduced him into a new training school. At Pharaoh's court he had learned much that was required to fit him for his vocation, as the deliverer and leader of Israel, as the mediator of the ancient covenant and founder of the theocracy, and also as a prophet and lawgiver. But his education there had been of a very partial character. He had learned to rule, but not to serve, and the latter was as necessary, if not more so than the former. He possessed the fiery zeal of youth, but not the circumspection, the patience, or the firmness of age. A consciousness of his vocation had been aroused within him when in Egypt; but it was mixed with selfishness, pride, and ambition, with headstrong zeal, but yet with a pusillanimity which was soon daunted. He did not understand the art of being still and enduring, of waiting and listening for the direction of God, an art so indispensable for all who labour in the kingdom of God. In the school of Egyptian wisdom his mind had been enriched with all the treasures of man's wisdom, but his heart was still the rebellious unbelieving heart of the natural man, and therefore but little adapted for the reception of divine wisdom, and by no means fitted for performing the works of God. And even the habit of sifting and selecting, of pondering and testing, acquired by a man of learning and experience, must certainly have been far from securing anything like the mature wisdom and steadfastness demanded by his vocation. All this he had yet to acquire. Persecution and affliction, want and exile, nature and solitude, were now to be his tutors, and complete his education, before he entered upon the duties of his divine vocation (8).

(1). On *Amram* and *Jochebed* see § 14. 1. Moses was not their first-born son. His brother Aaron was three years older than he (Ex. vii. 7); whilst his sister, whose name (*Miriam*, LXX. *Μαρίαμ*) we do not learn till afterwards (Ex. xv. 20), had evidently grown up before he was born (Ex. ii. 4). The following is the family-pedigree:

Levi.
Gershon, KEHATH, Merari.
AMRAM.
Miriam. AARON, MOSES.
Nadab, Abihu, ELEAZAR, Ithamar. Gershon, Eliezer.
PHINEHAS.

(2). The biblical record expressly mentions the striking *beauty* of the child, as leading to the mother's determination to conceal and, if possible, save it. Ver. 2: "And she saw the child that it was good" (כִּי־טוֹב, LXX., ἀστεῖος). It is true, that it is not an unusual thing for a mother to think her new-born child beautiful; but just because it is not unusual, the peculiar character of the sacred record leads to the conclusion, that in this case there must have been something more than usual. *Stephen* had this impression, for he expressly traces the connexion between the beauty of the child and God himself (*καὶ ἦν αστεῖος τῷ θεῷ*). Some message from God must have been communicated to the mother in a peculiar manner by the eyes of the child; she may have seen in them the intimations of an eventful future, which, with her faith in the promises made to the fathers, stood out before her mind in marked contrast with the oppressions, the sufferings, and the anxieties of the present.

This was also the view taken by the author of the epistle to the Hebrews (chap. xi. 23), for he extols the concealment of the child as an act of faith. The whole affair would be still clearer, if we could rely upon the Jewish tradition that Amram was a prophet. But there is nothing to warrant this; on the contrary the tradition itself appears to have been founded entirely upon the passage before us. If the birth of Moses had been attended by any direct revelations or predictions from God, the sacred record, according to its usual custom, would certainly have mentioned them. And in its silence in this respect we find a proof of its historical fidelity.—*Josephus* mentions the name of Pharaoh's daughter. In Ant. 2. 9. 5 he calls her *Thermuthis*. But there is no more reliance to be placed upon his account, than upon that of *Eusebius* (praep. evang. ix. 27) who calls her *Μέῤῥις*. The latter looks like a corruption of Miriam. —The *queen's daughter bathing* in the Nile causes great offence to *Herr v. Bohlen* (Genesis lxxxi.), who regards it as an evidence of the author's gross ignorance of Egyptian customs. However the "gross ignorance" falls back upon the critic. In Egypt there was nothing like the same restraint upon women as in oriental countries or even in Greece. On some of the monuments we meet with scenes, in which the women associate with the men with almost as much freedom as modern Europeans (*Hengstenberg* Egypt and Moses, p. 26). "That the king's daughter went to the Nile to wash (לִרְחֹץ) is explained by the Egyptian notion of the sacredness of the Nile. A representation of an Egyptian bathing scene—a lady with four female servants who attend upon her to perform various offices,—is found in *Wilkinson* iii. 389" (*Hengstenberg*, p. 86). The preparation of the little ark too (whose name תֵּבָה reminds one of Noah's ark), the papyrus of which it was composed, and the asphalt and pitch with which it was covered, all harmonize with the antiquities of Egypt (see *Hengstenberg*, p. 85).—Under the circumstances there is nothing surprising in the fact, that as soon as the princess saw the boy, she concluded that it must be a Hebrew child; and there is certainly no necessity for assuming with *Aben Ezra* and *Theodoret*, ὅτι ἡ περιτομὴ τοῦτο ἐδήλωσε.—We may introduce here a most sensible remark made by *Baumgarten* in his Theological Commentary (i. 1, p. 399): "In the fact, that it was necessary for the deliverer of Israel from the power of Egypt to be himself first delivered by the daughter of the king of Egypt, we find the same interweaving of the history of Israel with the history of the Gentiles, which we have already observed in the history of Joseph; and we may now regard it as a law, that the preference shown to Israel, when it was selected as the chosen seed, on whom the blessings were first bestowed, was to be counterbalanced by the fact, that the salvation of Israel could not be fully effected without the intervention of the Gentiles. This was the opinion of *Cyril of Alexandria*, which he expressed in his usual allegorical style by saying: the daughter of Pharaoh is the community of the Gentiles." In all the decisive turning points of the sacred history, whenever a new bud was about to open, some heathen power always came forward, as though summoned by the providence of God, to assist in bursting the fetters by which the bud was held, in order that it might open into a splendid and fragrant flower.

(3). The time of *weaning* is generally supposed (according to 2 Macc. vii. 27; 1 Sam. i. 23, 24; *Josephus*, Ant. ii. 9, 6), to have been at the end of the third year. As the princess was about to adopt the child and bring it up as her own (ver. 10), it is most likely that, according to a mother's rights, she gave it its *name*. If so, she would naturally select an Egyptian name. But the name מֹשֶׁה is certainly Hebrew (= one who draws out, the deliverer). We have here, however, without doubt, a similar case to that which we meet with in Gen. xli. 45, where the Egyptian name, Psomtomphanech, which Pharaoh gave to Joseph, is handed down in the form, Zaphnath-Paaneah, which admits of a Hebrew etymology (Vol. i. § 88, 2). And in both cases the *Septuagint* puts us upon the right track, by writing the name in a manner more closely resembling the original Egyptian form. Thus the name of Moses is always written, Μωϋσῆς, of which *Josephus* (Ant. ii. 9, 6), gives the correct explanation: τὸ γὰρ ὕδωρ ΜΩ οἱ Αἰγύπτιοι καλοῦσιν, ΥΣΗΣ δὲ τοὺς ἐξ ὕδατος σωθέντας. *Philo* explains it in a similar manner (de vita Mos. ii. 83, ed. Mang.): διὰ τὸ ἐκ τοῦ ὕδατος αὐτὸν ἀνελέσθαι· τὸ γὰρ ὕδωρ ΜΩΣ ὀνομάζουσιν Αἰγύπτιοι). In this *Clemens of Alexandria* (Strom. i. 251 ed. Sylb.), and *Ezekiel* the tragedian (Eusebius praep. ev. ix. 28) agree. The derivation here given is confirmed by our present acquaintance with the Coptic, in which *Mo* means *water*, and *Udshe* saved (*cf. Jablonski* opusc. i. 152 sqq.). Most modern authors adopt it; but though *Gesenius* will not actually reject it, he says in his Thesaurus that reputans sibi nominum propriorum apud veteres Aegyptios usitatorum, quae pleraque cum Deorum nominibus conjuncta sunt ratione (*e.g.* Amôs, Thuthmôs, Phthamôs, Rhamôs, &c.), he must prefer to trace the name to the Egyptian word Môs, a *son*, and to assume that the first part of the word, which contained the name of a god, was dropped in Hebrew usage. No one but *Lengerke* (i. 390), supports this explanation, and it will hardly meet with any further approval. Many of the earlier theologians made it, to a certain extent, a point of honour to affirm, that it was not Pharaoh's daughter, but the child's own mother, who gave it its name. Thus *Pfeiffer* (dub. vex., p. 214), following *Abarbanel*, renders ver. 10: "adduxit eum (sc. mater ipsius) ad filiam Pharaonis et factus est ipsi filius. Vocarat vero nomen ejus (sc. mater jam dudum) *Mose* (quod tum indicabat filiae Pharaonis), et dicebat: quia ex aqua educendum *curasti* eum," defending his translation on the ground that מְשִׁיתִהוּ not only *can*, but *must* be the second person feminine (since it is written *defective*, without י). But apart from every other consideration, we should in this case expect to find not מֹשֶׁה but מָשׁוּי. *Meier* also decides that the name was originally Hebrew (Wurzel-wörterb. p. 704). In his opinion the real name of

Moses was Osarsiph (meaning Osiris-sword), for which we have the testimony of *Manetho* (in Josephus c. Apion i. 26—28); and he received the name *Moses* (meaning the deliverer, leader, duke, *dux*) in connexion with the exodus from Egypt. We cannot adopt this explanation, since the Scripture-record attributes the naming to the princess, though under other circumstances it would commend itself; on the other hand, we do not hesitate for a moment to adopt the explanation given by *Josephus*, *Philo*, and *Clemens*, and based upon the *Septuagint*, since it meets all the requirements of the language and necessities of the case. The name did not suit the Hebrew organs of speech, and was therefore involuntarily changed by the Israelites into the form in which we have received it. At the same time this involuntary change became an unintentional prophecy, for he who had been *delivered* (taken out) actually became a *deliverer*. *Vox populi, vox Dei.*

(4). "*Moses was trained*," says *Stephen*, (Acts vii. 22), *in all the wisdom of the Egyptians*." These words are not founded upon a baseless tradition, or the creation of his own fancy, but form a just and necessary comment upon Ex. ii. 10: "and he became her son." The adopted son of the daughter of an Egyptian king *must* have been trained in all the wisdom of Egypt. This is also in harmony with the tradition reported by *Manetho*, which makes Moses a priest of Heliopolis, and therefore presupposes a priestly education. It was precisely this education in the wisdom of the Egyptians, which was the ultimate design of God in all the leadings of his providence, not only with reference to the boy, but, we might say, to the whole of Israel. For it was in order to appropriate the wisdom and culture of Egypt, and to take possession of them as a human basis for divine instruction and direction, that Jacob's family left the land of their fathers' pilgrimage, and their descendants' hope and promise. But the guidance and fate of the whole of Israel were at this time concentrated in Moses. "As Joseph's elevation to the post of grand-vizier of Egypt placed him in a position to provide for his father's house in the time of famine, so was Moses fitted by the Egyptian training received at Pharaoh's court to become the leader and lawgiver of his people." (*Baumgarten* theol. Comm. I. i. 399). There can be no doubt that the foster-son of the king's daughter, the highly-gifted and well educated youth, had the most brilliant course open before him in the Egyptian state. Had he desired it, he would most likely have been able to rise like Joseph to the highest honours. But affairs were very different now. Moses could not enter on such a course as this without sacrificing his nation, his convictions, his hopes, his faith, and his vocation. But that he neither would, nor durst, nor could. And hence it is with perfect truth that the author of the epistle to the Hebrews, when tracing the course of the history, says (ch. xi. 24—26): "by faith Moses, when he was come to years, refused to be called the son of Pharaoh's daughter, choosing rather to suffer affliction with the people of God, than to enjoy the pleasures of sin for a season; esteeming the reproach of Christ greater riches than the treasures of Egypt; for he had respect unto the recompence of the reward." *Winer*, who generally defends the historical character of our record against the attacks of the myth-loving critics, finds it difficult to explain "how it is that Moses should have been trained by an Egyptian princess, and yet is never represented as known to the court, when engaged in his subsequent negotiations with Pharaoh, whilst even in Ex. ii. 11, there is no allusion to his connexion with Pharaoh's daughter." (Realwörterbuch ii. 10). But for my part I cannot perceive the slightest difficulty in this. With regard to the former, a long series of years had passed since the flight of Moses from Egypt (Ex. vii. 7); the king who was reigning then had long since died (Ex. ii. 23); an entirely new generation had grown up; and we cannot therefore be surprised at the fact that Moses was no longer known at the court. But even supposing that he had been recognised, was there any reason why this should be specially noticed in the biblical narrative? Is there anywhere an express statement to the effect that he was not known? We believe that a negative reply must be given to both these questions. There is just as little ground for the second difficulty. The princess may have been dead when the event referred to in Ex. ii. 11 occurred, or, if not, it is just possible that as the attachment of Moses to his own people and his dislike of their oppressors became more and more apparent, there may have sprung up a growing estrangement between him and his foster-mother. And it is also probable that he may have begun to keep aloof from the court, meditating more upon the way to deliver his people, than upon the means of retaining the favour he had previously enjoyed.

Winer has made a good collection of the legendary tales associated with the early history of Moses in the Jewish mythology: "He is said to have been instructed in all the wisdom of the Egyptians, both by Egyptian and foreign teachers, including Greeks, Assyrians, and Chaldeans, and to have been remarkable as a boy for his enchanting beauty (*Philo* Opp. ii. 84, *cf. Clemens of Alexandria* Strom. i. 148. *Josephus* Ant. l. c. *cf. Justin* 36. 2). *Justin* says, '*Moses . . . quem formae pulchritudo commendabat.*' When he had grown up to be a young man, he led an Egyptian army against Ethiopia, and forced his way to Meroë, where he married the Ethiopian princess Tharbis, who had become enamoured of the fine manly youth, and had opened the gates of the fortress to his army (*Josephus* Ant. ii. 20)." The additional account given by *Josephus* (Ant. ii. 9. 7) is evidently copied from the legend concerning the elder Cyrus. He says the childless princess intended that the child should succeed to the throne, and endeavoured to win over her father, the aged king, to her plan. As a token of his consent, the king took the boy in his arms, hugged him, and put the royal diadem upon his head. But the child threw the crown upon the ground and stamped upon it. Upon this a scribe, who had formerly prophesied that a child would be born, who would be dangerous to Egypt, declared that this was the dangerous child, whose birth he had predicted, and requested that he should be put to death. But Thermuthis protected the child, and the king gradually forgot the occurrence, &c. The marriage of Moses with an Ethiopian princess was probably founded upon Num. xii. 1, where we read of his Ethiopian wife (Vol. iii. § 27. 3).

(5). The conduct of Moses towards the *offending Egyptian*, and the reply he received from the *insolent Israelite*, are very important, as helps to an acquaintance with his inner life at that time, his thoughts and imaginations, his hopes and fears. Here again, *Stephen* furnishes us with a complete, and well-founded explanation (Acts vii. 25): "he supposed his brethren would have understood how that God by his hand would deliver them; but they understood not." He is full of thoughts of deliverance, but does not yet know how he shall carry them out. He feels within himself that he is called to this work, he believes that this feeling is the voicc of God; but there is a carnal thirst for great achievements and worldly ambition mixed up with it, and these are unfitted for the ways of God. He is anxious to attract the attention of his brethren; he thinks that the adopted son of the king's daughter has naturally a right to stand at the head of his people; he has only to show that he is ready to do this, and all the people will acquiesce in a moment. But he was greatly mistaken. This, however, formed part of his training for his vocation; it was necessary that he should pass through some such experience as this, before he could be matured for his future work. He must find out the perverseness of his people, who would not observe and learn; and the perverseness of his own heart, whose courage and confidence were changed into cowardice and despair at the first failure that occurred. He must also discover the ways of God, who will not tolerate a man's self-confidence or self-elected ways. Still the love of Moses to his people was strong and noble, his vocation was true, and his aims were essentially godly. All these, however (as in the case of Joseph, Vol. i. § 84. 1), required a thorough purification and sanctification in the school of affliction and of humiliation, before God could use them to work out his designs. Hence, for the present, Moses could not succeed. Moreover, the weakness of his carnal mind and his natural pride was soon apparent, from the manner in which they gave way to despondency and cowardice at the very first failure. But this also was necessary to bring him into *the* school, whose discipline he still greatly needed, and in which his training was to be of a very different kind. The means which he thought most likely to rescue his people from misery, only led him into misery himself. And the events which seemed to carry him away from his vocation, were those which opened the right way for its accomplishment. Such are the ways of God. The Scripture-record does not blame him for killing the Egyptian, but leaves the Nemesis, which appeared in the consequences, to pronounce the sentence. The *Pentateuch* contains no information as to the age of Moses, when he fled from Egypt; but when he returned in obedience to the command of God, he is said to have been in his eightieth year (chap. vii. 7). *Stephen*, who most likely follows the Jewish tradition, says that he was forty years old when he fled (Acts vii 30). But when we consider that his sons must have been still young at the time of the exodus from Egypt (Ex. iv. 20, 25, xviii. 3), it does not seem probable that he remained with the Midianites so long as forty years. Still it is possible that his sons may have been born some years after his marriage.

(6). The *Midianites* were an Arabian tribe, descended, according to Gen. xxv. 2—4, from Abraham by his second wife Keturah. As early as the time of Jacob, we find them associated with the Ishmaelites in carrying on the caravan-trade between Asia and Egypt (Gen. xxxvii. 28, 36). This is in itself sufficient to indicate that their proper and original settlement was in the neighbourhood of the Elanitic gulf, which was the central point of such international commerce. And our conclusion is confirmed both by ancient and modern accounts and researches (*vid. Ritter's* Erdkunde xiii. p. 287). *Eusebius* says that the town of Midian was situated *ἐπέκεινα τῆς ’Αραβίας πρὸς νότον ἐν ἐρήμῳ τῶν Σαρακηνῶν τῆς ἐρυθρᾶς θαλάσσης ἐπ’ ἀνατολάς* (*Onomast.* s. v. *Μαδιάμ*) ; and in the middle ages the Arabian geographers *Edrisi* and *Abulfeda* (Arab. descr. p. 77 ed. Rommel) spoke of the ruins of this city as being found on the eastern side of the Elanitic gulf, five days' journey from Ailah. *Seetzen*, following these accounts, fixed upon a spot in the Wady Magne (Mukne) on the eastern side of the gulf, nearly opposite to Sinai, as the site of the town, of which at present no trace remains (*vid.* monatl. Corresp. xxv. 1812, p. 395). *Laborde*, on the contrary, thinks that he has proved that the city stood upon the western side of the gulf, near to the present harbour of *Dahab*, in the same latitude as Mount Sinai, with which it was connected by the Wady Zakal (es-Sa'l) : (comment. geogr. p. 6 sqq.). This opinion is expressed with great confidence, but it is fallacious in every respect, and destitute of the slightest foundation. Dahab is undoubtedly identical with the biblical Di-Sahab (Deut. i. 1) ; *cf. Ritter* Erdkunde xiv. 233 ; *Hengstenberg*, Balaam p. 225 ; *Ewald* ii. 326, 327. Towards the end of the Mosaic period, however, we meet with a numerous tribe of Midianites, who lived to the east of Canaan near the Moabites and Edomites, and who sustained a considerable defeat from the Israelites (Num. xxii. 4, 7 ; xxv. 6, 17 ; xxxi.). These Midianites had come into collision with the Edomites at an earlier period, and had been repulsed by them (Gen. xxxvi.

35). From the data thus obtained, we conclude that the Midianites spread northwards from Midian as far as the borders of Moab, but it is very doubtful whether they also spread westwards into Arabia Petraea. The sojourn of Moses in the land of the Midianites has been adduced as a proof that this was the case. For it seems more likely that Ex. ii. 15 sqq. refers to Arabia Petraea, where Moses would undoubtedly have been perfectly secure from discovery by Pharaoh, than to the more distant land on the other side of the Elanitic gulf. Moreover, when we read in Ex. iii. 1 that Moses led the flock of his father-in-law to the back of the desert, to *Horeb* the mountain of God, it can hardly be supposed that the fixed abode of the Midianitish Emir was so far from his flock, as it must have been if the settlements of the tribe were on the other side of the gulf. And again, the accurate acquaintance of Hobab the son of Reguel with the localities of Arabia Petraea (Num. x. 31) favours the conclusion, that his tribe had formerly dwelt in that district. But there are data, on the other hand, which render such an assumption a very doubtful one. The Israelites did not once meet with the Midianites during their journey through the desert; and when the father-in-law of Moses visited him during the encampment at Sinai, and brought him his wife and children, he evidently came from a great distance (Ex. xviii.). Now it is evident that there is no irreconcileable discrepancy between these different accounts. The only difficulty is to make a selection between the many possible solutions; and the Scripture-record does not supply us with data of sufficient certainty for this. The only precise information is that given in Ex. iii. 1: Moses led the flock of his father-in-law behind the desert (אַחַר הַמִּדְבָּר) to Mount Horeb. *Ritter* believes that this is quite in harmony with his assumption that Reguel's tribe also dwelt on the east of the Elanitic gulf (Erdk. xiv. 234). He explains אחר as meaning westwards: Moses drove the flock from the eastern coast of the gulf to the western. But if we regard this explanation as admissible, it seems to me that we must also assume that the whole tribe, of which Reguel was the head, went over to the eastern coast at the same time as Moses, for it is highly improbable that Moses went away alone to so great a distance with the flock entrusted to his care. However, "*westward to the desert*" is in itself a very questionable rendering of *achar ham-*

midbar. In any case it might be more advisable to abide by the natural translation, "*to the back of the desert*," from which it would follow that Moses traversed a barren tract of desert with his flocks, before he arrived at the pasture land of the mountains of Sinai. We should then have to look for the settlements of this tribe of Midianites somewhere to the east or north-east of Sinai, but still on the western side of the gulf. Subsequently, however, and *after* the call of Moses, they must have left this district and sought pasturage elsewhere, probably returning once more to the eastern side of the gulf. We are obliged to assume this, for the simple reason that the Israelites never met with the Midianites, and the father-in-law of Moses came from a distance to visit him (Ex. xviii., Num. x. 30). But whatever our decision may be, we must at all events regard the Midianitish tribe of which Reguel was the head as a nomadic branch, which had separated from the main body of the nation, and never united with the rest again; for whilst the great mass of the Midianites always maintained a hostile position towards Israel, the descendants of Reguel continued friendly to the last (Vol. iii. § 32. 2).

(7). A fresh difficulty arises from the different names given to the *Midianitish priest*, into whose service Moses entered, and to whom he became related. In Ex. ii. 18 sqq. he is called Reguel (רְעוּאֵל), and described as the *father* of Zipporah. But afterwards (in chap. iii. 1, iv. 18, xviii. 1 sqq.) he is called Jethro, and described as the *father-in-law* (חֹתֵן) of Moses. In Num. x. 29 we meet with him under the name of Hobab, where he is described as the *son* of Reguel, and the *Chothen* of Moses; and the same description occurs again in Judg. iv. 11. *Hartmann*, *De Wette*, and others regard these differences as attributable to differences and discrepancies in the genealogies employed. But in that case we should have to impute to the author, be he who he may, an amount of carelessness, which is really inconceivable (and this even *Winer* admits, ii. 310). The author, who wrote two different names so close together as in chap. ii. 18 and iii. 1), must certainly have been conscious of this difference, and if he had found any discrepancy in the two accounts, he would not have adopted them both. But if he saw no discrepancy, we are not justified in supposing that any really existed. The different notions conveyed by the word אָב, which meant both father and grand-father, and by חתן, which was used for brother-in-law and father-in-law, as well as the constant fluctuations in the use of names, justify us in assuming that the cause of the difference is to be sought in the one or the other. The most probable explanation is, that one of the names was a title of honour, given to indicate his priestly and princely dignity. *Lengerke* supposes the name Reguel (*i.e.* friend of God) to have been the official name (Kenaan i. 391). But he appears to me to be mistaken in his selection, since we should expect to find the proper name, and not the official designation, mentioned in connexion with his first appearance in Ex. ii. 21, and still more in the genealogical account in Num. x. 29. We prefer to ascribe to the name *Jethro (i.e. excellentia ejus)* the dignity of an official title, especially as we find it written in the form יֶתֶר in Ex. iv. 18. The three names would thus be reduced to two, and the only questions remaining would be: (1) whether we are to identify the Jethro of Ex. iii. 4, 18, with the *Reguel* in Ex. ii. 18, or with the *Hobab* in Num. x. 29 and Judg. iv. 11; and (2) whether we are to regard Reguel as Zipporah's *father*, or *grand-father*, and Hobab as the *brother-in law* or *father-in-law* of Moses. To the first question it seems to us that the only possible answer is that the Jethro, mentioned in Ex. iii. 4, 18, is the same person as the *Reguel* referred to in Ex. ii. 18; with regard to the second we are doubtful whether we are to consider the אב in Ex. ii. 18 or the חתן in Num. x. 29 as used indefinitely, *i.e.* whether the former is to be rendered *grand-father*, or the latter *brother-in-law*. *Ranke* (Pentat. ii. 8) decides in favour of the latter, and adduces Judg. xix. 4, 6, 9, to confirm the indefinite character of the word חתן; for in these passages, on account of the ambiguity of the word, which might just as well mean *brother-in-law* as *father-in-law*, the words "the father of the damsel" are added to point out what the meaning of the word really is.—So much, at all events, is clear: that *Reguel*, who was also called Jethro, was at the head of the tribe up to the period referred to in Ex. xviii. It is in Num. x. that we *first* meet with *Hobab* as the leader of the tribe, and on this account he is also classed genealogically as the son of Reguel. In the meantime, therefore, Reguel must have died. The father-in-law of Moses is held in veneration as a prophet, both in the Koran and among the Arabs, under the name *Shoeib* (which has arisen probably from an alteration of the name Hobab).

The description given of Reguel, that he was a *priest* of Midian, suggests the enquiry, what was the *religious condition* of that people? In seeking for an answer to this question, we must necessarily make a distinction between the different groups into which the Midianites were divided. We know nothing at all with regard to the religion of those who dwelt on the eastern side of the Elanitic gulf, and who, according to Gen. xxxvii. 28, 36, were a trading community mixed up with the Ishmaelites. On the other hand, we know that those who dwelt on the north, and were allies of the Moabites (Num. xxii. 25), had given themselves up to the abominable worship of Baal-peor, probably in consequence of their connexion with the Moabites. With reference to the third group, of which Reguel, and subsequently Hobab, were chiefs, we can safely assume, so much at least, that they were not worshippers of Baal-peor. Such a thing is absolutely inconceivable, when we consider the close association which was constantly maintained between them and the Israelites (Vol. iii. § 32. 2). Their nomadic isolation from the rest of the tribe renders it probable (and the earlier the separation took place the greater the probability would be), that in general they had preserved the theism, which they inherited from Abraham (see Ex. xviii. 9 sqq.). Still, we must not form too exalted a notion of the purity and genuineness of their theism, since Moses evidently refrained from communicating much to Jethro respecting the divine revelations which he had received. And the obstinate refusal of Zipporah to allow her sons to be circumcised (Ex. iv. 25) indicates a feeling of contempt for the religion of the Israelites.

(8). The house of the Midianitish priest was, doubtless, a severe but salutary *school of humiliation and affliction*, of want and self-denial, to the spoiled foster-son of the king's daughter. We can understand this, if we merely picture to ourselves the contrast between the luxury of the court and the toil connected with a shepherd's life in the desert. But we have good ground for supposing that his present situation was trying and humiliating in other respects also. His marriage does not seem to have been a happy one, and his position in the house of his father-in-law was apparently somewhat subordinate and servile. The account, given in Ex. iv. 24 sqq. (§ 21. 3), shows us clearly enough the character of his wife. *Zipporah* is there represented as a querulous, self-willed, and passionate woman, who sets her own will in opposition to that of her husband, who will not trouble herself about his religious convictions, and, even when his life is evidently in danger, does not conceal the reluctance with which she agrees to submit, in order to save him. We might be astonished to find that a man of so much force of character as Moses possessed, could ever suffer this female government. But the circumstances in which he was placed sufficiently explain them. He had arrived there poor and helpless, as a man who was flying from pursuit. A fortunate combination of circumstances led to his receiving the Emir's daughter as his wife. It is true he could not pay the usual dowry. But the remarkable antecedents of his life, his superior mental endowments, his manly beauty, and other things, may have been regarded at first by his chosen bride and her relations as an adequate compensation for its omission. But if the character of Zipporah were such as we may conclude it to have been from Ex. iv. 24 sqq., we can very well imagine that she soon began to despise all these, and made her husband feel that he was only eating the bread of charity in her father's house. Nor does he seem to have been admitted to any very intimate terms with his father-in-law; at least we might be led to this conclusion by the reserve with which he communicated to Jethro his intended departure, and the little confidence which he displayed (Ex. iv. 18). Thus he was, and continued to be, a foreigner among the Midianites; kept in the background and misunderstood, even by those who were related to him by the closest ties. And if this was his condition, the sorrows arising from his exile, and his homeless and forlorn condition, must have been doubly, yea trebly severe. Under circumstances such as these, his attachment to his people, and his longing to rejoin them, instead of cooling, would grow stronger and stronger. There is something very expressive in this respect in the names which he gave to the sons who were born to him during his exile (Ex. ii. 22; xviii. 3, 4). They enable us to look deeply into the state of his mind at that time, for (as so frequently happened) he incorporated in them the strongest feelings and desires of his heart. The eldest he named *Gershom*, which means a stranger there, "for," he said, "I have become a stranger in a strange land;" and when the second was born, he said, "the God of my father has been my help, and has delivered me from the hand of Pharaoh," and he called him *Eliezer* (God is help). We may also call to mind the miserable style in which he set out to return to Egypt (Ex. iv. 20): his wife and child he placed upon an ass, and he himself went on foot by their side.

THE CALL OF MOSES.

*Vid. Die Berufung Moseh's (*by *Hengstenberg?)* in the *Evangelische Kirchenzeitung* 1837. No. 50—51.

§ 20 (Ex. ii. 23—iv. 17).—The oppression of the Israelites in Egypt still continued. The king died, but the principles of his government were carried out by his successor. The change of rulers appears to have excited hopes in the minds of the Israelites, which were doomed to disappointment. Their oppression was not only perpetuated, but rendered increasingly severe, and their disappointment added to their sufferings. But the first signs of a powerful agitation were just appearing among the people, an agitation which was to ripen them for freedom. It was not a resolution to help themselves, or a plot to overthrow the existing government, which grew out of these disappointed hopes, but a movement of a much more powerful character, namely a disposition to sigh and mourn and call upon Him who is an avenger of the oppressed, and a friend of the miserable. And this movement attained its object; God heard their complaint and remembered his covenant with Abraham, Isaac, and Jacob. The hour of their redemption was drawing nigh. Moses, too, who was destined to be the saviour of Israel, had passed through the chief school of his life, the school of humiliation and affliction, and was now ready to obey his call. This call was now for the first time distinctly made known to him as the voice of God. He was feeding the flock of Jethro in the fertile meadows of Mount *Horeb* (1), when there appeared to him one day a miraculous vision. He saw a *bush* in the distance *burning* with brilliant flames, and yet *not consumed* (2). As he was hastening to the spot to look at this wonderful phenomenon more closely, he heard a voice calling to him and saying, "put off thy shoes from off thy feet, for the place whereon thou standest is holy ground." This was a *voice* which had been silent for 400 years, the voice of the angel of God, in whom God had so often appeared to the fathers of his people (Vol. i. § 50. 2). Moses was not left for a moment in doubt as to the Being who was addressing him, for the voice continued: "I am the God of thy father, the God of Abraham, Isaac, and Jacob." On hearing this, Moses hid his face, for he was afraid to look on God (3). The word of God, which was then addressed to Moses by the angel of the Lord, contained the key to a right understanding of the vision: Jehovah had seen the affliction of his people in Egypt, had heard their sighing and their cries, and had come down to deliver them out of the hands of the Egyptians, and bring them into the land of promise. "Come, now, therefore," he said, "I will send thee unto *Pharaoh*, that *thou* mayest bring forth my people out of Egypt." Moses was directed to go to Egypt, and having assembled the elders of Israel, to introduce himself to them as a messenger of God sent to effect their deliverance. He was then to go with them to Pharaoh, and first of all demand of him, in the name of Jehovah, the God of the Hebrews, that he would let the people go a three days' journey into the wilderness and sacrifice to their God. It could be foreseen that such a request would be strongly opposed by the king; in fact, this was expressly foretold him by God: but with this prediction there was coupled the assurance, that the almighty hand of Jehovah would open the way before him by means of signs and wonders (4).

How did Moses act when he heard the words of God announce this divine commission, and beheld the representation of its object in the miraculous sign? He had become a different man in his exile. Formerly he had burned with eager desire to appear as the deliverer of his people, and had offered to effect it of his own accord; but now he sought in every way to excuse himself from the divine command, by which he was called and equipped for the task. The training he received at Pharaoh's court had borne its fruit, and this fruit was essential to the fulfilment of his vocation; but it also gave birth to pride, false confidence, and a trust in his own power, which were unsuitable for the work. The discipline of his desert-school had broken down this pride and taught him humility, and had made him conscious of his utter weakness. His false confidence in his own power and wisdom had vanished, but he still wanted that true and proper confidence in the power and wisdom of God, by which the weak can be made strong. Not that he had any doubt as to the power of God; but he doubted *his own* fitness to serve as the organ of this power, although God himself had called him: and in these doubts there was just as much *false* humility, as there was false pride in the confidence he felt before. Still, excessive humility is always nearer to the proper state of mind than pride, that knows no bounds. And this was the case with Moses. With inexhaustible patience God follows the windings of his false humility, meeting his difficulties with promises and assurances of strength, and his refusals with mildness, but with firmness also (5). "Who am I," said Moses, "that I should go unto Pharaoh, and that I should bring forth the children of Israel out of Egypt?" To this Jehovah replies: "I will be with thee," and places the issue of his mission in the most striking manner before his mind, by telling him that on that very *mountain* the people should sacrifice to God, when they had been delivered out of Egypt. The *altar* for the sacrifice was already built. So certain was God that it would be offered, and so important was the sacrifice in the estimation of God, that when he founded the world, he had prepared the place on which it was to be presented. Thus Sinai itself was a pledge of success, a monument, and a witness of the call of Moses and the promises of God. The scruples of Moses were at length removed, at least for a time. He began to grow familiar with the thought, that he was to appear before the people as the messenger of God, and to reflect upon the manner in which he should introduce himself to them. It was now four hundred years since the God of the fathers had manifested himself. Hence it appeared the more important, that this God should be announced to the people by a name, which would clearly and definitely express the character of the new revelation. It was requisite that the *name* of the God, who appeared to deliver, should contain in itself a pledge of success, if it was to excite any confidence at all. Moses, therefore, asked for some name, which he might hold up before the people, as the banner that was to lead them to victory, and which he might use as the watchword of the coming conflict. His request was granted. God communicated to him *the* name, which from the very first had expressed his relation to the sacred history, the name *Jehovah ;* but by the explanation, which He gave of that name, He made Moses feel that it was a name, whose fulness would not be exhausted, till the eternal counsels of salvation had been fulfilled and exhausted by the events of history, and which therefore, whatever might be its age, would still be always new (6). Moses then raised another difficulty: " Will they believe me, when I appear before them as the messenger of God?" Jehovah met this difficulty by giving him a *threefold miraculous power*, by which to attest his mission both before the people and Pharaoh (7). There was still one obstacle remaining: his slowness of speech, his want of eloquence. But Jehovah replied: "Did not I create man's mouth? Go and I will be with thy mouth, and teach thee what thou shall say." The difficulties in Moses' path were now all removed, and his *reasons* for refusing were exhausted; so that we naturally expect to find him cheerfully yielding obedience to the will of God. But no; faint-hearted and froward, praying and doubting at the same time, he exclaimed: "O my Lord, send I pray thee whom thou wilt send!" This showed at once all that was in his mind, and the festering unbelief, which had been hidden, unknown to him, beneath the outward covering of humility, now came to a head. But this is the way to a cure, first softening applications, then the sharp lancet of the physician. "Then," says the record, "the anger of the Lord was kindled" (8). But this anger was still attended by the love which assists the weak. Moses was told that Aaron, his brother, should be sent by Jehovah to meet him, and should stand by his side to assist him in his arduous task. The eloquence of Aaron would thus hide his brother's want of the gift of speech, and supply the deficiency. "He shall be thy mouth, and thou shalt be his God. And now take the rod in thy hand, with which thou shalt work miracles, and go." And Moses went (9).

(1). The name *Horeb* is applied in the Bible to the whole of the mountains in the peninsula; *Sinai*, on the other hand, is the name of the particular mountain, on which the law was delivered. (See Vol. iii. § 8. 1). The fact that the mountain, on which God appeared to Moses, is here called "the mountain of God," is a proof that the call of Moses took place on the very same spot which was afterwards to be the scene of the calling of the people, the conclusion of the covenant, and the giving of the law. Even now it was holy ground (chap. iii. 5); when Israel departed from Egypt to offer sacrifice to the Lord in the desert, they had a definite spot in view, and one which had been already appointed by God. And in this consecrated spot they were to gain the assurance, that as the call of Moses, which had previously taken place there, had been attested by signs and wonders in Egypt, so their own call would be attested by signs and wonders in the land of Canaan. As the spot, on which Moses was then feeding the sheep, was one from which Sinai could be seen (chap. iii. 12), it must have been either one of the side valleys (Wady Leja and Wady Shoeib), which form the eastern and western boundaries, or the broad plain of Sebayeh to the south of the mountain. This will appear from the further descriptions given at Vol. iii. § 6. The testimony of tradition is in favour of the Wady Shoeib (*i.e.* Jethro-valley) on the eastern side of Sinai.

(2). It is very evident that the *vision of the* BRIER, *which burned but was not consumed*, was not merely a θαυμαστόν, but was especially designed to be σημεῖον; in other words it was not merely intended, that the fresh manifestation of God, which was thus introduced, should be attested by the extraordinary character of the phenomenon, and its evidently supernatural cause, but also that the meaning of the revelation to be made should be symbolically represented by the phenomenon itself. Had the former been the only design, any extraordinary appearance would have answered the purpose quite as well, and in that case the selection of this, out of the one or two thousand means which the Almighty had at command, would be perfectly arbitrary and unimportant. But where God works there can be nothing arbitrary.

The divine miracle was followed by the divine address. They must have stood in close connexion with each other. The address served to explain the meaning of the symbol; the symbol was a visible representation of the substance of the address. Not that the two necessarily cover and exhaust each other. A symbol often reaches beyond the substance of the address which accompanies it. In a symbol the whole revelation is made simultaneously, it represents in one complete picture, and from one single point of view, all that is to be revealed. An address, on the contrary, is a successive revelation, the substance of which is gradually unfolded and explained. This we shall find to be the case here. The sign which Moses beheld was a picture of the whole of that stage in the progress of revelation, which was then about to commence. And it retained its validity to the end of that stage, whereas the words addressed to Moses in connexion with the sign, merely referred to the circumstances and necessities belonging to the very commencement of this stage of revelation.

There are two things requiring explanation in connexion with this sign: viz. the bush and the fire. In the BRIER we have a symbol of the people of Israel. From this time till the cursing of the fig-tree, which had no fruit on it but only leaves, the chosen people of God are frequently and variously referred to, under the figure of a bush or tree. Here they are represented as a low, contemptible brier, in contradistinction to the tall majestic trees, which proudly rear their heads to the clouds, and are gazed at and admired by the world. Hence the brier was symbolical of Israel, as a people despised by the world. The FIRE is always used in the Scriptures as a symbol of divine holiness. And this is the case here; for the record expressly says that the presence of God was made known in the fire: "the *angel of the Lord* appeared to him *in* the flame of fire out of the midst of a bush" (chap. iii. 2); God spake "out of the midst of the bush" (ver. 4); Moses had to take off his shoes, because the place on which he stood was rendered *holy* by this appearance (ver. 5); he "hid his face, for he was afraid to look upon *God*" (ver. 6). The burning brier, therefore, was a symbol of the community of God, in which the holiness of God had its abode. The brier was burning in the fire, but it was not consumed, although from its nature it deserved to be consumed, and could easily be so. It was a miracle that it was not consumed. And thus was it also a miracle of mercy, that the holiness of God could dwell in a sinful community without consuming it. But in the midst of the thorns of the natural life of the community there was hidden a noble, imperishable germ, namely, the seed of the promise, which Jehovah himself had prepared. It could not, indeed, be set free without the pain of burning, but by that burning it was made holy and pure. There was also another fact of great importance represented by this symbol, viz., that the fire of divine holiness, which burned in Israel, without consuming it, served also as an outward defence. Hitherto, every one who passed by might ridicule, injure, or trample on the insignificant bush, but henceforth whoever touched it would burn his own fingers. "I will be unto her a *wall of fire round about*," said the Lord by the prophet Zechariah (2. 4), " and will show my glory *in the midst of her*." Pharaoh was soon to find this out.

The brier had not always been surrounded by the flames. The time had only just arrived, when the holiness of God was about to condescend to dwell in Israel. The words afterwards spoken by God (ver. 12), informed Moses of the time when the events, symbolized by that vision, would occur. On that very mountain, on which the miraculous *sign* appeared, (and which for that reason was called the *mountain of God*, chap. iii. 1), the great miraculous *fact* which it represented was to be afterwards accomplished, and to that end Moses was to bring the people out of Egypt and conduct them thither.

This explanation of the vision is different from the traditional one; and only *Hofmann* supports it (Schriftbeweis i. 335). Most commentators (including even *Baumgarten* i. 1, p. 406), explain the vision as referring not to the future condition of Israel, into which it was to be first brought at Sinai, but to the circumstances in which Israel had hitherto been placed, and was still living in Egypt. The fire is generally supposed to represent the affliction, which the people were enduring from the oppressions of Pharaoh, an affliction in which Israel was burning, but was not consumed, was suffering pain, but for its good, since by that means it was being purified and fitted for its future destiny. This explanation is borne out by Deut. iv. 20, where the sufferings of the Israelites in Egypt are compared to an iron furnace. But as fire is introduced in innumerable passages, in fact every where else, as a symbol of divine holiness, and as the fire which envelopes the bush is expressly described in this passage as a manifestation of God, we must either give up the reference to the afflictions in Egypt, or else identify those afflictions with the fire of the holiness of God. The only way in which the latter conclusion is arrived at, is by regarding the afflictions as sent by God for the purification of Israel, although in one sense they proceeded from Pharaoh.

But there are very grave difficulties in the way of this interpretation. The attempted combination of the two points of view is so forced and complicated, that even if there were nothing else in the way, we should on that account alone feel very great hesitation about giving it our support. But, when we look at the design of the vision, it becomes at once apparent that it is absolutely inadmissible. The vision was intended to bring distinctly before the mind of Moses, not something present, but something future; from the nature of the case it must have been so, moreover it is expressly intimated in ver. 12. Moses did not require to be reminded of the misery of his people in Egypt, or to have the afflictions which they were enduring brought visibly before him; that fire was burning perceptibly enough in his own life. And even the information that Israel was not to be consumed, but purified by this fire, had but little connexion with the call, which Moses was about to receive. What he actually required was, now that God had come down to deliver his people, to have their future condition set before him. There was no occasion to set before him the circumstances that had led to his mission; the design and issue of that mission were what he wanted to know.

(3). The command *to take off the shoes*, was in accordance with oriental views and customs which are still in force. As shoes are worn in the East as a protection, not against cold, but against dirt, the necessity for wearing them ceases where a place is clean. Moreover, as the shoes *are* already defiled with dust, if not with mud, by walking on the streets and roads, not only would it be unnecessary to wear shoes in a clean place, but the clean place itself would be thereby defiled. *Here*, of course, the removal of the shoes had a symbolical meaning. The respect due from the feet to the clean place, represented the reverence, with which the inward man should approach the Holy One. As soon as Moses perceived that God was in the fire, he *hid his face;* a sinful man cannot look openly and freely at the self-revealing holiness of God, and therefore he shuts or hides his eyes.

We must not overlook the fact that in this, the first revelation that had been made for 400 years, God announced himself as the *God of Abraham, Isaac, and Jacob*. This name was, as it were, a bridge, which spanned the vacant interval of 400 years between Jacob and Moses, a bridge, by which the past was linked with the present, and which gave to the present a hold upon the promises, the lessons, and the results of the past. The lives of Abraham, Isaac, and Jacob, constituted the first stage in the history of the covenant. In each of these patriarchs the subjective, human side of the kingdom of God under the ancient covenant was unfolded in a peculiar manner, and hence in each of them, the presence and power of God were peculiarly attested. Together, they formed a distinct and complete whole. The distinction, however, was merely temporary, the completeness merely relative. It was only in the form of a family, that the development of the covenant had attained distinction and completeness. As soon as the family had grown to a nation, the covenant entered afresh upon its course of development, with the same powers and tendencies as before, but on a larger scale and with more abundant materials. All that God had effected and promised during the first stage of the covenant, was summed up in the name, "the God of Abraham, Isaac, and Jacob." This name was the inscription on the portal of the historical development of the covenant in the form of a nation, and it continued to be the seal of that covenant, till the Old Testament expanded into the New, till the covenant with one nation gave place to the covenant with all, and the God of Abraham, Isaac, and Jacob, became the God and Father of our Lord Jesus Christ ; in other words, until the time arrived, in which Abraham ceased to be the rock whence the people of the covenant were hewn, and Sarah the hole of the pit whence they were digged (Is. li. 1, 2), and the new Israel found in Christ the author and finisher of faith, and in the Spirit of God the fountain of life (*vid.* Vol. i. § 48. 1).

(4). In ver. 12 we read: "This shall be a token unto thee, that I have sent thee ; when thou hast brought forth the people out of Egypt, *ye shall sacrifice to God upon this mountain.*" We have already pointed out in the paragraph above, how an event, which was not to happen till afterwards, was a token of *assurance* to Moses even at that time. But we cannot for a moment doubt, that something *more* was intended than merely to give to Moses a sign which should strengthen his faith, that there was a highly important end to be answered by this *sacrifice.* Even in the presence of Pharaoh, the motive to be assigned for the departure of the people was, that they might sacrifice to their God in the desert. Every sacrifice presupposes an interruption of communion with God, which is to be restored and renewed by the sacrifice offered. Now, even though it is possible that sacrifices may have been continually offered by individuals during the sojourn in Egypt, it is certain that the community as a whole, as a united body, had *never* sacrificed for 400 years (§ 17). Hence this sacrifice which was represented as the immediate object, the first-fruits of the deliverance, was the first offering ever presented by the community, *the first national sacrifice.* But every first thing contains potentially the fulness of all that comes afterwards; it is the type, representative, and pledge of all that succeeds. And if Israel, henceforth, was related to God as a community, as an entire body, if it entered into that communion with God, which is the object of sacrifice, and persevered in such communion; the foundation of all this was laid by the sacrifice at Sinai. Hence, from the very outset we must ascribe to this sacrifice an extraordinary meaning, a meaning such as cannot be ascribed to any of the sacrifices, which were subsequently offered during the same stage of the history of revelation;—in a word, this was to be the sacrifice, by which Israel was to become the people of God, the covenant-sacrifice, by which the covenant, made with Abraham, Isaac, and Jacob, was to be renewed, that it might be more fully and gloriously developed. And if, in accordance with the vision which Moses saw, we expect to see Israel exalted to be a nation, in which the holiness of God resided without consuming it, it could only be by means of *this* sacrifice, that such an end would be attained.

In ver. 18 we read that Moses was directed to say to Pharaoh: "Let us go *three days' journey* into the wilderness, that we may sacrifice to Jehovah, our God." In these words we naturally suppose that Sinai was the place intended, for according to ver. 12 it was there that the sacrifice was to be offered. But the geography does not permit of such a supposition. For Sinai is 150 miles from Suez, and even a caravan, with the greatest possible speed, would be unable to accomplish so much as this in three days. It must, therefore, be admitted that permission was to be asked of Pharaoh, not to go to Sinai, but merely to cross the borders of Egypt. The request was represented to him as indispensable, on the ground that the sacrifices of the Israelites were an abomination to the Egyptians (chap. v. 3, viii. 25 seq.). Still God had already expressed his will, that the sacrifice should not be offered in any place which they might choose in the desert immediately bordering upon the frontier of Egypt, but on Sinai, which was at least seven days' journey from the Egyptian boundary. Moreover, we learn from ver. 8, that God had come down to lead Israel altogether out of Egypt, and bring them back to the land of their fathers' pilgrimage, that he might assign them this land as a permanent possession and abode; and yet Moses was merely to request permission of Pharaoh to go away to a distance of three days' journey at the most; a request, which of course tacitly involved a promise to return when the sacrifice had been offered. Are we then to regard this as an intentional and dishonourable deception of Pharaoh on the part of both God and Moses? By no means. It would certainly have been a deception, or would have become one, if Pharaoh had acceded to the request in good faith, and had given them permission to go a three days' journey and no more; and if, in spite of their promise, and without further permission, they had marched away to Canaan, or even if from the first they had intended this. But such was not the case. Pharaoh, as was foretold in ver. 19, did not accede to the request of Moses. And as he annulled the *request* by his refusal, so *eo ipso* he annulled the *promise* implied in that request. It is true that he afterwards permitted the Israelites to depart; but he was forced to do so by the plagues, with which the God of Israel smote him, and his permission had no connection, therefore, with their friendly petition. Hence, when once Jehovah had placed himself in a hostile relation to Pharaoh, the amicable negotiations having entirely failed, the first limited and conditional request had no longer any force, and even Pharaoh himself, when utterly defeated, gave an unconditional and not a conditional permission to depart (*vid.* § 35. 2). God could foresee that this would be the ultimate issue of the whole transaction; and on the ground of this foreknowledge he announced to Moses, that Sinai would be the spot on which the appointed sacrifice would be offered, and Canaan their final and permanent destination. But although at the very outset he made known to his friend and servant Moses the whole of his design, it did not follow that it was necessary to exhibit it to Pharaoh also. And it was *mercy* towards Pharaoh, which dictated a different course. If Pharaoh had known at the outset the whole of the divine demand, which was eventually to be made upon him, it would have been an infinitely more difficult matter for him to prepare his heart to obey the will of God, than it was under the plan adopted by the wisdom and condescension of God which commenced by making known but a small part of the entire demand. If we could imagine Pharaoh yielding to the first, partial request made by Moses, of course the second and more unpalatable half of the demand would still remain to be made. But as faithfulness in little prepares for faithfulness in much; so would the obedience manifested to the will of God in that which was least, have been a stepping-stone to obedience in something greater, and God would have given him grace to overcome the resistance and selfishness of his own will.

(5). The *conversation between God and Moses*, this divine "*I will not let thee go*," forms a counterpart to the human "*I will not let thee go*," which is first met with in its original force in the case of Jacob (Gen. xxxii. 26). The whole conversation is related in such a manner, that it carries in an eminent degree the pledge of its own authenticity within itself. Where can we find a legend in the whole range of mythology, which will bear comparison with the narrative before us? Where does fiction present a similar psychological picture, with such striking scenes and yet such deep, psychological truth? The only object of a myth is to glorify the hero; he is without a flaw, a hero from top to toe, from the cradle to the grave, with superabundant force and confident elation at the beginning, with irresistible, irrepressible power in the middle, and triumphant power at the end. We never meet with a myth, which imputes to its hero such timidity and despondency, as was here displayed by the greatest, the most celebrated, and the most powerful hero of the people of Israel. Look, for example, at the full and elaborate way in which the weakness and faint-heartedness of Moses are here described, the evident interest with which the author dwells upon them, as though he could hardly turn away from the subject. How striking and unusual, and yet how deep and true are the lineaments of this picture from life! Observe the progressive steps in the psychological development, how appropriate they evidently are! When Jehovah first gives him the commission, Moses shrinks from the weight of the enormous burden, which is laid upon his shoulders, and which he is not strong enough to bear. But he does not unconditionally decline the commission, on the contrary he fancies the future already present, transports himself to the circumstances in which he will be placed; and counts the cost, to see whether he has enough to meet the demands. Then doubts and fears, wants and weaknesses, start up on every side. But Jehovah has an answer to every doubt, a promise for every fear, an inexhaustible supply for every want, and divine strength for every human weakness, which he lays in the scale. We look with heartfelt joy at the manner in which one fear after another is taken away from the trembling Moses. And when at length his fears are all exhausted, and he has no more excuses left, we expect to find him yield, and to hear at length his "yea and Amen." But no, hitherto his refusal has been conditional, but now it is unconditional. All that God has spoken, and promised, and done, appears to be thrown away, and to have been utterly in vain.

How unexpected and improbable does this turn *appear*; and yet how true and necessary it really *is!* So long as the refusal of the flesh could give reasons for its opposition to the demands of the spirit, it appeared to be somewhat justifiable. And an inexperienced observer might fancy that as soon as every doubt had been entirely removed, and every fear had been set aside, the will would yield itself captive, that in fact it could not do anything else. But whoever has looked into the dark recesses of the human heart, knows well that it is there that the most severe and violent opposition commences, namely, the opposition of capriciousness and self-will. "*I cannot*" then comes forward openly and without reserve in its true character, as an unvarnished "*I will not.*" And thus the most improbable of all the steps is that in which there is the deepest *truth.*

Hitherto, when all that had been said was, "I cannot," the mercy of God replied to the objection, with inexhaustible long-suffering and patience, "thou canst in my power," but now it meets the insolent "I will not," with a determined "thou shalt," and *this* is the moment of decision. As faith has to constrain God, (Gen. xxxii. 26; Matt. xi. 12; Luke xi. 8, xviii. 6 sqq.), that the mercy of God may break forth from righteousness; so does God put restraint upon man, that the germ of faith, which is imprisoned in unbelief (Mark ix. 24), may be set free and expand in all its glory. And this was necessarily the case with Moses. That which was least expected was just the most *necessary* step of all.

A certain amount of enthusiasm, which is sometimes regarded as inspiration, the confidence of self-assurance, a superabundance of power, insensibility to difficulties, and the boldness which rushes headlong into dangers, all these befit the hero of this world. But for men, who are to do the work of God, the heroes who are to fight in His cause, not only are they unsuitable, but they actually disqualify them for their vocation. Forty years before, Moses possessed them all in rich abundance. But at that time, God found him unfitted for the work he was destined to perform. Modesty and circumspection, humility and self-abasement, consciousness of one's own weakness and insufficiency, are the indispensable conditions of all employment in the kingdom of God, for they are the vehicles of divine inspiration and wisdom, of divine power and strength. Therefore it is that the apostle says: "when I am weak, then am I strong" (2 Cor. xii. 10). To this *weakness* Moses had been trained in the desert-school. But the perverseness of human nature was again manifested in the fact, that even in this he went to the extreme. He overstepped the boundary between negative weakness, which places no confidence in self, and is the weakness that God desires and demands, and that positive weakness, which not only renounces all self-confidence, but cherishes a want of confidence in the power of God. Thus he went from one extreme to the very opposite. But the discipline of God can reclaim a man from his wandering to the right, just as well as when he wanders to the left.

We may see how necessary it was, that all the weakness and faint-heartedness, the incredulity and unbelief which Moses displayed, should be brought out and overcome before he entered upon his mission, when we consider how serious and dangerous the slightest manifestation of it at a later period would have been, whether in the presence of Pharaoh or of the people. Then the reproach of Moses would have been the reproach of God, and his fall would have ruined his work. It was necessary that he should stand before God weak and fainthearted, despairing and of little faith, in order that he might be strengthened by God to stand firm before Pharaoh and the people, as a divine hero possessed of undaunted courage and unshaken confidence.

(6). It is apparent from the etymology, that אֶהְיֶה אֲשֶׁר

אֶהְיֶה, the name which God announced to Moses (in ver. 14), as *the* name by which he was to make him known to the people as their deliverer, was merely an alteration and explanation of the name *Jehovah*, which was so well known to the people; and this is put entirely beyond the reach of doubt by Ex. vi. 1 sqq. We have already entered into some explanation of these two passages in § 5, and in vol. i. § 13; but we have a few additional remarks to make here. The different theories respecting the origin of this name, which have been based upon the assumption that it is of foreign origin, have been for ever refuted by *Tholuck* (über den Ursprung des Namens Jehova, in his vermischte Schriften i. 377—405), and *Hengstenberg* (Pentateuch, vol. i. p. 235 sqq. *cf. Gesenius*, thesaurus, p. 576 sqq.), such, for example, as that the Egyptians invented the name of the deity by combining together the seven vowels *Ιεηωουα*; that the name originated in Phoenicia, India, or China; and again that it was identical with Jupiter, Jovis, &c.—The derivation of the name from הוה = היה, which is based upon Ex. iii. 14, is still firmly maintained by every commentator, in spite of *Ewald's* curious objections (Vol. i. § 13. 1).—The superstitious fear with which the Jews refrained from uttering the quadriliteral word יהוה (calling it in consequence שֵׁם הַמְּפוֹרָשׁ = *nomen separatum*, ὄνομα ἄῤῥητον), was founded upon Lev. xxiv. 16: וְנֹקֵב שֵׁם־יְהוָה מוֹת יוּמָת. Now, the verb נקב, by itself, does not mean to *blaspheme*, to *curse* = קבב, but to *utter*; still the context shows plainly enough what kind of utterance is referred to, viz., utterance in the way of blasphemy (*Buxtorf*, lex. talmud. p. 1847, and *Hengstenberg*, Pentateuch vol. i. 245). The frequent occurrence of the name of Jehovah in the composition of proper names is a proof that no such fear existed in Old Testament times. It probably arose shortly after the return from the captivity. For it must have prevailed as early as the date of the *Septuagint* translation, where the name Jehovah is always rendered by *Κύριος*. Even the apocryphal writers of the Old Testament do not venture to use the name. *Philo* (de vita Mos.) describes it as a name, ὅ μόνοις τοῖς ὦτα καὶ γλῶτταν σοφίᾳ κεκαθαρμένοις θέμις ἀκούειν καὶ λέγειν ἐν ἁγίοις, ἄλλῳ δ' οὐδενὶ τὸ παράπαν οὐδαμοῦ, and *Josephus* says (Ant 2. 12. 4): Ὁ Θεὸς αὐτῷ σημαίνει την ἑαυτοῦ προσηγορίαν . . . . περὶ ἧς οὐ μοὶ θέμις εἰπεῖν. In the *Talmud* we read: Etiam qui pronunciat Nomen suis litteris, non est ei pars in seculo futuro. *Maimonides* (*More Nevochim* i. 61) states: "Nomen hoc non pronunciabatur nisi in Sanctuario et quidem a sacerdotibus Dei sanctificatis in benedictione sacerdotali et a pontifice ipso die jejunii. But after the death of Simeon the just even this custom was abolished, and henceforth the substitution of אֲדֹנָי was required in temple, as it had long been outside, ne illud nomen disceret homo, qui non esset honestus et bonae existimationis" (Jad chasaka, xiv. 10).

In consequence of the substitution of אדני (Κύριος) for יהוה, which had taken place as early as the date of the *Septuagint*, the vowel points of the former were attached to the latter, when points were introduced into the text, and thus יְהֹוָה became a *Keri perpetuum*. But notwithstanding the fact that this is evident and indisputable, there have even been Christian theologians, who have maintained with great pertinacity that יְהֹוָה was the correct and original pointing. The last who asserted this was *J. F. v. Meyer* (Blätter für höhere Wahrheit xi. 306; *cf. Stier* Lehrgeb. d. hebr. Grammatik i. 327). *Reland* made a collection of all the treatises on the subject that were in existence in his time, whichever side they advocated; (see his *Decas exercitationum de vera pronunciatione nominis Jehova.* 1707). In the preface to his work he gives it as his opinion, that we have not the original pointing. *Hengstenberg* has, since then, published a most elaborate defence of the correct view, (Genuineness of the Pentateuch, Vol. i., 231 sqq. trans.). The following points are especially decisive: 1, Wherever אדני is joined with יהוה, the latter has the pointing of אֱלֹהִים; 2, whenever יהוה occurs with the prefixes ב, ל, כ, ו, מ, they do not take *chirek* (בִּיהוָֹה &c. like בִּיהוּדָה) but *patach*, except in the case of מ which takes *Zere*, just as they do when joined to אדני; 3, when יהוה is followed by one of the letters ב ג ד כ פ ת, the latter takes a *Dagesh lene*, though it could not take it with the reading יְהוָה, which ends in a *litteraquiescens*. Moreover the word יְהֹוָה is beyond the reach of any admissible etymology; for the favourite explanation, given by the earlier theologians, that יְהֹוָה is a composite word formed from the future, present, and preterite of the verb הוה (יִ representing the future, הֹו the participle, וָה the preterite), can only be regarded as a curiosity belonging to the childish days of Hebrew Grammar. Such an etymology is impossible according to the rules of the language, and is not sustained, as some suppose, by Rev. i. 4. We subscribe to the opinion that the description there given of God, ὁ ὢν καὶ ὁ ἦν καὶ ὁ ἐρχόμενος, is as a paraphrastic rendering of the name *Jehovah;* but the ground for deducing all three from the word *Jehovah*, is to be found solely and exclusively in the imperfect formation of the name, and this reason is amply sufficient.—יהוה is, no doubt, formed by changing the first person imperfect אֶהְיֶה (Ex. iii. 14) into the third person; or rather the explanatory word אהיה in Ex. iii. 14 is merely יהוה changed into the first person. We have already observed in Vol. i. § 13. 1, that among the different pointings that are grammatically admissible the reading יַהְוֶה is most probably the genuine one. And this agrees with *Theodoret's* remark in reference to Ex. vi.: καλοῦσι δὲ αὐτὸ Σαμαρεῖται μὲν Ἰαβὲ, Ἰουδαῖοι δὲ Ἀϊά. The latter form is merely יהוה written in Greek characters. It is well worthy of observation that the name יהוה is founded not upon the more modern form היה, but upon the form הוה which was already antiquated when the *Pentateuch* was composed; for it follows from this, that if the *Pentateuch* dates from the time of Moses or Joshua, יהוה must have been an ancient, pre-Mosaic name. If we consider the dignity of the verb היה (essence, being) on the one hand, and that of the imperfect sense on the other (*cf. Ewald's* ausführl. Lehrbuch § 136), it will be at once apparent that the name יהוה means *the existing one*, whose operations have commenced and still continue, and who permits us continually to look for more glorious manifestations of his existence (*vid.* Vol. i. § 13. 1).

We have already entered into a thorough discussion of Ex. vi. 3, 4, at § 5. 1, 2, and have shown that the words, "I am *Jehovah*, and appeared unto Abraham, Isaac, and Jacob as אֵל, שַׁדִּי, but by my name *Jehovah* was I not known to them," do not imply that the name Jehovah was altogether unknown till then. And we have also fully explained there, why, even assuming the correctness of the supplement-theory, the author of the so-called ground-work naturally avoided the use of the name,

Jehovah until Ex. vi., whilst the writer of the supplementary parts employed it without hesitation. As proofs of the incorrectness of *Ewald's* opinion, that the name was first made known to the people by Moses, but was used proleptically by the (sole) author of Genesis, we may mention the following arguments, in addition to those already adduced: 1. The name occurs in pre-Mosaic surnames. Among these we reckon the name of the mountain *Moriah* in Gen. xxii. 14 (*cf.* Vol. i. § 65. 1), that of Jochebed, the mother of Moses, Ex. vi. 20, and the name *Bithjah* (1 Chr. iv. 18 *cf.* § 15. 3). The comparatively rare occurrence of the name Jehovah, in the proper names of the Mosaic and pre-Mosaic times, may be explained simply enough, from the fact that the name had certainly not taken so firm a hold of the minds of the people, and was not so fully understood before the time of Moses as afterwards (this is evidently implied in Ex. vi. 3). 2. The name *Jehovah* is formed from the verb הוה, which was an obsolete form in the time of Moses (Ex. iii. 14). 3. The name occurs in Jacob's blessing, which is a pre-Mosaic document (Gen. xlix. 18). 4. The words of God in Ex. iii. 14 do not by any means indicate that he was about to reveal a new name, which had hitherto been quite unknown, but are evidently explanatory of a name already known, whose depth and fulness had never yet been completely exhausted. The latest attempt to explain the passage has been made by *J. Richers* (die Schöpfungs-, Paradieses-, und Sündfluths-geschichte, Lpz. 1854, p. 453 sqq.), but we cannot adopt his explanation. He thinks that the frequency with which the passages are altered in Hebrew, the same pronouns and verbs being employed with reference to entirely different persons in one and the same discourse, justify the conclusion that לָהֶם, "to them," in the words of God, "but by my name *Jehovah* was I not known *to them*," does not refer to the patriarchs and the families mentioned in the book of Genesis at all, but to the Israelites who were groaning beneath the Egyptian yoke. In his opinion God was known to the patriarchs as Jehovah, but their descendants in Egypt lost both the name and the knowledge of Jehovah; they neither knew him, nor lived under his especial protection and rule.

The name Jehovah *was* (or rather *became*) undoubtedly a new one then, but only in the sense in which Christ said (John xiii. 34): "A *new* commandment give I unto you," whereas he merely repeated one of the primary commandments, which we find in the Old Testament, and meet with on every hand in the laws of Moses. It was a commandment, however, the fulness and depth, the meaning, force, and value of which were first unfolded by the gospel. And just as the greatest act of love, which the world ever witnessed, provided a new field for the exemplification of this command, in greater glory than was possible under the law, and thus the old commandment became a *new* one; so did the new act of God, in the redemption of Israel from Egypt, furnish a new field in which the ancient name of God struck fresh and deeper roots, and thus the ancient name became a new one.

(7). *Three miraculous signs* were given to Moses, by which he was to legitimate himself as a messenger from God, first before *the people*, and then in the presence of *Pharaoh*, in order that it might be evident that he had *not come empty*, but had been sent by God and filled by Him with divine power. It was not stated at the outset, that the signs were meant for Pharaoh as well as for the people (this first occurs in chap. iv. 21). All that was required here was to remove the difficulty suggested in chap. iv. 1: "The people will not believe me." Moreover, there were only two of the signs, which were intended and adapted to be performed in the presence of Pharaoh. The gift of miracles was also communicated with especial reference to Moses himself. His doubts and incredulity were to be overcome by the consciousness, that he was the possessor of such powers; and the miracles themselves were of such a nature, as to furnish a type and guarantee of the progress and success of his mission. *Baumgarten* (Comm. i. 1, p. 415) is on a wrong track, when he looks upon the three signs, which Moses received, as intended *for the people* alone, and denies that they had any reference *to Moses* himself. It is not true that the sign referred to in chap. iii. 12, which could only be witnessed in the future, was sufficient for Moses, and that the people required more visible signs, because they were more completely under the dominion of the senses than he was. Had this been the case there would have been no necessity for the signs to be performed, as they now were, before the eyes of Moses. But the continued refusal of Moses proved, that the sign referred to in chap. iii. 12 was too spiritual for him, and that he needed a present sign of a more tangible character.

The miracle of the burning bush was, as we have seen, not merely a θαυμαστόν, but also a most significant σημεῖον (note 2); and in the same manner the three miracles which Moses was commissioned to perform, must have been significant signs, intended as revelations from God, adapted to the senses of all concerned, of Moses as well as of the people of Israel and the king of Egypt, and designed to convey to the heart of each, just so much as he required to know. For this reason, as *Baumgarten* has well observed, we find in chap. iv. 8 *voices* ascribed to the signs.

The *first sign* was as follows. At the command of God, Moses threw his *staff* upon the ground. The *staff* became a serpent, and Moses *fled* from it. But Jehovah told him to take it by the tail; and on his doing so, it became a staff in his hand once more. The staff was the *shepherd's crook*, with which he had hitherto conducted the flock of Jethro. Hence it represented his vocation as a *shepherd*. This he was to throw away, *i.e.*, he was to give up his calling and follow a new one. But *the staff which he had thrown away became a serpent, and Moses fled before it.* His vocation hitherto had been a poor and despised one; but it was also quiet, peaceful, and free from danger. When this was given up, he was to be exposed to dangers of such magnitude, that even his life would be threatened. Moses could foresee all this, and hence the obstinacy with which he refused to enter upon his new vocation. But at the word of God he laid hold of the snake, and it became a staff in his hand once more. This showed that, by the power of God, he would be able to overcome the dangers that would surround him, when he relinquished his present calling. By overpowering the snake he recovered his staff, but it was no longer *his* staff; it was the *rod of God* (iv. 20), and with the staff thus altered he was to perform the work entrusted to him (iv. 17). It was still a shepherd's staff, and his new vocation was a shepherd's calling. From being a shepherd of Jethro's sheep he was to become the shepherd of God's sheep, the leader and lawgiver of the people of God. And he became so, by overcoming the dangers which intervened between these two different employments. We must also observe, that this was the rod with which he was to bring the plagues upon Egypt; and therefore it was the retributory counterpart to the rod with which the Egyptian taskmasters had beaten the Israelites (chap. v. 14; *vid. Hengstenberg*, Beiträge iii. 523). As soon, then, as Moses appeared before the people and performed this sign, it showed them, first, that the dangers to which the mission of Moses would expose them—dangers which they soon experienced (chap. v.)—would be overcome; and secondly, that the staff of shepherd and ruler, with which Moses was to lead and govern them, was not assumed without authority, but given to him by God, and therefore the question could not be asked, as it was before, "who made thee a prince and a judge over us" (chap. ii. 14)? He afterwards performed the same miracle *in the presence of Pharaoh* (vii. 10 sqq.). We shall see by and by what was its meaning then (§ 24. 2).

We come now to the *second sign*. Moses put his hand into his bosom, and it became LEPROUS *like snow*. But as soon as he put it into his bosom a second time, it became CLEAN *and whole as before*. The *bosom* is the place into which the hand is put, to shelter it from cold and other evils; in the warmth of the bosom it is protected and cherished as in a mother's lap. But behold in that very place, in which we expect it to be protected and warmed, the hand of Moses became leprous. *Leprosy* was impurity in its worst possible form; and for this reason the leper was put away and banished from the society of his fellow-citizens. These data, which are indisputable, are amply sufficient to explain the sign. That which happened to the hand of Moses was a picture of what had happened, and was still to happen, to the people of Israel. By going down to Egypt, the Israelites had been preserved from the injurious influence of Canaanitish customs (§ 1. 7). Through the favour of the first Pharaohs, Egypt was undoubtedly a hiding-place, in which the family of Jacob had been cherished and preserved, when it was distressed both in body and mind. But there had been a change in both the men and the times, and Israel was enslaved, despised, and held in abomination in the land of Egypt. When Israel departed from Egypt, he was like a homeless leper. But Jehovah led him once more to a hiding-place, where he was cleansed from the leprosy which he had brought with him from Egypt, and where he was set apart as a holy people and a priestly nation (xix. 6). It is very easy to explain why *this* sign was not exhibited before Pharaoh as well as the others (chap. vii.). The thing signified was of too internal and spiritual a nature, it was too closely connected with the counsel of God concerning his people to be appropriately displayed to Pharaoh. The objects indicated by this sign, were such as could only be treated of between God and his people.

The *third sign* belonged to a totally different sphere. It is clearly and expressly separated and distinguished from the other two. Moreover, it was not manifested on this occasion, even to the eyes of Moses; it could only be seen in Egypt. Moses was directed, in the event of the people disbelieving the first two signs, to take some *water out of the stream* (the Nile), and pour it upon the earth, where it would be turned to blood. The reason that in this instance Moses only received a command and a promise, and that the sign was not actually accomplished before his eyes, is not to be sought in the fact that there was no water near. The whole force of this sign depended upon the water being taken *from the stream*, *i.e.* from *the Nile.* This sign, as will afterwards appear, Moses was also to perform in the presence of Pharaoh, but on an incomparably larger scale. It was then not merely a handful of water which was affected, but the whole body of water, when touched by the rod of Moses, was turned into blood (vii. 17 sqq.); for in this instance it was not merely a sign for *Pharaoh*, but a *plague* and judgment upon the land of Egypt (*vid.* § 26. 1). To enter fully into the meaning of this sign, we must remember what the Nile was to the Egyptians. It was the source of all the wealth and fertility of the land of Egypt, and was therefore worshipped as a god. With the same rod, which God had placed in the hand of Moses, that he might tend and lead the Israelites, he was also to overthrow the gods of Egypt, and demonstrate their utter weakness in comparison with the power of the God he worshipped. That which brought a blessing to Egypt was turned by Moses into a curse; and that which had been the object of veneration and worship was made an object of disgust and abhorrence. Moses was first of all to show to his people that he possessed this power, by taking a handful of water out of the Nile; but to Pharaoh he was to demonstrate that God was in earnest in his determination to smite the gods of Egypt, by corrupting the whole of the river.

Thus did God furnish his servant with *three* signs. They were all related, each in its own way, to the work which he had to perform. The number *three* is the mark of perfection and completeness, an indication that the process of development is at an end, and the idea embodied is fully manifested. Hence the three miraculous signs were proofs, that the miraculous power of God would be put forth in all its fulness through the instrumentality of Moses. A fourth sign Israel could not and durst not demand, without convicting itself of obdurate unbelief. But in this case it was necessary, that with the triple sign there should also be progress and a culminating point in the manifestation of the idea. And our interpretation of the signs exhibits this. There were *three* objects upon which the power of God was to be exerted: *Moses*, who was appointed to be the leader and shepherd of Israel; *the people of Israel*, who were to be cleansed from their leprosy and made a holy priestly nation; and *Egypt*, which was now to discover the impotence of its gods. There was a special sign, containing a distinct intimation, for each of these; and thus the idea was actually exhibited in the whole course of its development. Even the order in which the signs occurred, is thus shown to have been both significant and necessary. No other order would have harmonized with the natural development and the idea.

There is one more point, and a very essential one, to which we have still to refer. It has already been frequently observed, that Moses was the first prophet sent by God, and more especially the *first* worker of miracles in the history of the world. We refer the reader to our remarks in § 6. 1, on the absence of the gifts of prophecy and miracles during the pre-Mosaic age. In accordance with those remarks, we discover here an essential advance in the course of the kingdom of God upon earth, and the arrival of a turning-point in its history. Hitherto the covenant works of God and man, though constantly related to each other, had been kept more distinct the one from the other. The union of the divine and human, which is the true characteristic of the history, and the perfect realization of which in the person of the God-man was foreshadowed from the first, had never before been displayed in such a way through the instrumentality of a man, expressly called and fitted for the purpose of manifesting the word and power of God. In this respect, therefore, Moses was the first type of Christ, the God-man.

(8). When Moses complained of his *heavy tongue* and his want of eloquence,—faults the removal of which he could see no reason to anticipate, notwithstanding the call that he had received,—the words do not, in our opinion, denote that he *stammered*, as some commentators suppose. Moses said that he was not a man of words, his mouth was not fitted for addressing others. And Jehovah replied, "Did not I make man's mouth?" This answer implied a promise, that a gift of grace should make up for the lacking gift of nature; and the subsequent history of Moses' career contains the proof that the promise was fulfilled (compare ver. 15, "I will be with *thy* mouth and with his mouth, and will teach *you* what ye shall do"). But notwithstanding this promise, Moses replied: "send, Lord, whom thou wilt send." The difficulty arising from the fact that he was not a man of words, was removed by the promise of God; but Moses could not fully enter into the meaning of the promise, or place implicit confidence in it. Jehovah, therefore, referred him to the eloquence of his brother Aaron, who was to be always at his side when engaged in the duties of his office. I cannot persuade myself that *M. Baumgarten* (i. 1 p. 418) and *O. v. Gerlach* (i. p. 213 seq.) are right in the inference which they draw from the passage before us: viz. that it was the original intention of God that Moses should stand alone in delivering, conducting, and organizing the people, and should even hold the office of High Priest as well; but that he forfeited the honour by his refusal, and had now to share with Aaron both the office and the glory. There is no reference at all to the high priesthood here. Aaron was to be Moses' interpreter; this is stated in ver. 16, "he shall be thy spokesman unto the people, and he shall be thy mouth, and thou shalt be his God," and in chap. vii. 1, 2, "Aaron shall be thy prophet, and shall speak *unto Pharaoh* all that I command *thee*." There is nothing else referred to here. And this office of Aaron does not seem to have lasted long. It served for a time to remove the doubts and difficulties of Moses, and to sustain and help him in his *first* appearance before the people and Pharaoh. But when once Moses had discovered that the grace and gifts of God rendered him mighty in word and deed (Acts vii. 22), he stood no longer in need of another's mouth; henceforth he was his own. It was at a later period that Aaron received the call to the priesthood, and even then the arrangement was not of such a nature as that Moses relinquished anything in consequence of his brother's call, or in any way shared his office and his honours with Aaron. Even in the capacity of high priest Aaron was *under* Moses, and did not stand in an independent position by his side. Moses still continued to be Aaron's God, and Aaron the interpreter of Moses. It was this which constituted the peculiarity of Moses' position, a position which has no parallel in the Old Testament (Vol. iii., § 33. 4); he stood entirely alone as the founder and mediator of the ancient covenant just as Christ was alone as the founder and mediator of the new; though Moses and the ancient covenant were but feeble and imperfect types and copies of Christ and the new (*cf.* § 11).

The expression in ver. 16, "*he thy mouth, thou his God*" scarcely requires an explanation. As the prophet stood in such a relation to God, that he only spake what God put into his mouth, so was it to be with Aaron and Moses. Moses was the inspiring God of Aaron's prophetic activity. Aaron was the organ and representative of Moses, as Moses was the organ and representative of God. Compare chap. vii. 1, 2, "behold, I have made thee a God to Pharaoh, and Aaron shall be *thy prophet*," &c.

Simultaneously with the call of Moses, a violent agitation took place among the people, by which they were prepared for his mission. Here was an exhibition of the great and secret power of sympathy, a vague presentiment, in many perhaps a conscious anticipation, that a turning point was approaching, and that the time of deliverance was at hand. The divine promise to Abraham (Gen. xv. 13, 14): "They shall afflict them *four hundred years*, and afterward shall they come out with great substance," was probably associated with this presentiment, and served to explain it. The fact that there was such a movement among the people, we infer from the people's fervour in prayer, to which reference is made in chap. ii. 23 and iii. 7, but more especially from the impulse which constrained Aaron to go and seek out his brother in his exile (iv. 14); probably for the sake of consulting him, possibly to urge him to return to Egypt. It is not likely that the people in Egypt had entirely forgotten Moses, or that they had altogether relinquished the hopes, which the marvellous events of his life had apparently justified them in cherishing. It is even possible that Aaron may have been charged with a commission from a select body of men from among the people, who had already drawn up plans of escape, and were desirous of seeing Moses at the head of their enterprise.

FIRST APPEARANCE OF MOSES IN EGYPT.

§ 21. (Ex. iv. 18—31).—Moses at once obtained leave of absence from his father-in-law. He said nothing to him about what had occurred at Horeb (§ 19. 8); but merely expressed a desire to visit his relations in Egypt. He then set off upon his journey with his wife and children.—The intercourse between God and Moses was uninterruptedly maintained, after the first appearance of God at Mount Horeb. Even on the road Moses was not without divine encouragement. He received further instructions as to his future interview with Pharaoh. "*Israel is my first-born son,*" said Jehovah. Upon *this* Moses was to found his demand upon the king of Egypt, and also the threat, that Pharaoh's refusal should be punished with the death of *his* first-born son (1). He was further reminded once more, that he had to expect the most obstinate resistance on the part of the king (*vid.* chap. iii. 19, sqq.). It would be to no purpose, that he would perform before him all the miracles which Jehovah had commissioned him to work. Nevertheless there was no reason why Moses should be afraid of this opposition on the part of the king, for Jehovah had already taken it into account in his counsels. In fact Jehovah had *willed* this resistance, and was bringing it about as a judgment upon Pharaoh, for the greater glory of his own name, not only in the sight of the Israelites and Egyptians, but in that of all the nations round about. "But *I*," said He to Moses, "*but I will harden his heart*, that he shall not let the people go" (2). Thus Jehovah prepared himself for judgment. It was right, however, that judgment should begin at the house of God (1 Pet. iv. 17). The demand of Jehovah upon Pharaoh was founded upon the fact, that Israel was his *first-born son.* Israel had become so by the election of Abraham, and by the covenant with Abraham, Isaac, and Jacob, which was now to be renewed and extended, and to enter upon a higher stage in its development. Circumcision had been instituted as the sign of this covenant. And yet Moses, who was to contend for the covenant, had broken it himself, for his youngest son was still *uncircumcised.* In the weakness of his heart he had yielded to the haughty spirit of his wife, who from a false maternal tenderness, and a disregard to the religious institutions of Israel, had refused her consent to the bloody operation. As Moses was now returning from his exile to enter upon the duties of his vocation in Egypt, he found himself in similar circumstances to those in which Jacob was placed, when he returned from his exile in Mesopotamia to the land of Canaan, to enter upon the work of his life (vol. i. § 80, 4). The relation in which Moses stood to God was, like Jacob's, not a pure one. In his case there was also a disturbing element, which had to be removed, before Jehovah could acknowledge him entirely and without reserve. And it was necessary that this should be removed, before Moses entered the land of Egypt, in which he was called to labour. Jehovah, the friend and protector of Moses, still found in him reasons for anger and enmity. When Moses, therefore, was stopping at an inn on the road, Jehovah met him and was about to kill him. Moses at once discovered the cause, either because his own conscience accused him, or else from some intimation of his guilt, with which the hostile encounter of Jehovah was accompanied. Zipporah then took a stone-knife, and circumcised her son, and in the excitement of passion threw the foreskin at her husband's feet. (3) The way was now clear, and the first intimation that the favour of God had been restored, was the arrival of Aaron, who had been sent by Jehovah, and met his brother at Horeb, the mount of God. It was probably from this spot that Moses sent back his wife and children to his father-in-law (Ex. xviii. 2) (4).

The two brothers then proceeded to Egypt, where they called together the elders, and having performed the miracles in attestation of their mission, announced to them the words of God. The people believed, and bowed their heads and worshipped (5).

(1). *Israel is Jehovah's first-born son* (iv. 22). *M. Baumgarten* justly complains, that commentators have shown such a disposition to explain these words away, and to regard them as indicating nothing more than the preference of God for Israel, which resembled the love that a father generally bears to his first-born son, above all the rest of his children. "If," as he truly says, "Jehovah calls himself the father to Israel, we must understand these words as referring to some *fact*, which is to be regarded as the generation of Israel. But we cannot possibly concur with him in his opinion, that this fact is to be found in the physical generation of Isaac, which resulted, not from the power of nature, but from the power of grace. Shortly afterwards (i. 1 p. 425) *Baumgarten* correctly observes, that the expression, first-born son, has reference to the Gentiles; since the term *first*-born implies a contrast to those who are born *afterwards*, and by the latter we must necessarily understand the Gentile nations. He seems, however, to have been quite unconscious, that by this explanation, which is undoubtedly the correct one, he entirely upsets his previous theory. For, if the term *first*-born can only be fully justified, by our tracing its origin to a *physical* generation through the grace of God, the same rule must also apply to those who are born *afterwards;* and if those who regard the former as indicating merely a spiritual relation are to be charged with explaining the words away, the same charge must certainly be brought against those who do precisely the same with the latter. And where could such a generation be found in the case of the Gentiles?

It cannot be disputed that the notion, contained in the term son of God, requires some concrete act of generation on the part of God. We cannot discover this in creation; for here there was no difference between the Gentiles and the Israelites. Nor can it be found in their organization as a distinct people, that is to say, in the multiplication of the descendants of the patriarch to such an extent as to constitute a nation, in consequence of the blessing pronounced by God (Gen. i. 26, ix. 1) ; for in that case Israel could not be the first-born, but the youngest of the nations. Moreover, in either of these cases it would not be *Jehovah*, but *Elohim*, who would be described as the father. All nations are sons of Elohim from the very first, for they all owe their origin, their existence, to the creative, world-sustaining, and superintending operations of God. But only those, who are begotten according to the counsel of salvation, can be called sons of *Jehovah*. The generation of Isaac was undoubtedly of this kind. But Israel was not called a *son of Jehovah* merely (he is never called the *only son)*, but the *first-born son*, who would therefore be followed by other sons, begotten in the same manner. Hence, as we understand the words, we are shut up to the spiritual explanation ; and the generative act of God, which constituted Israel his first-born son, cannot have been any other than that one act, by which Israel received its peculiar character, as a people distinguished from all other nations on the earth, by which *the* seal of Jehovah was stamped upon it, and which was to make a perpetual distinction between the Israelites and other nations, until the time arrived when these also should be described as sons of a later birth. This act was the election of Abraham, with all the consequent leadings, and promises, the blessings and chastisements, which had made Israel what it then was ; that is to say, all the dealings of God with Abraham and his seed, from the first call out of Ur in Chaldæa to the summons to the mountain of God in Midian, which are thus brought into a focus and placed in one single point of view. This also serves to explain the reason, why the seed of Abraham could not be designated as the son of Jehovah until now (a fact which *Baumgarten's* views will not allow him to explain) ; for the birth of this son was only completed by the exodus from Egypt. Till then, the Israelites had no individual and independent existence.

The idea of sonship embraces both the act of begetting on the part of the father, and essential likeness on the part of the son ; for generation is the transmission of being, and the nature of the father must also be that of the son. If, then, *Jehovah* had begotten Israel, there would necessarily be a *Jehovistic* nature in Israel. But the Jehovistic nature of God relates exclusively to His operations in the development of the plan of salvation. Israel, therefore, if begotten of Jehovah, must have had imparted to it some living germ and divine call, in connexion with that plan.

Moreover, the idea of sonship involves both paternal and filial rights. The son owes to the father obedience, confidence, reverence, and love, and the father is bound to render to the son sustenance, protection, and education. Thus, in the name " son," there is involved the duty of faith in Jehovah on the part of Israel, and the pledge of constant and immediate training on the part of Jehovah.

But the idea of sonship is still further defined by that of primogeniture ; and the first-born has peculiar rights and privileges apart from the rest. He has already enjoyed the father's discipline and care, long before the others begin to participate in it. His education is complete, when the training of the others is still in progress, and therefore, for a time at least, he has essential advantages over them. Moreover he also takes part in the father's supervision over the rest, and assists in their training. And at all times he is the first and most natural representative of the father. If, then, Israel was actually the *first-born* son of Jehovah, all this must have been manifest in their case. The very consciousness of being the first-born in the house of Jehovah, was therefore a prediction for Israel of the future history of that house ; it was a repetition in a more concrete form of the original prediction: " In thee and in thy seed shall all the nations of the earth be blessed." When Israel knew that he was the first-born, he must also have known that the rest of the nations were destined to be born at a later period, and therefore that they also were called to inherit and share the possessions of the father's house. And this knowledge determined the duties of Israel with regard to these nations, in both the present and future course of their history.

Such a sonship as this, though it may be the result of spiritual generation, is quite as real as that which proceeds from physical generation. In consequence of this generation, Jehovah could no more forsake the Israelites, than a father neglect his son (*cf.* Jer. xxxi. 20 ; Is. xlix. 15, &c.), for he had made Israel an actual partaker of his own nature. Moreover, the character which was imparted to Israel through this generation, and which for the time distinguished it from every other nation, was a thoroughly real one, which had been implanted and had taken shape in the flesh and blood, as well as in the spirit and soul of the people. The call and election of Israel were something more than a mere idea, which floated like a vapour above the people, and could be driven away by the first wind that blew. It had become the soul, the national soul of Israel, and continued to fill with true life all its healthy, normal, vital functions, so long as it did not touch its own existence with suicidal hands.

(2). On the *hardening of Pharaoh's heart* we have an excellent dissertation by *Hengstenberg* (Pentateuch, vol. ii. p. 380), which, though not entirely free from partial views, has rendered essential service towards the elucidation of this subject. The earlier Lutheran theologians went so far in their opposition to the doctrine of predestination, as to maintain that the sinner always *hardened himself*, the part performed by God being limited to *permission* alone. "The rationalistic theology appropriated the rationalism of the orthodox all the more readily, since in the estimation of the former the co-operation of God even for good does not extend beyond permission." In this, however, both the orthodox and the rationalist were at variance with the Holy Scriptures, which so frequently and distinctly represent the hardening of man as the result of an actual interposition on the part of God. But the Scriptures were regarded by orthodox theologians as the word of God, and therefore they endeavoured to show that the discrepancy was merely apparent, and explained the Bible according to *their* notion of what hardening is. Thus, for example, in *Pfeiffer's Dubia vexata*, p. 229, the *decisio* respecting the case before us is as follows: "*Deus dicitur cor Pharaonis indurare permissive, permittendo scil. justo judicio, ut ille, qui se emolliri non patiebatur, sibi permissus durus maneret in propriam perniciem.*" The rationalistic theology, on the other hand, was not fettered by any doctrine of inspiration, and therefore candidly acknowledged the discrepancy, and even exaggerated it to such an extent as to affirm that the Scriptures made God the author of sin. *Hengstenberg* has defended the authority of the Scriptures in opposition to both of these, with especial reference to the hardening of Pharaoh.

The first thing to be decided is, what interpretation the writer of the Scriptural record intended us to put upon the whole transaction? With regard to this, *Hengstenberg* has at the outset very properly laid stress upon the fact, that the Scriptural account represents the hardening of Pharaoh's heart, no less frequently and decidedly as Pharaoh's own act, that is as self-hardening, than as an act of God of which Pharaoh was the object. He finds on examination that there are *seven* passages in which Pharaoh is said to have hardened his own heart (Ex. vii. 13, 22, viii. 15, 19, 32, ix. 7, 34), and also *seven* in which God is said to have hardened Pharaoh's heart (chap. iv. 21, vii. 3, ix. 12, x. 1, 20, 27, xi. 10). In his opinion the number seven is significant. "It indicates," he says, "that the hardening rested upon the covenant of God with Israel, of which this number was the mark." But in this we must differ from him. From the point of view referred to, we can easily understand why the hardening should be represented seven times as *an act of God*, for His covenant with Israel was the cause of all that He did. But it is impossible to apply the same explanation to the fact that Pharaoh is also referred to seven times, as hardening himself by his own voluntary act. Where Pharaoh acted *freely*, he cannot be regarded as having in any way acted in subservience to the covenant. It is only where his actions appear as the result of what God had done, that such a reference is admissible. However, as *Baumgarten* has already shown, *Hengstenberg* has not reckoned correctly. It is not seven times but *ten times* that God is said to have hardened Pharaoh's heart, and *ten times* also that Pharaoh is said to have hardened himself.[1] At the commencement of the narrative the hardening is attributed twice to God (chap. iv. 21, vii. 3), then seven times to Pharaoh (chap. vii. 13, 14, 22; viii. 15, 19, 32; ix. 7), then again once to God (chap. ix. 12), twice to Pharaoh (chap. ix. 34, 35), four times to God (x. 1, 20, 27; xi. 10), once more to Pharaoh (xiii. 15), and, lastly, three times to God (xiv. 4, 8, 17). This considerably alters the state of the case. *Ten* is the sign of completeness, being the last number of the *decad*, in which every possible numeral appears. If, then, the Scriptural record sets

[1] If the article in the evang. Kirchenzeitung 1837 was written by *Hengstenberg* himself, (and we have no reason to doubt it), it appears an inexplicable thing, that he should have given the right number "ten" in that article (p. 496), and then afterwards have altered it to seven.

before us the hardening of Pharaoh's heart ten times as his own act, and also ten times as the act of God, we may conclude from the equality in the numbers, that the two aspects are to be placed side by side as of equal importance, and that neither of them is to be sacrificed to the other. On the other hand, from the fact that the number *ten* is used on both sides, we may infer that each of the two processes, which were the determining causes of Pharaoh's hardness, ran its own course both freely and fully, and that in his case they were both of them exhausted and completely fulfilled. For this reason we are also unable to subscribe to *Hengstenberg's* opinion, when he says that " the equality in the numbers denotes, that the hardening attributed to Pharaoh stood to that ascribed to God, in the relation of effect to cause." We might reverse the sentence with quite as much propriety, or rather impropriety. If the author had desired to convey the idea, which *Hengstenberg* imagines, he would certainly have arranged the two causalities in such a manner, that every instance in which the hardening was ascribed to God, should be followed by another, in which it was attributed to Pharaoh. Even in the fact that, " in the introduction and the summing up the hardening is attributed to God," and therefore " the part performed by Pharaoh is surrounded by that of God," we can not discover any evidence of an intention to represent " the former as determined by the latter." The announcement of the obduracy of Pharaoh was necessarily made from that point of view, in which it appeared as the work of God ; for this was the only light, in which it could awaken confidence or give the intended pledge. In making such an announcement, God could not possibly refer exclusively or even prominently to the fact that Pharaoh would harden *himself*; for that would have implied, that the people would be left to the caprice and hard-heartedness of an enemy. On the other hand, there is certainly truth in the further remarks of *Hengstenberg*, that " there are also marks of design in the fact, that the hardening at the beginning of the plagues is attributed, in a preponderating degree, to Pharaoh, and towards the end to God. The higher the plagues rise, so much the more does the hardening of Pharaoh assume a supernatural character, and so much the more obvious is the reference to a supernatural cause." But after the divine causality has been placed in a most decided manner in the foreground, the self-hardening is again brought prominently forward; and in this we discover an evident intimation, that the two causalities are to be regarded as working side by side to the very close. We will here merely observe in passing, that *Hengstenberg* really contradicts himself, one of his assertions completely upsetting the other. In one place he says that the hardening attributed to Pharaoh stood to that ascribed to God in the relation of effect to cause; and yet afterwards he says, that in the circumstance that the hardening is at first chiefly attributed to Pharaoh, and towards the end almost exclusively to God, we have a proof that at the outset the human causality predominated, and subsequently the divine, in other words that the former was the effect, the latter the cause (the former the cause, the latter the effect—*Tr.*). If the two causalities really stood to *each other* in the relation of cause and effect, it is evident from the fact just referred to, that the human has the stronger claim to be regarded as the cause. But we deny that this is the relation in which they are to be placed, and we found our denial upon the Scriptural narrative itself. Each contains its own *cause*. within itself, the one in the evil will of man, the other in the holy will of God; the *effect* of the one is the hardening of a man to his own destruction, that of the other the hardening of man to the glory of God. Still each of these forms of hardening is determined by the other, and the one can never take place apart from the other. Had Pharaoh not received that testimony from God, of which the narrative before us speaks, he would not have hardened himself; and had not Pharaoh's sinful will resisted the divine decree, God would never have hardened him.

To prevent mistake, however, we must enter in several respects into a fuller explanation of what we have said above. Both the expressions employed and the facts themselves lead to the conclusion, that hardening can only take place, where there is a conflict between human freedom and divine *grace*. It is the ultimate result which is sure to ensue, when the human will continues to maintain a negative attitude towards the will of God, after the latter has positively announced itself, in accordance with the plan of salvation, as having no pleasure in the death of the wicked, but that the wicked turn from his way and live (Ezek. xxxiii. 11, *cf.* 2 Pet. iii. 9; 1 Tim. ii. 4). Hence such a thing can only occur within the sphere of revelation, or where the tidings of salvation have been received. In the case of a heathen, then, the hardening referred to is impossible, so long as he continues fixed in a purely heathen state. It is only when he comes into contact with a special message from God, in connexion with the history of salvation, as was the case with the king of Egypt, that the hardening of the heart becomes possible for him.

Again, it follows from the notion of hardening, that it can only result from a *conscious* and obstinate resistance to the will of God. It cannot take place where there is either ignorance or error. So long as a man has not been fully convinced that he is resisting the power and will of God, there remains a possibility that as soon as the conviction of this is brought home to his mind, his heart may be changed, and so long as there is still a possibility of his conversion, he cannot be said to be really hardened. The hardening of the heart commences from the moment in which a man becomes clearly conscious that he is resisting God, and it increases in proportion as this consciousness becomes stronger and clearer, and the testimony of God comes home to his mind with greater vividness and power. The course of Pharaoh's history will show how truly this applied to him.

Hardening, then, cannot even commence till some manifestation of *God* has been brought home to a man, with the express declaration and proof that it *is* such a manifestation. As God desires that all men should be saved, the first manifestation of God to a man must necessarily be one of mercy, designed to lead to his salvation. Even in the case of Pharaoh, the demand of God that he should let Israel go, supported as it was by signs and wonders, was an act of mercy; for it afforded him the opportunity and means of coming to a knowledge of the true God, of assisting in the development of the history of salvation, and of participating in its blessings. If, after this, a man rejects the mercy offered, and steels himself against its *saving* influence, this fact, which is the commencement of the hardening of his heart, proves that there was already an ungodly disposition within him, and that this disposition is still maintained in spite of such manifestation of the will of God. There is, it is true, an ungodly disposition in every man, because all are sinners. There was such a disposition in Moses as well as in Pharaoh; and even Moses refused for a time to acquiesce in the will of God. In *his* refusal, too, there was contained the *possibilty* of his becoming hardened, but this possibility did not become a *reality*, because the refusal was not an absolute one, and because his disposition *to depart from God* was counteracted by a leaning towards God, which eventually became victorious under the assistance of God himself. When Moses exclaimed, "Lord, send whom thou wilt," he had probably approached very nearly to the boundary of absolute refusal, which is the commencement of hardness ; but even these words did not contain an absolute refusal, since there was still a tone of petition heard by the side of the refusal. The divine summons then became a stern command, and the only alternative was either unconditional submission to God, or unconditional rebellion against him. Had Moses chosen the latter then, we fear that such a choice would have proved that his heart was hardened.

With Pharaoh it was altogether different. His ungodly disposition was already determined and unconditional, at the time when the manifestation of God was made to him. His refusal, therefore, was from the very first an absolute one, and had no counterpoise in a leaning towards God, as was the case with Moses. And, therefore, from the very first, his opposition was also a hardening of the heart. Pharaoh hardened himself on the first approach of God. He did not recognise the mercy contained in that approach, as being really mercy. But he could not conceal from himself the fact, that it was God who was approaching him, and that by his refusal he was fighting against God. Not that this is the only way in which the heart can be hardened. *Judas* obeyed the call of the Lord, tasted the delights of fellowship with the Lord, and yet was given up to hardness of heart.

No one would think of describing the refusal of *Moses*, reprehensible as it was, or *Peter's* denial, however grievous the sin, or lastly, the fury of *Saul* against the church of the Lord, bitterly as he afterwards condemned himself for it, as a hardening of the heart, or even as the commencement of hardening. The commencement of hardening is really hardening itself, for it contains the whole process of hardening potentially within itself. This furnishes us with two new criteria of hardening ; (1) before it commences, there is already in existence a certain moral condition, which only needs to be called into activity, to become positive hardness ; and (2) as soon as it has actually entered upon the very first stage, the completion of the hardening may be regarded as certain. The ossification of the heart may progress, but it can never return to a *status integer;* its course may possibly be checked or retarded for a time by the removal of everything that is likely to promote it, but nothing can prevent it from becoming complete in the end.

In what relation, then, does God stand to the hardening of the heart? Certainly His part is not limited to mere permission. *Hengstenberg* has proved, that this is utterly inadmissible on doctrinal grounds; and an impartial examination of the Scriptural record will show that it is exegetically inadmissible here. No. God *desires* the hardening, and, therefore, self-hardening is always at the same time hardening through God. The moral condition, which we have pointed out as the pre-requisite of hardening, the soil from which it springs, is a man's own fault, the result of the free determination of his own will. But it is not without the co-operation of God, that this moral condition becomes actual hardness. Up to a certain point the will of God operates on a man in the form of mercy drawing to himself, he desires his *salvation;* but henceforth the mercy is changed into judicial wrath, and desires his *condemnation.*

The *will of God* (as the will of *the Creator*), when contrasted with the will of *man* (as the will of *the creature*), is from the outset irresistible and overpowering. But yet the will of man is able to resist the will of God, since God has created him for freedom, self-control, and responsibility; and thus when the human will has taken an ungodly direction and persists in it, the divine will necessarily gives way. Hence, the human will is at the same time dependent on the divine will, and independent of it. The solution of this contradiction is to be found in the fact, that the will of God is not an inflexibly rigid thing, but something living, and that it maintains a different bearing towards a man's obedience, from that which it assumes towards his stubborn resistance. *In itself* it never changes, whatever the circumstances may be; but in relation to a creature, endowed with freedom, the manifestation of this will differs according to the different attitudes assumed by the freedom of the creature. *In itself* it is exactly the same will which blesses the obedient and condemns the impenitent,—there has been no change in its *nature*, but only in its *operations*,—just as the heat of the sun which causes one tree to bloom is precisely the same as that by which another is withered. As there are two states of the human will, obedience and disobedience; so are there two corresponding states of the divine will, mercy and wrath, and the twofold effects of these are a blessing and a curse. Even when the divine will yields to the human, it maintains its absolute supremacy, for in yielding it merely proceeds to manifest itself in another form, in accordance with the conditions of human freedom. But the human will is free, only because the living action of the divine will endows it with the power of self-determination. And it is also under restraint, since it can never escape from the will of God; for when it withdraws itself from one form of the divine will, by that very act it yields itself captive to another. Whenever a man obstinately refuses to submit to the will of God, who desires his happiness, God yields at length to the will of the man himself, who is seeking for happiness in an ungodly way, and working out his own condemnation. The ultimate result, then, is that described in Ps. cix. 17: "He wished for cursing, and it will come to him; he had no wish for blessing, and it will be far from him." The hardening of the heart, so far as it is permitted by God, is nothing but a recognition of human freedom, even to its utmost abuse; but God does more than permit this hardening, he wills it and even promotes it, when once a man's own sin has gone to such lengths, that all the pre-requisites of hardness are already there, and nothing more is required than an opportunity or inducement for putting it in practice. The occasion is then created by *God*, and it is this which constitutes the co-operation of God in the hardening of a man. Under these circumstances hardness of heart *must necessarily* ensue. Still the necessity for such an issue does not arise from the measures adopted by God, but is contained in the man's own moral condition, a condition brought about entirely by himself. Hence, it is not the will of God, which forces him into hardness, but his own ungodly will. The message from God, which furnishes the occasion for the entrance, continuance, and consummation of the process of hardening, is in itself as much a means of sanctification as of condemnation. The very same divine manifestation, which furnished the occasion for the hardening of the king of Egypt in his peculiar moral condition, would have been the means of leading him to salvation, if his moral condition had not been what it was. Whenever a message from God is received by a man, it urges him forward, either to salvation or condemnation. And his condition must become better, or worse than it was before. Such a message cannot pass over him without effect, for it is not dead but living.

If the message from God had not been delivered to Pharaoh, his heart would not have been hardened. If God had not brought the matter to a climax by fresh manifestations of a more and more striking character, the hardening would not have been completed. But neither his present nor his future condition would have been improved thereby. For the soil, from which this hardness sprang, would have been just the same, and a period in his history would have been sure to arise, as in that of every other man, when a decision, an actual, absolute decision must necessarily have been formed. Such decisions are a necessary part of the moral, that is the divine, government of the world. As the judgment is the end and aim of the history of the world, and of the life of every individual, it is necessary that such a decision should be formed by every individual as a prerequisite of judgment. It makes no difference to the individual himself, whether the decision is hastened or delayed. Nor are the reasons of such acceleration or delay to be found in the man himself, but in the position which he occupies in the world. The history of the world is woven of innumerable threads, and He who sits at the loom takes every thread just as it suits his purpose, but sooner or later he is sure to take them all. In the present instance it is easy to discover why this was just the moment, when it was requisite that Pharaoh's decision should take place. A new thread was about to be introduced into the plan of salvation and the history of the world, and that decision could not be dispensed with.

(3). The *occurrence in the inn,* which is narrated in chap. iv. 24—26, is in many respects difficult to understand. In ver. 24 it is said that "it came to pass by the way in the inn, that Jehovah met him, and wished to kill him." As the words stand here, they cannot possibly be referred to any one but Moses. But several commentators have assumed that it was not Moses, the father, but the uncircumcised son, who was *threatened with death. E. Meier* has lately adopted this explanation (Wurzelwörtb. p. 402). The opinion that it was Moses' son, whom

Jehovah wanted to kill, is derived from a reference to ver. 23. Jehovah is supposed to threaten, that he will punish Moses for neglecting to circumcise his son, in the same manner as he had already declared that he would punish Pharaoh for his disobedience. As such an interpretation is impossible with the present reading, *Meier* pronounces the passage mutilated; and supposes a verse to have been omitted between vers. 23 and 24, which contained the information that the first-born of Moses had not been circumcised, and concluded with the threat, "if thou do not circumcise him, I will slay thy son, even thy first-born." The similarity between this conclusion and that of ver. 23 is said to explain the accidental omission of the conjectural verse. But this is merely one of the arbitrary and unfounded assumptions, to which we have become accustomed, from such men as *Meier*. There is no reference at all to the *first-born* son of Moses. As we find from chap. iv. 20 that he had more than one son (according to chap. xviii. 3, he had *two*), and as only *one* of these is said to have been uncircumcised, it is scarcely possible to come to any other conclusion, than that it was the *youngest*. If it had been the *first-born*, this would certainly have been stated in ver. 25. But there is another difficulty, of greater importance than *Meier's* foundling, which induced some of the earlier commentators to refer the threat, contained in ver. 24, to the son rather than the father. In Gen. xvii. 14 it is the neglected son, not the negligent father, who is threatened with being cut off, if circumcision should not be performed. But even in this passage the destruction of the child is intended to be set forth, as primarily and chiefly a punishment for the parents; and what is of more importance still, the threat applied to the period when the covenant was in full force. When the covenant was suspended, or had been almost lost sight of by the parties concerned, as was the case now, after the Israelites had been in Egypt for 400 years without any revelation (*cf.* ii. 24), the threat contained in Gen. xvii. 14 lost its relentless severity. Still it was an act of sinful weakness and perverseness on the part of Moses, to give way in this matter to the self-will of his wife, a weakness which became the more conspicuous, now that he was about to come forward as the hero of God, and a perverseness, which seemed all the greater, as he was on his way to Egypt to renew the covenant, whose provisions he had himself neglected. The anger of God was therefore directed immediately against Moses himself, and not, first of all, against his son; that is to say, it threatened his own life and not his son's. This occurrence was, at the same time, a fresh and striking proof of the holy and inviolable manner, in which the covenant was to be henceforth maintained; but the threat also involved a promise and a pledge, that Jehovah would display as much vigilance and zeal in his defence, as he now displayed on behalf of the covenant. In the place of "Jehovah," the *Septuagint* reads ἄγγελος Κυρίου in ver. 24, a reading which is certainly justified by the sense (*cf.* chap. iii. 2, 4). It is doubtful, however, whether we are to think of a visible appearance, on the part of Jehovah or his angel, or merely of some act performed by Jehovah, which threatened to put an end to his life. The brief and indefinite notice in the Bible appears to favour the latter conclusion. Moses was probably suddenly attacked by some mortal disease, in which he could clearly discover the hand of God.

It is stated plainly enough in the history, that Zipporah was chiefly to blame for *the omission of the circumcision.* Whether her maternal feelings had led her to set herself against his being circumcised at all, or whether, from her contempt for the Israelitish rite (Gen. xvii. 12), and her preference for the custom of the Ishmaelites (Gen. xvii. 25), she wished to delay the circumcision of her sons till their 13th year, cannot possibly be decided. Her accurate acquaintance with the mode of performing the operation, which is presupposed by ver. 25, might perhaps be regarded as favouring the latter conclusion. Her use of a *sharp stone* for that purpose, is in harmony with Josh. v. 2. The agreement between these two passages seems to imply, that in the earlier times stone knives were generally employed in the operation. If this was the case, the explanation of such a custom is not that metal knives were as yet but little used, but that on symbolical grounds stone knives, which are a simple product of nature, were preferred to metal knives, which had been prepared by human art and were in general use every day. Even in heathen countries stone knives were employed in operations of a religious nature, *e.g.* in the preparation of the mummies in Egypt (Herod ii. 16: λίθῳ Αἰθιοπικῷ ὀξέϊ παρασχίσαντες παρὰ τὴν λαπάρην), and in the emasculation of the priests of Cybele (Catull. xliii. 5: *devolvit acuto sibi pondera silice).*

Zipporah cut off the foreskin of her son וַתַּגַּע לְרַגְלָיו. How are we to understand these words? Do they refer to Moses' feet, or his son's? This depends upon the interpretation to be given to הִגִּיעַ. If the meaning of this word will admit of its being referred to Moses, there is no grammatical difficulty to prevent our referring the suffix to him also, as he is mentioned in the previous verse. *Meier* renders the passage "she cut off the foreskin of her son, and smeared his feet (with it, or with the blood)," on the ground that the Hiphil of נגע is also used in Ex. xii. 22, with reference to the smearing of the lintel and door-posts with the blood of the paschal lamb, and moreover (listen and admire his acuteness!) the smearing of the blood on the lintel and door-posts exactly corresponded to the blood on the feet and the place of the wound (!!!) We do not think it worth the trouble to reply to such nonsense. נגע means to *touch*, the Hiphil to *cause to touch*, to bring into contact. This may be done in a slow and quiet way, or in a violent and angry manner. The passionate excitement of the woman, which is apparent enough from the history, justifies us in giving the preference to the latter. The words would then mean: she threw down at his feet. Smearing the feet of the child with the blood of the wound would be thoroughly senseless and without any analogy. And there would have been just as little sense in her throwing the foreskin at the boy's own feet. But the whole scene is intelligible enough, if we refer the words to Moses. It is her husband's fault, she thinks, that she is obliged to perform this bloody operation against her will. In her ill-humour she throws the foreskin at his feet, which was as much as to say: "now you have what you want."

If רגליו is to be thus referred to Moses, there can be hardly any doubt that Zipporah's exclamation, "a *blood-bridegroom* art thou to me," was addressed to the husband and not to the son. But in this instance also *E. Meier* has displayed most remarkable wisdom. "חֲתַן דָּמִים," he says, is "an expression borrowed from the consummation of marriage, and therefore points out the newly circumcised child as consecrated, entrusted to God. The act of circumcision bears this resemblance to the consummation of marriage, that in both cases there is an offering of blood to the great Deity of nature, and by this offering, this symbolical sacrifice to the God of life, in either case there is a self-consecration to the Deity of life, by which the right of existence is first obtained." *Spencer* also is of opinion that the meaning of Zipporah's words is: Ritu illo Deo et ecclesiae nostrae, quasi conjugii foedere copulatus es (p. 61 ed. Pfaff.). There is no doubt whatever that circumcision may be regarded in the light of a marriage or union with God, and the use of the Arabic word ختن for circumcision shows that it has been so interpreted. But the idea of there being any reference to connubial intercourse is nothing but a colossal absurdity, and even *Spencer's* explanation is inadmissible, for Zipporah says: "A blood-bridegroom thou art *to me*." *Aben-Ezra* and *Kimchi* say that the Jews were accustomed to call a newly circumcised child Chatan (though this cannot have been a universal custom, for we find no reference to it anywhere else; *Bodenschatz*, for example, does not once refer to it): but if so, this only proves that the later Jews gave this interpretation, or rather misinterpretation, to the passage before us.—On the other hand, the whole is perfectly clear, if we understand the words as referring to Moses. Moses had been as good as taken from her, by the deadly attack which had been made upon him. She purchased his life by the blood of her son; she received him back, as it were, from the dead, and married him anew, he was in fact a bridegroom of blood to her. In ver. 26 we read: "She said 'blood-husband' because of the *circumcisions* (לַמּוּלֹת)." The plural in this case must either be regarded as an abstract, according to the well-known custom in Hebrew (referring to circumcision in general as a religious rite, which Moses had wished to observe, but which *she* had hitherto obstinately refused, and not to the particular concrete act), or we may take it as a concrete, and refer it to the circumcision of the two sons.

(4). We learn from Ex. xviii. 2 that Moses *sent back* his *wife* and *children* to his father-in-law. This probably occurred now. The event in the inn had convinced him, that Zipporah was by no means in a proper state of mind to encounter all the dangers which threatened him in Egypt, with equanimity and faith. His brother Aaron's advice may also have led him to adopt this resolution.

(5). It was a most important thing both for Moses and the people, that the latter should *believe God* on the first interview with Moses. *M. Baumgarten* has the following apt remarks on this subject: " The text exhibits this declaration of feeling, with which the entire nation responded to the first message from God, as a most important commencement. . . . By faith Israel now proved itself to be the son of Jehovah (ver. 22), for the son believes the father. And the commencement thus made by the seed of Abraham, as a nation, answered to the disposition manifested by their father Abraham himself (Gen. xv. 6). Thus, whatever might be the course henceforth pursued by the nation of Israel, it enjoyed this honour, that its first mental act was *faith*, in which, though still suffering the severest oppression and hardship, it looked upon the redemption of Israel as already secured.

§ 22 (Ex. v., vi.).—A good beginning was made; the *people* believed and worshipped. But when Moses and Aaron appeared before Pharaoh, and in the name of Jehovah, the God of Israel, requested him to allow the Israelites to go a three days' journey into the desert (*cf.* § 22. 4), that they might celebrate a festival in honour of their God, they met with nothing but ridicule and insult (1). "Who is Jehovah," said Pharaoh, "that I should obey his voice? I know nothing of your Jehovah, and will not let Israel go." The king of Egypt, from his heathen point of view, looked upon Jehovah as merely the national god of the Hebrews, who in his estimation was as powerless and contemptible in comparison with the gods of the Egyptians, as the enslaved Israel when compared with the despotic and powerful Egypt. Like people, like god, was his notion; and, to show his contempt of both, he contemptuously increased the oppression under which Israel was groaning. The people, he thought, have too easy a life of it, and hence the wish for liberty is growing up among them; he therefore ordered their task to be doubled, that he might thoroughly eradicate any such desire. Hitherto they had had the material for the work brought to them; but henceforth they were to get it for themselves, and yet produce as many bricks as before (2). This was beyond their power. They fell into arrears with their deliveries, and their *shoterim* (or scribes, § 16) were beaten in consequence. They complained to the king of such inhuman proceedings, but their complaints were disregarded. And now the weakness of the people's faith became at once apparent. They heaped reproaches upon Moses and Aaron, for having brought them into deeper misery instead of bringing them relief, and refused to listen to their consolations and promises any more (3). But this only afforded the occasion for a display of the ability of Jehovah both to overcome the incredulity of the people, and break down the opposition of Pharaoh.

(1). The *request*, that Pharaoh would let the people go a three days' journey into the desert to celebrate a festival, does not seem to have struck the Egyptians as anything surprising. This may possibly be explained on the ground that the Egyptians were in the habit of making similar pilgrimages from time to time. *Niebuhr* discovered a mountain, called Surabit-el-Khadim, in the desert between Suez and Sinai, the whole plateau of which was covered with fragments of statuary, and pillars overturned, evidently the ruins of a temple, the pillars being crowned with the head of Isis. All the walls, pillars, and fragments, that were left, were covered with Egyptian hieroglyphics, symbols, and representations of priests offering sacrifice. *Lord Prudhoe* supposes this to have been a sacred spot, to which pilgrimages were made by the ancient Egyptians. The supposition is well founded, though *Robinson* has expressed a different opinion (Travels, vol. i. 112—116).

(2). The tributary service referred to here, consisted of the making of *bricks* for the royal buildings (*vid.* § 14. 5). Up to this time the straw that was required had been supplied to the Israelites; but henceforth they were ordered to go into the fields and gather it for themselves. The bricks, most extensively used by the Egyptians, were not burnt (as *Luther's* translation erroneously implies), but dried in the sun. The clay was mixed with chopped straw to give it the greater consistency. *Rosellini* brought some bricks from Thebes with the stamp of King Thothmes IV., the fifth king of the 18th dynasty, upon them. On examination, it was found that they were always mixed with straw. *Prokesch* (Erinnerungen ii. 31) says: "The bricks (of the pyramids at Dashur) are made of the fine mud of the Nile mixed with stubble. This mixture gave to the bricks an inconceivable durability." *Hengstenberg* (Egypt and Moses, p. 79 transl.), has properly laid stress upon this, as a proof that the author of the account before us possessed a most accurate acquaintance with the customs of Egypt.

L. de Laborde (comment. géogr. p. 18), has the following comment upon this passage: "J'ai assisté aux travaux du canal, et les moyens comme le résultat m'ont semblé en tous points répondre aux versets de l'Exode. Cent mille malheureux remuaient la terre, la plupart avec les mains, parceque le gouvernement n'avait fourni en nombre suffisant que des fouets pour les frapper; les pioches, les pelles et les couffes manquaient. Ces paysans, hommes infirmes, vieillards (les jeunes gens avaient été réservés pour l'armée et la culture des terres) femmes et enfants venaient principalement de la haute Egypte, et étaient répartis sur le cours présumé du canal en escouades plus ou moins nombreuses. L'entreprise était dirigée par des Turcs et des Albanois, qui avaient établi parmi les paysans des conducteurs de travaux responsables de la tâche imposée á chaque masse d'hommes. Il faut dire, que ces derniers abusaient plus que les autres de l'autorité, qu'ils avaient reçue. Tout ce monde de travailleurs était censé recevoir une paie et une nourriture, mais l'une manquait, depuis le commencement des travaux jusqu'à la fin, l'autre était si précaire, si incertaine, qu'un cinquième des ouvriers mourut daus cette misère sous les coups de fouet, en criant vainement, comme le peuple d'Israel (v. 15, 16), &c."

(3). By modern critics, who suppose that chap. vi. formed part of the original document, and that the previous chapters (iii.—v.) are supplementary, the two passages are regarded as different accounts of one and the same event, whereas according to their present position they form different parts of a continuous narrative. Undoubtedly nearly all the particular details of the call described in chapter vi. are also found in chaps. iii.—v.; and hence one might be tempted to regard the former as an earlier, more concise, and summary account of the same event. But it is also conceivable that, after the failure of the first mission to Pharaoh, the same doubts and fears may have arisen again in the mind of Moses, which he had already expressed at Horeb, and hence it may have been necessary that the call should be renewed, with a repetition of the consolations and promises by which they had once before been allayed. But at any rate, even if the two sections must be regarded as different accounts of the same event, there is sufficient progress in the second section to justify the editor in placing the summary account, contained in chap. vi., after the more detailed narrative in chap. iii.—v. This progress consists in the change from the strong faith, evinced by the Israelites at the outset (iv. 31), to the incredulity, manifested by them immediately upon the failure of the first attempt (vi. 9).

(2), On the CAPITAL *of the King of Egypt* at that time, see § 1. 5, and § 41. 2.

THE SIGNS AND WONDERS IN EGYPT.

Vid. Lilienthal, gute Sache ix. p. 31 sqq.—*S. Oedmann* vermischte Sammlung aus der Naturkunde zur Erklärung der heiligen Schrift. Aus d. Schwed. v. Gröning, Rost. 1786 sqq.—*Rosenmüller*, altes und neues Morgenland, vol. i. *Hengstenberg*, Egypt and the books of Moses, p. 95—125, Eng. transl.—*L. de Laborde*, comment. géogr. p. 22 sqq.—*J. B. Friedreich*, zur Bibel, Nürnberg 1848, i. 95 sqq.

§ 23 (Ex. vii. 1—7).—Pharaoh had contemptuously rejected the word of God, and therefore God spoke to him in deeds. The instrumentality of Moses was also employed in the *deeds*, as it had formerly been in the *word*. The fruitless negotiations were followed first by a declaration of war, and then by war itself. Moses, the shepherd and leader of Israel, was opposed to Pharaoh, the King of Egypt. But Moses was the messenger and representative of Jehovah, whom Pharaoh despised, so strong was his confidence in the superior might of his own deities. Hence the contest, which was now about to commence, was essentially a *war on the part of Jehovah against the gods of Egypt* (1). For that reason, Moses did not conduct the armed hosts of his people against the horses and warriors of Pharaoh; it was not to the secular power of the Egyptian monarch, but to his *gods*, that the gauntlet was thrown down. It was in the domain of miracles that the battle was to be fought—a domain in which Egypt regarded itself as peculiarly strong—for it was in Egypt, the land of conjurors and magicians, of interpreters of dreams and signs, that *magic*, that mysterious life-blood of heathenism, had put forth its marvellous power in its most fully developed forms (2).

(1). The whole of the ancient church was most fully convinced of the *reality of the heathen gods*. Idolatry in its esteem was devil-worship in the strict sense of the term. The fathers of the church had no more doubt than the heathen themselves, who still adhered without the least misgiving to the religion they had inherited from their fathers, that the gods and goddesses of mythology were real beings, and had a personal existence, and that the worship with which they were honoured was not only subjectively directed, in the minds of the worshippers, to certain supernatural beings, but actually reached such beings and was accepted by them. The fathers of the church undoubtedly lived in an age, when the original power of heathenism was broken; but even this shattered heathenism, the *disjecta membra poetae*, still produced upon their minds the powerful and indelible impression, that there was something more in all this than the empty fancies or foolish speculations of idle brains; that there were actually supernatural powers at work, who possessed a fearfully serious reality. The impression thus produced upon their minds, by their own observation of the tendency of heathen idolatry, was confirmed by their reading of both the Old and New Testaments; and the greater the confidence with which they looked upon the salvation they had experienced in Christ, as something real and personal, the less doubt did they feel, as to the reality of the powers of evil by which it was opposed in heathenism. In a word, the gods and goddesses of heathenism were in their estimation the destructive powers of darkness, the fallen spirits, the principalities and powers that rule in the air, of whom the Scriptures speak. It is not to be denied, that in this they went farther than the Bible authorised them to go. But it must be maintained, on the other hand, that they had laid hold of the substantial truth contained in the Bible; whilst their error was merely formal, and confined exclusively to their doctrinal exposition of that truth. But modern theology, both believing and sceptical, by denying all objective reality to the heathen deities, and pronouncing them nothing but creations of the imagination, has departed altogether from the truth, and rendered it impossible to understand either heathenism itself, or the conflict which is carried on by the kingdom of God against the powers of heathenism. We find *Hengstenberg* still following this false track (Beiträge iii. 247 seq.). In the zeal, with which he has so worthily contended against the rationalist foundling of a national God of the Hebrews, he has persuaded himself that he may safely assert, that the Bible does not once attribute even a sphere of existence to the gods of heathenism, much less a sphere of action. On the other hand, the theologians of the present day are again beginning to discover the true solution of the problem. Among others we may refer to *J. T. Beck* (Einleitung in das System der christl. Lehre p. 102 seq. and christl. Lehrwissenschaft i. 259), *Rodatz* (luth. Zeitschrift, 1844), *Delitzsch* (biblisch-prophetische Theologie p. 81), *M. Baumgarten* (Commentar i. 1. p. 469; i. 2 p. 351, &c.), *Hofmann* (Weissagung und Erfüllung i. 120; Schriftbeweis i. 302 sqq.), *Nägelsbach* (Der Gottmensch Nürnb. 1853. 1. 244) &c. There were others of earlier date who held the correct view, for example: *G. Menken* (Homilien über die Geschichte des Elias., 2 A. Bremen, 1823, p. 107 sqq.), and still farther back *Chr. A. Crusius* (Hypomnemata ad theol. proph. i. 129 sqq.).

Crusius maintains with perfect justice: Sacrae literae a Mose usque ad Novum Testamentum constanter docent, Deastros esse daemones. Quorum etsi *Deitas* negatur, non ideo *entitas*, ut ita dicam, negari censenda est, cum potius contrarium aperte pateat. What impartial expositor can possibly deny that such passages as Ex. xii. 12, xv. 11, Num. xxxiii. 4, Deut. x. 17, Ps. lxxxvi. 8, xcv. 3, xcvi. 4, xcvii. 9, cxxxv. 5, cxxxvi. 2, seq., &c., attribute to heathen deities not merely a "sphere of existence," but a "sphere of action" also? In Ex. xii. 12 Jehovah promises: "I will pass through the land of Egypt this night . . . . . and against all the gods of Egypt I will execute judgment, I Jehovah." In his song of praise (Ex. xv. 11) Moses sings: "Who is like unto thee, O Lord, among the gods?" In Ex. xviii. 11 Jethro confesses: "Now know I that Jehovah is greater than all gods!" Even on the gods whom Israel served in the desert Jehovah executed judgment (Num. xxxiii. 4). In Deut. x. 17 Moses declares to the people: "Jehovah, your God, is the God of gods and the Lord of lords." The Psalmists describe Jehovah as highly exalted above all gods (Ps. xcvii. 9, cxxxv. 5), as a great king above all gods (Ps. xcv. 3), as to be feared above all gods (Ps. xcvi. 4), whilst there is none like him among the gods (Ps. lxxxvi. 8). In the prophets the judgments of God on heathen powers are spoken of, as a victory on the part of God over the heathen deities, and a judgment inflicted on them. Now who would suppose the theocratic lawgiver, the poets, or the prophets, capable of such absurdity, as to think that the best way of convincing the people of the absolute power and supremacy of Jehovah, was to demonstrate continually that he was stronger than *nothing*, more exalted than a mere fancy, greater than what had no existence at all, victorious over something which had no sphere of operation or of life, ruler over that which was not, and judge of that which had never been? *Cervantes* makes the knight of La Mancha fight against windmills; but the prophets would have done something worse than this, if they had made their Jehovah attack, conquer, and execute judgment upon something, of which they were convinced that it never existed at all.

Let us see what reply *Hengstenberg* has to make to this. He proves the non-existence of the heathen deities, first of all, from what the *Pentateuch* says of Jehovah: "Jehovah is Elohim, the God of Israel is also *the* deity; *quidquid divini est*, is contained in him." But the gods of the heathen are also Elohim, and are so called. "Jehovah is the God of the spirits of all flesh (Num. xvi. 22, xxvii. 16); He is the creator of the heavens and the earth; the heaven and the heaven of heavens are His, the earth also and all that therein is (Deut. x. 14); He feeds and clothes the strangers (Deut. x. 17, 18); from Him proceeds the blessing, which is to flow through the posterity of the patriarchs to all the families of the earth; He is the judge of the whole earth (Gen. xviii.). What is there then that is left for the idols, seeing that all is preoccupied by Jehovah? They can be nothing but λεγόμενοι θεοί, 1 Cor. viii. 5."—This is all perfectly correct; they cannot be gods in the same sense as Jehovah, or on an equality with Him, of the same essence, and with equal power. But *Hengstenberg* asserts more than this. He maintains that "they *cannot exist* at all, since they have neither *a sphere of action* nor *a sphere of existence*." What a leap in the demonstration! *Hengstenberg* might have said, with *quite* as much propriety, that angels, and men, and animals cannot exist at all, because they have neither a sphere of action, nor a sphere of existence. The sphere of existence belonging to the heathen deities is within the limits of creation, though it is super-terrestrial; and their sphere of action is simply heathenism, viz., that which is estranged from God, and has rebelled against Him, who alone is God.—*Hengstenberg* further demonstrates the non-existence of the heathen deities from the terms employed to designate them in the *Pentateuch*: "They are called אֱלִילִים *nothings* (Lev. xix. 4), לֹא אֵל and הֲבָלִים "not God" and "vanities" (Deut. xxxii. 21), לֹא אֱלֹהַּ (*ibid.* ver. 17), גִּלּוּלִים *stercorei* (Lev. xxvi. 30; Deut. xxix. 17)."—And from this we are to infer, that they have no existence at all, that they are merely "creatures of fancy!" And this is to overthrow the strong testimony, afforded by the passages cited above, to the objective reality of those powers, which the heathen worshipped as their gods! Does it follow that, because the λεγόμενοι θεοί of heathenism are *not-God*, therefore they are *nothing*, have no existence at all? Does it follow that, because they are powerless in relation to Jehovah, they are also powerless in relation to man? Does אֱלִיל mean *that which has no existence whatever*; does it not rather mean that which is not what it pretends to be, or is supposed to be? (*cf. Gesenius, s.v.*, p. 103. 1) adj. *qui nihili est, vanus, inanis, debilis*). Did *Job* (xiii. 4) mean to say, that the friends who came to comfort him, had no existence, when he called them רֹפְאֵי אֱלִיל (no physicians?). Or was it with reference to something that did not exist, that *Zechariah* said, "Woe to the רֹעִי הָאֱלִיל (the worthless shepherds) who neglect the flock?" (xi. 17).—The same remarks apply to the passages in which the deities are called הֲבָלִים. Was it the opinion of Eve that her second son had no existence whatever, when she called him הבל? When Job exclaimed כִּי הֶבֶל יָמָי (vii. 16), did he mean to say his days had never been? Or do the words of the Preacher, כֹּל הָבֶל (all is breath, vain), deny that anything exists?

Lastly, it is with peculiar emphasis that he says: "If a man should say that Christ conquered the gods of Greece, is there any one in the world who would infer from this, that he believed in the existence of those gods, especially if at other times he had repeatedly and expressly stated, that he looked upon them as merely creatures of the imagination?" Most certainly if the latter were the case, the former could only be regarded as a mode of speech, intended to be poetical. But in no other case; that is to say, we could only interpret the words as poetical, if it had been first of all established that the gods of Greece were vain, non-existent creatures of fancy; and this, as we have seen, cannot be proved on Scriptural grounds.

We adhere, therefore, to the opinion expressed by the excellent *Crusius*: Quorum etsi *deitas* negatur, non ideo entitas negari censenda est. We cannot otherwise explain the fact, that in the same breath the sacred Scriptures maintain the reality, and also the nothingness of the strange gods; *e.g.*, Ps. xcvi. 4, 5; 1 Chr. xvi. 25, 26; compare also 1 Cor. viii. 4, 5, and x. 19, 21.

We cannot at present enter upon the enquiry, to what extent the doctrine respecting the *daemonia* had been developed in the age of the *Pentateuch*. So much, however, we may safely affirm, that if there was no daemonology before, it must have arisen from the prevalence of the views referred to respecting the heathen deities. If Jehovah is the one and only God, the supreme and absolute Deity, and if, on the other hand, the so-called gods of the heathen are real, super-terrestrial beings and powers (Elohim), which are objects of fear and reverence to the heathen on account of their power,[1] but from their weakness are אלילים and הבלים in the estimation of Jehovah and his people; which were created by Jehovah, but have resisted him, though he has defeated and judged them; we have here the necessary data for determining the daemonology of the Bible.— The ungodly and rebellious nature of these powers, whom the heathen worshipped as *Elohim*, is apparent from the judgment executed upon them by Jehovah; and it was expressed in the name שֵׁדִים (from שׁוּד *violenter egit, vastavit)*, *i.e.*, destroyer, devastator, ungodly daemons (Deut. xxxii. 17; Ps. cvi. 37).

We may see from the *Septuagint*, that in the opinion of later Jews the heathen deities were representatives of daemoniac powers. The words שֵׁדִים in Deut. xxxii. 17, and Ps. cvi. 37, אלילים, in Ps. xcvi. 5, (*πάντες οἱ θεοὶ τῶν ἐθνῶν δαιμόνια*), גַּד (god of fortune = Baal), in Is. lxv. 11, and other similar expressions, are all rendered *δαιμόνια.*

This view is also fully sanctioned by the New Testament. The description of the spirit by which the girl at Philippi was possessed (Acts xvi. 16), as a *πνεῦμα Πύθωνος*, may be adduced as a proof of this. And Paul says most clearly and indisputably in 1 Cor. x. 20, 21, "the things which the Gentiles sacrifice, they sacrifice to *daemons* and not to God. And I would not that ye should *have fellowship with daemons* (*κοινωνοὶ τῶν δαιμονίων*). Ye cannot drink the cup of the Lord and the cup of *daemons;* ye cannot be partakers of the Lord's table and *the table of daemons.*" *Ex quo statim apparet*, says *Crusius* (l. c. p. 133), *Apostolum daemonia illa pro naturis existentibus habere, non pro figmentis cerebri humani. Alioqui communione cum illis interdicere non poterat, quia non entis nulla praedicata sunt.* In this passage the apostle expresses his conviction of the actual personal connexion between daemoniac powers and the worship of the heathen, in a manner so clear and decided that no other explanation is possible. It is true that in another passage (1 Cor. viii. 4, 5), he brings forward the other side of the ques-

[1] I have been convinced by *Delitzsch* (Genesis ed. ii. 1. 31 and ii. 171 sqq.) that the opinion, formerly expressed by me (Vol. i. § 13. 1) that אלהים is derived from איל, *to be strong*, is untenable, and that it must be traced to the Arabic Aliha = *stupuit, pavore correptus est;* transitive: *coluit, adoravit Deum.*

tion in the same manner as the *Pentateuch* does: "as concerning the eating of those things that are offered in sacrifice to idols, we know that an idol is nothing in the world (*ὅτι οὐδὲν εἴδωλον ἐν κόσμῳ*, *cf.* x. 19, and vii. 19), and that there is none other God but one." But it is evident from what follows, that the apostle does not mean to deny that the idols have any real, objective existence (a statement which would be directly at variance with chap. x. 20, 21). In ver. 5 he says, "for though there are really so-called gods (*καὶ γὰρ εἴπερ λεγόμενοι θεοί*), whether in heaven or on the earth, as there are actually many gods and many lords (*ὥσπερ εἰσὶ θεοὶ πολλοὶ καὶ κύριοι πολλοί*), yet *we* have one God, the Father," &c. The apostle first of all introduces the statement, that the *λεγόμενοι θεοί* really exist, in a *hypothetical* manner, as a mere supposition, an opinion generally entertained; but he afterwards guards against any doubt that might arise, as to his own agreement with that opinion, by introducing the parenthetical clause, *ὥσπερ εἰσι θεοὶ πολλοὶ*, &c., which contains a most distinct assertion, that the popular opinion is perfectly true.

The Scriptures do not anywhere affirm, that the mythological world of heathen deities exactly corresponds to the objective world of daemons, that is to say, that every individual god in the heathen worship is to be personally identified with an individual daemon, or *vice versâ*, that each particular daemon is represented by some heathen deity, so that we can say that Osiris and Isis, or Jupiter, Mars, Venus and others, are all representatives of particular personal daemons, and that the same name always denotes the same daemon. On the contrary, they merely affirm that the worship of the heathen has respect to real objects; that all the homage paid to a heathen deity reaches some existing, personal, supernatural power, and is accepted by that power; and that, as the heathen devotes himself to some such power by the worship which he presents, so does that power come near to him, and enter into living fellowship with him. "The things which the heathen sacrifice," says Paul, "they sacrifice to daemons,"—they think they are offered to a god, but they only reach a daemon, a being opposed to God and not God; and he who sacrifices enters thereby into fellowship with daemons, as the Christian, when he comes to the table of the Lord, enters into fellowship with Christ.

The relation between the mythological world of deities and the world of daemons may be thus explained: The commencement of all heathenism was a departure from the personal, holy, spiritual, and supermundane God, who had become burdensome and troublesome to the desires of the heart. The fulness of life, apparent in all nature, the rich variety of its forms, the energy of its powers, its inexhaustible resources of enjoyment, the charm of its mysteries, and other things, then became objects of veneration. The first object of adoration was the *ἕν καὶ πᾶν*, the force of nature which gives to every thing its life, and shape, and motion;—thus the primary form of heathenism was *pantheism*. But the various methods in which the common force of nature manifests itself, the different functions and instrumentalities employed, and the manifold spheres in which it developes itself, caused the *ἕν καὶ πᾶν* to be grasped not merely in its unity but in its diversity also; and thus pantheism was shaped and developed into *polytheism*. It was from the contemplation of nature, and the play of speculation and fancy, that mythological systems sprang. The names and forms of the various deities, and the peculiar powers and functions attributed to them, were purely creations of the fancy, empty and airy phantoms. There was no living personal object by whom the worship could be received. But this did not continue to be the case. The empty forms, which the fancy had created, were soon filled with something real. For the *ἐθελοθρησκεία*, which turned away from the one living God, and sought for other objects besides him, afforded to the spirits, described in Eph. vi. 12 as *ἀρχαί* and *ἐξουσίαι, κοσμοκράτορες τοῦ σκότους τούτου* and *πνευματικὰ τῆς πονηρίας ἐν τοῖς ἐπουρανίοις*, a sphere of action on the earth, such as they had never had before, and supplied them with empty forms, which they were able at once to fill with their substance. Powers of magic and augury were now put forth, on the basis and in the ceremonies of this *ἐθελοθρησκεία*; and these powers attested the presence of real supernatural agencies, and tended to confirm and enslave the heathen in their errors. Hence it is equally true that the heathen deities are vain creations of the fancy, and that they are real and personal powers, that the *εἴδωλον* is a *nothing*, and yet that it is *something* possessed of power. And it is no more true, that they have in *themselves* any personal reality, than that they have continued to be merely airy phantoms in the concrete development of heathen life.

(2). "As there is no nation without religion, so there is also none without *magic*, which clings like a shadow to religion in all its forms." (*Georgii* in Pauly's Real-encyclopädie iv. 1377). Magic, according to the notions prevalent among those who placed implicit confidence in it, is a power, acquired or inherited, which enables the possessor, by means of some secret art or science, to employ at will the forces of a supernatural world of spirits or deities, either for the purpose of finding out what is naturally hidden from human knowledge *(augury)*, or of performing things, beyond the natural power of the human will *(magic)*. Three different methods have been proposed, for explaining those examples of magic which are found recorded in well-authenticated history. The *first* method treats the whole affair as fraud and trickery on the one side, and superstition or excessive credulity on the other. This is the explanation suggested by the modern enlightened schools of Deism and Rationalism, in which *Balthazar Becker*, with his "Enchanted world," first led the way (De betoverde Weereld, 1—4. Amst. 1691—93. 4). But these things are already looked upon as antiquated. The *second* method admits the credibility of those accounts of magic, which have been handed down, and regards the feats described as actually performed by supernatural powers, either good or evil. A distinction was frequently made between black and white magic. White magic was referred to God or his angels and saints, whose assistance was supposed to have been obtained by means of prayers, asceticism, by word or sacrament. Black magic was attributed to Satan and his angels. With this explanation, the magic of heathenism was of course set down as exclusively the work of Satan. From the time of the Rabbis and Fathers till the days of modern "enlightenment," this was *the* explanation adopted in both the synagogue and the church. The *third* method suggested, for explaining the enigmatical data, traces the whole to natural powers, which are either acquired or inherent in the human mind, and which are only secret so far as they have not been thoroughly investigated by science. These powers are said to consist, partly in the control possessed by the human mind over nature itself, and partly in the connexion between one mind and another. In the ordinary every-day life these powers lie dormant in the depths of the soul, shut up and confined by the bolts and bars of the outward life of sense. But there are occasions and circumstances, sometimes unsought (as for example in certain diseases, or at the approach of death), at other times induced at will by some influence from without, in which these bolts are drawn back, the veil of the Psyche is lifted, and the hidden dormant power of the mind wakes up and moves, free and unfettered, in regions of light and knowledge, of will and action, from which it is entirely excluded when the bodily life, the life of sense, is in its healthy normal condition. This explanation has found many friends and supporters, since mesmerism has thrown some light upon the mysteries of somnambulism, and induced phenomena, corresponding in many respects to those displayed by the magicians of old. Thus not only are the phenomena of magic, of which we have received accounts both from antiquity and from the middle ages, regarded as something more than a mere delusion, a mournful aberration of the human mind; but they are even supposed by some to be a continuous series of profound anticipations of a science, the first letters of whose alphabet have been but recently learned, and not only this, but anticipations of that state of activity, to which the human mind will first fully attain, when it has entered upon the perfect life of the future state, where it will no longer be encumbered by an outward corporeal frame. The work of *Joseph Ennemoser* (Geschichte der Magie, being the first volume of his Geschichte des Magnetismus ed. 2. 1844) is founded upon this hypothesis.

With regard to the phenomena of *heathen magic*, to which our plan requires that we should confine our attention in the further discussion of this mysterious subject, we feel obliged to maintain at the outset, that neither of the methods described above is equally applicable to all the cases which present themselves. On the contrary, sometimes one will furnish a satisfactory explanation and sometimes another; most frequently it is necessary to combine two of them together; and there are cases, which it is perhaps impossible to explain without uniting all the three. It is very seldom indeed that the mysterious phenomena of magic can be set down as mere trickery, a clever attempt to deceive; and the farther we go back into antiquity the more rarely do such cases occur. On the other hand, the farther we depart from the period, in which we find heathenism in its most simple state, towards a period of abstraction and reflexion, the further we leave behind us the age, in which it flourished with unbroken power, and approach the times in which its power was shattered and its end was at hand—so that instead of the vigorous breath of life, we become more and more sensible of the odour of dissolution and decay—the less hesitation do we feel in assuming that there is some trickery, even if the whole is not fraud.

In addition to what we have already said, with reference to the nature of the heathen deities and heathen worship, we have still the following remarks to offer, in explanation of the enigmatical phenomena of heathen magic. There are three different sources, from which extraordinary and miraculous knowledge and power may be obtained: life in God and with God, the fellowship of daemons, and a magical power acquired by the mind over both nature and spirit. The third is a middle-sphere, capable of serving as the channel of both divine and daemoniac knowledge and power. And in itself it is undoubtedly sufficient, under certain circumstances and within certain bounds, to confer the ability to look beyond the limits of time and space, and to will and perform things which in our ordinary every-day life are absolutely impossible. Yet this never occurs in the present life, when the body and soul are in a healthy and normal condition, but only during some temporary, and more or less violent and unnatural disturbance of their proper relation to each other. When our present life is in a sound and natural state, such faculties as these are suppressed and hidden, and merely exist as dormant potentialities bound up in the inmost recesses of the soul. There can be no more doubt, that, when they were first implanted in man, they were intended to be unfolded and put forth in this present life, than that they will still, by virtue of the counsel of redemption, attain to full development and activity, though this may only take place in the future state. For the present, however, they are *shut up* and *restrained* according to the gracious will of God, because their exercise in connexion with the sinfulness of humanity could only be injurious and ungodly, and therefore unnatural also. Hence every arbitrary and self-willed attempt to burst the fetters by which they are bound, and so to loosen or snap the intimate connexion which exists between the body and the soul as to open up a chasm from which they may come forth, is from the very first ungodly and unnatural. It matters not whether this be accomplished by means of the stupifying vapour from the cavern at Delphi, or the intoxicating poison of the toadstool; by fixing the eyes upon a tin plate, or looking intently at the navel, the method adopted by the Omphalopsychi; by the magnetic influence of a physician, who goes beyond the laws of medicine in the performance of unwarrantable experiments, or by any other means by which the outward, clear, self-consciousness is forcibly suspended. And where this is accomplished, let no confidence be placed in the revelations which may be made, or in the morality of the power that may be at work; the prophecy is as likely to be false as true, and the power at work is just as likely to be injurious and destructive, as beneficial and saving: for a so-called natural magic is essentially unnatural and ungodly. Such experiments are doubly dangerous, for a man thus lets the sceptre of self-consciousness and spontaneity fall from his hands, and knows not whither the emancipated, will-less *(willenlose)* power of the *psyche* may hurry him, and to what strange, dark, and hostile powers it may thus be laid open and become a helpless prey. In the case of many heathen oracles, and also of many of the revelations made by modern sleep-walking Pythiae, it is indisputable that some wicked, mischievous, intentionally deceptive intelligence has been at work *(cf. G. .H v. Schubert, Zaubreisünden* p. 38); and the Angekoks of Greenland acknowledged, after their conversion to Christianity, that "much of their conjuring had been nothing but trickery, but in a great deal of it there had been some spiritual influence, which they now abhorred, but could not describe" *(Crantz*, Hist. v. Grönland i. 273).

There is an essential difference, however, between this natural magic and the exertion of miraculous power, either divine or daemoniac; although it may serve as the channel for either. Of the former we have not to speak here. But with regard to *the latter* it was clearly the conviction of the biblical writers, not only that it was within the bounds of possibility, but that it was actually put forth in heathenism (in the period of its power and glory), and will once more be displayed in the final conflict, which has yet to take place between the kingdom of God and the kingdom of darkness (2 Thess. ii. 9; Matt. xxiv. 24; Rev. xiii. 13). Even if we leave out of sight the feats performed by the Egyptian magicians, so far as they were successful; the earnestness and emphatic manner, in which the law prohibits every kind of heathen witchcraft, forces the conviction upon our mind, that the lawgiver did not regard the practice as mere superstition, a foolish fancy, or delusive trickery. And if we turn to a later period; who can read the history of the contest between Elijah and the priests of Baal on Mount Carmel, and doubt for a moment that they actually and confidently expected signs and wonders from their god? And how can this confidence be explained, unless on a former occasion they had received some such proof of his power?

But if, according to the Scriptures, we must thus attribute to daemoniac powers the ability to perform signs and wonders,—except so far as they were prevented by a divine interdict,—we maintain just as firmly, according to the Scriptures, that they were only *σημεῖα καὶ τέρατα ψεύδους* (2 Thess. ii. 9). They were *lying* signs and wonders, because they proceeded from a lie, and their aim was falsehood. They were lying, because they represented the *λεγόμενοι θεοί* as *ὄντες θεοί*; whereas in spite of all the signs and wonders the latter were only אלילים and הבלים, an *οὐδὲν ἐν κόσμῳ*. They were lying, because the powers employed by those who performed them were stolen and abused; and because they were the means of perpetuating error, falsehood, and destruction; in a word, they were lying because they gave themselves out for what they were not, and whilst professing to do good were really the cause of evil.

It is still a point in dispute, whether the feats performed by the Egyptian magicians were examples of a natural or daemoniac magic, or of both together. But to our mind there can be no doubt, that what they did was not effected without the co-operation of those powers, which they worshipped as gods. The whole scene from first to last is described as a contest between Jehovah and the gods of Egypt. The conjurors, we may be sure, left nothing untried which they thought likely to bring the gods to their help, and the gods of Egypt, that is the daemoniac powers, who were here engaged, will assuredly have endeavoured to maintain the appearance of power, as long as they possibly could. The scriptural record says with regard to the magicians (Ex. vii. 11, 22; viii. 7, 18 [iii. 14]): "they also did in like manner with their enchantments (בְּלַטֵּיהֶם)," but it does not inform us what these enchantments were.

§ 24 (Ex. vii. 8—13).—Moses and Aaron went to Pharaoh again a second time at the command of God. Aaron, as the prophet of Moses, carried the rod with which the gods of Egypt were to be defeated (§ 20. 7). When Pharaoh demanded a miraculous proof of the power of their God, Aaron threw down his rod before the king and his servants, and immediately it became a serpent. Thus was the contest commenced in the very territory in which the magic of Egypt was strongest, that of snake-charming. Pharaoh imagined that he was certain of victory here. He therefore sent for his *wise men* and *sorcerers* (1) (*Charthummim*, *cf.*, vol. i., § 88. 1), that they might frustrate Aaron's power by their secret arts. They appeared and threw down their rods. These also became serpents, but were swallowed by Aaron's rod (2). Thus was the first decisive victory gained by the power of Jehovah over that of the gods of Egypt. This was so clear and unmistakeable, that even Pharaoh could not deny it. Yet he would not acknowledge it. His sorcerers had produced the same effects by their conjuring as Moses and Aaron; and the unfortunate result might, perhaps, have been merely attributable to accident, or the carelessness of his sorcerers. At all events, instead of yielding to the impression which he ought to have received from this manifestation of divine power, he hardened his heart against it and persisted in his refusal.

(1). The Apostle Paul has, no doubt, followed the Jewish tradition when he calls the sorcerers, who withstood Moses, *Jannes* and *Jambres* (2 Tim. iii. 8). This tradition is also found in the Targums and Talmud. Jannes and Jambres are represented in the latter as sons of Balaam (Num. xxii. 22), *cf. Buxtorf*, Lex. Chald. Talmud, p. 945 sqq. The same names also occur in a fragment of the Pythagorean Numenius (about

150 A.D.) quoted by *Eusebius* praep. evang. 9. 8, where they are said to have been the most distinguished magicians of their age, and to have been summoned by the Egyptians on that account, that they might resist Moses, whose prayers to God had been especially powerful; and that they actually succeeded in counteracting and driving away all the plagues, which Moses brought upon Egypt.—The Arabian tradition calls the leaders of the magicians, who contended against Moses, *Sabur* and *Gadur*, and does not relate anything remarkable concerning them. (*Vid. Herbelot* orient. Biblioth. s.v. Mussa Halle 1789, iii. 588 sqq.). The account given by *Pliny* in his *hist. nat.* 30. 1 (2), "*est et alia Magices factio, a* MOSE *etiamnum et* LOTAPEA *Judaeis pendens, sed multis millibus annorum post Zoroastrem tanto recentior est Cypria,*" has no connexion with this subject. (The reading *a Mose et Janne et Jotapa* [*Jochabele*] *Judaeis pendens* is corrupt). There is great plausibility in the conjecture offered by *Fr. C. Meier* (Judaica. Jena 1832, p. 24, n. 16): "*An designaverit Noster quosdam de circumeuntibus Judaeis exorcistis, quorum mentio fit in Act.* xix. 13, *quorum princeps fuerit in Cypro Judaeus, nomine Lotopeas?*" Compare especially *J. A. Fabricii. Cod. pseudepigr. V. T.* i. 813 sqq.; where the ancient accounts and modern opinions are most diligently collected together.

(2). One of the principal branches of Egyptian magic from the earliest times has been SNAKE-CHARMING; and even to the present day there are relics of this secret art, the astounding results of which no European observer, however incredulous, has been able to deny. The earlier accounts of snake-charming, to which reference is made in Ps. lviii. 6 and Jer. viii. 17, have been collected by *Bochart* (hierozoicon iii. 161 sqq. ed. *Rosenmüller)* and *Calmet* (Biblical Researches); the later by *Hengstenberg* (Egypt and the Books of Moses p. 98).—The modern snake-charmers or *Psylli* form a separate hereditary guild, their principal occupation being to attract from their hiding-places any poisonous snakes, that may have concealed themselves in a house, and thus to clear the house of them. The manner in which they handle the most venomous snakes, without having extracted their poisonous fangs, is almost incredible. In the learned work of the Franco-Egyptian expedition (vol. xxiv., p. 82 sqq.), it is stated that "at religious festivals the *Psylli* appear nearly naked, with snakes coiled round their neck, their arms, and other parts of their body. They allow them to bite and tear their breast and abdomen, defending themselves against them with a kind of frenzy and appearing as though the snakes were about to devour them alive. According to their own account they can *turn* the Haje, the snake generally selected for their experiments, *into a stick*, and compel it to look as if it were dead. When they want to produce this effect, they spit into its throat, force the mouth to, and lay the snake upon the ground. Then, as though giving it their last commands, they lay their hand upon its head, when the snake immediately becomes rigid and immoveable, and falls into a kind of torpor. They wake it up, whenever they wish, by laying hold of its tail and rubbing it quickly between their hands."

To these accounts *Hengstenberg* appends the following remarks, which are certainly correct: "It deserves to be noticed that the present condition of the *Psylli* in Egypt is entirely one of decay. It is torn away from its natural connexion, the soil of natural religion, from which it originally sprang. It exists in a land in which modern illumination has already exerted its influence in various ways, and thus fettered its freedom. Accordingly nothing was more natural, than that very much that is artificial should be associated with the ecstatic condition produced, and that much charlatanry should creep in" (p. 102 Eng. transl.).

We decidedly agree with *Hengstenberg*, that there is a close connexion between the events before us and this relic of the early Egyptian order of the *Psylli*; but we cannot assent to the manner in which he and others (*O. v. Gerlach*, *Hävernick*, *&c.*) dispose of the matter. *Hengstenberg* says: "Moses was furnished with the power to produce those effects, on which the Egyptians especially prided themselves, and on which they chiefly founded their authority," and *O. v. Gerlach* most näively copies from him this *quid pro quo*. But according to the account contained in the *Pentateuch* the problem was not to turn snakes into sticks, and then revive the snakes which had become as rigid as poles; but to turn a dry stick into a living snake, and then change it into the substance and condition of a dry stick again. And these learned men are certainly open to the charge of having missed this problem altogether; whilst their oversight gives a certain colour to the remark of *v. Lengerke* (i. 406), which is intended for wit: "The serpent of Moses is still a flying one, for it flies away at the appearance of criticism."

That Moses, according to the biblical account, changed a real stick, a piece of wood, into a real living snake, and then turned this again into a piece of wood, is a fact which must be admitted without hesitation or disguise. No twisting and turning, no passing over in silence, will avail anything here. If, then, Moses was empowered by God, as *Hengstenberg* says, to produce those effects on which the Egyptian magicians especially prided themselves, it will be necessary to admit with equal candour that the Charthummim are also said to have turned dry wood into living flesh. But in that case the remarks which he makes (at p. 102 transl.), are calculated to mislead. He there says: "Were the thing so simple as it is generally considered to be, were it either common jugglery or something really miraculous, performed by the permissicn of God through Satanic influence, then the author of the *Pentateuch* would not, it may be presumed, fail to express an opinion upon it. But, since the ground on which these things rest—a very dark and difficult one—is still but little explored by science even in its most advanced state, it is better to content ourselves with the outward effects produced, without attempting to penetrate into their actual nature."

It cannot, however, have been a matter of so little importance in the estimation of the writer, as *Hengstenberg* supposes, whether the feats performed by the Charthummim were effected by the indifferent laws of nature, or some ungodly daemoniac power; for the worth of the victory could only be righly estimated, when it was known over whom it was gained. If, then, the author has made no express declaration on this subject, the reason must have been, not that he felt any doubt or uncertainty himself; or that he thought it possible to leave his readers in doubt, but that he assumed that his readers would naturally understand how the matter was to be explained. The whole of the legislation of the *Pentateuch*, in which all such magical arts are treated as an abomination to the Lord, as rebellion against Jehovah, is based upon the assumption that daemoniac, ungodly powers were actively connected with heathenism (§ 23. 1). And up to the time when the destructive infidelity of Sadduceeism prevailed, we may be sure that no Israelitish reader put any other interpretation upon the narrative, than that there was an active co-operation on the part of ungodly, that is of daemoniac powers. And just as confidently are we prepared to assert, that the Charthummim themselves had no doubt whatever, that they were assisted by such powers—the only difference being that the Charthummim from their point of view regarded the assistance as coming from some *godly* daemoniac power, whilst the Israelites believed the power to be that of an *ungodly* daemoniac character. It is only when considered in the abstract, that the sphere of magic can be looked upon as a natural and indifferent one; in its concrete form it is filled with ungodliness. In its *earliest stages*, it did not arise from any influence exerted upon nature by daemoniac powers; but by the violence done to nature a breach was made in the natural boundaries which God had established, and through this breach daemoniac powers rushed in with irresistible force and obtained the supremacy. The Scriptures regard the practice of magic as *already actually* belonging to the spiritual powers of darkness.

Let us look, then, at the problem, without disguise or fear of disagreeable results! It is impossible to determine *a priori* to what extent magic, when its power is greatest, can penetrate into the sphere of miracles, and whether the ability to turn a stick (either really or apparently) into a snake is to be regarded as within the boundaries of that sphere. All depends upon whether the scriptural record says that it is to be so regarded. What we there read is: "The magicians of Egypt, they also did in like manner (as Moses had done) with their enchantments; they cast down every man his rod, and they became serpents, but Aaron's rod swallowed up their rods." Were the rods, then, which the sorcerers brought with them, wooden rods, or snakes which they had rendered as rigid as rods by their incantations? We cannot at once affirm that the latter was not the case. The sorcerers who were not summoned till Moses had already performed his miracle in the presence of Moses and his servants, knew beforehand for what purpose they were summoned. They could therefore make their preparations before they went, and take with them such rods as they would be able to turn into serpents. Moreover, such was the sacredness of the snake in Egypt, and so highly was the magical art of its Charthummim esteemed, that it is quite conceivable that the latter may have carried such snake sticks as the symbol of their order. The rod was employed from the very earliest times, as one of the *insignia* of office. We cannot, indeed, subscribe to the statement made by *Scholz* (Einleitung in die heil. Schriften. Köln 1845. 1. 399), to the effect that, " in countries, where suitable wood was so scarce that they were obliged to make use of whatever material was at hand, it is possible that they may not infrequently have made a kind of rod from snakes (? !). For the finest, such as shepherds' crooks and rulers' sceptres, they seem to have used the larger *horned-vipers* or *cerastes* (?). The sceptres which the Pharaohs are represented as holding in their hand, in the paintings on early Egyptian monuments, *e.g.* in the temples at Thebes, are always of this shape, with the head and neck bent forwards. It frequently occurs in the hieroglyphics, in very different connexions, as a symbol of supremacy. The wooden staffs carried by the shepherds of Arabia are still made in the same form, the magical power referred to above having long been lost." The form of a snake, which was adopted for ordinary staffs, may certainly point to actual snake-staffs (*i.e.* rigid snakes), from which they were copied, though the latter may have been carried by the adepts of magic alone. Still, we feel some hesitation in giving the preference without reserve to such an explanation. According to the biblical record, the superiority of Moses, and the victory gained by him over the Charthummim, were evidently displayed in the fact, that *his* snake swallowed up *their* snakes. But if the interpretation just given be correct, was there not another point of superiority worthy of being recorded, viz. that Moses was able to turn a real staff into a serpent, whilst the staffs of the sorcerers were only staffs in appearance after all? However, we do not regard this difficulty as conclusive. The biblical record merely describes with objective calmness what took place before the eyes of the spectators. It does not concern itself with any arts which the Charthummim may have previously employed, to get possession of staffs which they could turn into serpents. It is enough that the result of the whole gave to the miraculous power of Moses a most brilliant victory over the magical arts of the sorcerers. A brilliant victory it certainly was. The sorcerers were disarmed, the symbols and insignia of their calling and art were not only taken from them, but completely annihilated, and thus their art itself was shown to be completely defeated and annihilated too.

But, as we have already said, we are not afraid of the result, should any one press the letter of the scriptural narrative (as we think, unwarrantably), and try to force us to confess that the staffs of the Charthummim were also of wood. The Scriptures speak of *σημείοις καὶ τέρασι ψεύδους*, which are wrought *κατ' ἐνέργειαν τοῦ Σατανᾶ.* And should the explanation here given not be received, we do not hesitate to admit, that even in the present instance it is quite possible, that such *lying* signs and wonders may have been performed by jugglery, and by means of daemoniac agency.

§ 25. From the fruitlessness of the first sign, it was evident that Pharaoh would not learn wisdom, till he had been made to suffer. The great judgments and strong arm of the Lord therefore began at once to be manifested. Signs gave place to *plagues;* but the plagues still continued to be signs, which demonstrated the weakness of the gods of Egypt, and the complete supremacy of the God of Israel. The peculiarity of these plagues was, that they possessed at the same time a natural and a supernatural character; and therefore the way was left perfectly open for the exercise of either faith or unbelief: the more so as even that which was supernatural, when compared with the similar efforts of the Egyptian sorcerers, might be set down by unbelief as the result of ordinary magic (1). The first two plagues were repeated by the sorcerers; but their weakness was manifested in the fact, that they could only increase the evil, and were unable to remove the plagues, or render them harmless. But at the third plague their magical art was entirely exhausted, and they were unable to continue even their miserable imitations. When the plagues had reached the significant number ten, the victory of Jehovah over the gods of Egypt was at length complete, and the judgments of God on Pharaoh's hardened heart were brought to an end. The whole of the plagues appear to have been inflicted within the space of two months, commencing at the early part of February, and terminating at the beginning of April (2).

(1). A closer acquaintance with the physical condition of Egypt has shown that the *plagues*, which preceded the deliverance of Israel, were plagues peculiar to the country, and that they frequently occur there; though never in the same force, or to the same extent, or so rapidly one after the other, as on the occasion before us. This fact has been used by both deists and rationalists, for the purpose of bringing the whole series of signs and wonders, down to the level of purely natural and fortuitous occurrences. The natural basis, upon which the events described as miracles rests, places it in their opinion beyond the reach of doubt that all the incidents described in the scriptural account, which cannot be brought within the category of merely natural phenomena, were nothing more than mythical embellishments. The English deists led the way (see *Lilienthal's* reply to them); and after the great French expedition, *Du Bois Aymé* and *Eichhorn* went still farther in the same direction, the former in the *notice sur le séjour des Hébreux en Egypte*, in the 8th volume of the "Description de l'Egypte, ou recueil des observations qui ont été faites pendant l'expedition française," the latter in his article *de Egypti anno mirabili* (in the comment. societ. Gott. rec. IV. hist. p. 35 sqq.). *Hengstenberg*, on the other hand, has undertaken to show, that the natural basis, on which the plagues were founded, furnishes a proof of their miraculous character. The principal points in his argument are: "(1) The design of all these occurrences was, according to chap. viii. 22 (18), to prove that Jehovah was the Lord *in the midst of the land.* But this could not be thoroughly demonstrated by a series of altogether unwonted terrors. All that could follow from these would be, that Jehovah had obtained a temporary and external power over Egypt. But if the events, which happened every year, were shown to be dependent upon Jehovah, the best proof would thus be afforded that he was God in the midst of the land, and by the judgment inflicted upon the imaginary (?) deities, which had been put in his place, those deities would be completely banished from the very territory, which had been hitherto regarded as peculiarly their own. (2). The tendency of later fiction would be, to disturb the connexion between the natural and the supernatural, from a notion that such a connexion impaired the dignity of the latter, and obscured the omnipotence of Jehovah, and his love to Israel. It would aim at representing the plagues inflicted upon Egypt, as a number of terrors of the most extraordinary kind." (Egypt and the Books of Moses, p. 97).

The last argument, however, can hardly be considered a conclusive one ; and in our opinion it would have no force at all, unless directed against opponents, who denied that there was any historical basis for the narrative to rest upon, and pronounced the whole account a legendary fiction, or an invention of the author. The first argument, on the contrary, we regard as perfectly conclusive. The plagues were so arranged, as to embrace both the natural and the miraculous: *the natural*, so far as the same or similar phenomena occurred in Egypt in the common course of events,—*the miraculous*, inasmuch as they were unparalleled in their extent, their amount and their force. They probably occurred at an unusual season of the year. They came at the immediate command of Moses, they disappeared again at his bidding or his prayer, and lastly (and this is the most important feature of all) the land of Goshen and the Israelites were entirely free. The miraculous character of these plagues was therefore forcible and evident enough to strike any one, who was willing to see ; but it was not so unmixed and irresistible, as to render it impossible for determined unbelief to overlook it.

In this we see *one* of the reasons, why the miraculous power of God was associated with natural phenomena, which were in themselves by no means unusual. A *second* reason may be discovered in the fact, that the Egyptians looked upon the powers of nature as deities. For these very deities of theirs were compelled by Moses' rod, to bring misery and destruction upon their own worshippers. A large portion of the plagues were plagues of animals, and Egypt was the land of animal-worship. The author of the Book of Wisdom (chap. xi. 15 seq.) has given due prominence to this feature: " for the foolish thoughts of their unrighteous conversation, by which being deceived they worshipped irrational worms and contemptible beasts, thou didst send among them swarms of irrational beasts for vengeance, that they might learn that *by whatsoever any man sins, by the same shall he be punished.*"

A *third* feature, which explains to us the choice of such plagues, is this: that the contest of Jehovah against Egyptian heathenism was thus transferred to that particular sphere, which the latter regarded as its stronghold, and therefore the victory was complete. The Egyptian sorcerers were able to perform the same feats by the power of their gods, as Moses by the power of Jehovah. It is true that their arts were exhausted at the third plague. But we must not infer from this, that they had only power to conjure up frogs, but could not act upon gnats or flies. They tried their skill in the third plague also, and they did so, no doubt, with the full confidence that they would be quite as successful, as with the first and second. But they failed and confessed, "this is the finger of God!" *vid.* § 28. 2.

(2). *The season of the year at which the Egyptian plagues commenced, and the period of their duration*, can only be approximatively determined. The last plague occurred on the 14th of Nisan, that is, about the beginning of April. The first is frequently supposed to have taken place when the water of the Nile was high; on the assumption that the turning of the water into blood was the same thing as the reddening of the water, which very often takes place when the river overflows. This ordinarily happens in the month of July. If such a theory be correct, the divine controversy must have been carried on for a period of nine months. *Hengstenberg* thinks there was a special reason for its lasting so long (p. 106): "it must have had a peculiar significance, if Jehovah went through *an entire revolution, as it were*, with the Egyptians, and for once displayed his miraculous power in connexion with the ordinarily recurring circle of natural phenomena." But apart from the fact, that this identification of the miraculous change of the Nile with the reddening of the river, which is customary at the time of the overflow, is still very problematical, and in my opinion inadmissible (§ 26. 1); apart too from the probability, that the plagues were intentionally arranged, so as to happen at an unseasonable period, in order that their miraculous character might be clearly stamped upon them: there are other data, which render it highly probable that the whole course of this miraculous chastisement occupied a very limited period of time. It is expressly stated in chap. vii. 25, that only seven days intervened between the first and second plagues. When the seventh plague (that of hail) occurred "the barley was ripe and the flax was bolled" (chap. ix. 31 seq.). Now, in Egypt, barley and flax ripen in March (*Hengstenberg*, p. 121); so that there were not more than three weeks, between the seventh plague and the tenth. The interval, therefore, between the last four plagues must have been the same as that between the first and second, namely, one week. If we suppose this to have been the average interval, the whole period must have occupied about nine weeks, which would have to be reckoned from the commencement of February to that of April. Moreover, in spite of the reasons adduced by *Hengstenberg* to the contrary, we believe that a quick succession of plagues must have been incomparably more impressive and effectual, than the same plagues could possibly have been, if spread over the whole of the year.

§ 26. (Ex. vii. 14—25).—As Pharaoh was going one morning to the Nile, probably to offer sacrifice, or perform his religious ablutions, Moses and Aaron met him. At the very moment when the king came to present his homage and his worship to the Father of life, the Father of the gods (as the Egyptians designate the Nile), he was forced to look on while the messenger of Jehovah smote him in the face, till it became bloody. Aaron, that is to say, smote the stream with the rod of God, and all the water in the river, as well as in the canals, the trenches, and the ponds, which were connected with it, and even the water which had been previously taken from the Nile, and was set by to settle in wooden or earthen vessels, was turned into blood. The fishes died in the river; the water became corrupt and stank, so that no one could drink it; and the Egyptians were obliged to dig in the sand, to obtain water of a different kind. The Egyptian sorcerers did the same with their enchantments (2). But Pharaoh hardened his heart again.

(1). The first plague was the *turning of the water of the Nile into blood.* When we enquire into the natural groundwork of this miracle, the first thing that presents itself is the fact, that nearly every year, when the Nile overflows, the water is rendered more or less turbid and red, by the blood-red marl which is brought down by the stream from the more elevated districts. "Le Nil," says *Laborde*, p. 28, "en se répandant sur les rives et les terrains cultivés, entraîne des amas d'herbes sèches et de saletés qui obligent les habitants à boire son eau dans cet état de malpropreté ou à se contenter de celle dont ils ont fait provision. Plus tard, après ce premier passage, elle enlève une première couche de limon, qui mêlé à quelque terre rougeâtre, qui descend des régions les plus élevées, lui donne une couleur rouge, qui annonce, qu'elle devient potable, et alors les pots et les jarres de terre dans lesquelles on la laisse déposer la rendent bientôt aussi claire qu'en toute autre saison."

This phenomenon, which is so far from causing surprise or annoyance in Egypt, that it is rather anticipated as something desirable, is generally regarded as the groundwork of the miracle; and even *Hengstenberg* advocates this view. But whilst *Eichhorn* and his followers look upon the unusual concomitants, the impurity and stinking of the water, the dying of the fishes, &c., as the unnatural exaggerations of the legend, *Hengstenberg* considers these to have been the result of the miraculous interposition of God.

We feel obliged to reject this association as inadmissible. (1). It is at variance with the time when the plague occurred; for, unless we are entirely mistaken, the plague happened at the beginning of February (§ 25. 2), whereas the Nile does not turn red till July. (2). This phenomenon is only conceivable at the period when the Nile overflows; but there is not the least indication of an overflowing in the whole of the narrative before us; on the contrary, there are several things which lead us to an opposite conclusion: for example, Pharaoh *walks* to the *brink* of the river, and the Egyptians dig *round about the river* for water to drink. (3). The fact that the water became putrid, was an indication of fermentation and decomposition, and this again of stagnation. But overflowing and stagnation exclude each other. (4). The effect of what Aaron did was immediate, it extended at once to all the canals, and trenches, and pools, which were connected with the Nile, and even to the water which had previously been taken from the river and was put by in wooden and earthen vessels to settle. *Hengstenberg*, it is true, endeavours to explain the passage in such a manner as to leave out the latter, which was evidently the most miraculous part of the whole ; but he is far from being successful. When Aaron was directed (ver. 19) to stretch out his rod over all the streams (the arms of the Nile), the canals, the trenches, &c., that they might become blood, and that there might be blood in the vessels of wood and of earth also ; the latter could only refer to the water, which was in the vessels before the change took place, and not, as *Hengstenberg* explains it, to water which was taken from the Nile, and placed in the vessels after it had been turned into blood. In this we are supported by the well-known manners and customs of Egypt *(cf. Hengstenberg*, p. 107). The water of the Nile is *generally turbid*, and is filtered and clarified before it is brought to table. "Chaque habitant a sa provision d'eau, qu'il puise dans le Nil, s'il habite sur le bord du fleuve, ou dans les caneaux dérivés, qui l'amènent dans les villes. Cette eau est toujours trouble quand on la puise ; mais versée dans de grandes jarres de terre, elle dépose son limon avec rapidité" *(Laborde)*. Not only was this the most miraculous part of the whole miracle, but it was in reality the point of greatest importance. It was not intended that all the water of Egypt should be affected, but that all the water of *the Nile*, the source of health and of blessing, the chief of the deities, should be turned into blood, wherever it might be found. And when even the water, which had already been taken from the river, was thus changed, there could no longer be any doubt, that it was to the Nile *as such* that the miracle applied. The rest of the water, which was not connected with the Nile, remained unaffected, as verses 22 and 24 most clearly show. Quod olim superstitio voluit, sub ipsum baculi in Nilum protenti ictum omnes Nili ejusque canalium, rivorum et stagnorum aquas ruborem induisse, as *Eichhorn* says, is in our opinion an exegetical necessity, and therefore we adopt it. But for this very reason we cannot admit, that the miracle had any connexion with the redness of the water at the time of the overflow, which could not only be foreseen, but was a very gradual process. (5). The ordinary redness does not render the water unfit for use ; on the contrary, it cannot be used until it turns red (see the passage quoted from *Laborde)*, and this phenomenon has no injurious influence upon the fish in the river. There is not a single instance on record, in which the water was unfit for use when it was in this condition. *Hengstenberg* quotes Abdollatiph to the effect that "in the year 1199 the increase of the Nile was less than had ever been known, and about two months before the first indications of the inundation, the waters of the river assumed a *green* colour, which went on increasing until it had a foul and putrid taste," but the most cursory glance will suffice to show that this phenomenon has no connexion with the case before us. The green colour of the water, at a time when it was stagnant and putrid, was undoubtedly caused by the decomposition of vegetable matter, which in some way or other had accumulated in the river. *Hengstenberg*, it is true, regards the putridity and smell of the water, as well as the death of the fishes, as a proof that the natural elements in the phenomenon were miraculously intensified. But even this is inadmissible. The admixture of the water with marl, though increased to the utmost extent, would never produce such effects ; and the state of the water, from which such results ensued, must be regarded not as a heightening of the phenomenon referred to, but as a complete *alteration*, which rendered the water entirely different from what it was before, and brought the whole occurrence within the range of another series of the processes of nature.

We must therefore look elsewhere for the natural groundwork of the miracle. *C. G. Ehrenberg* (in Poggendorfs Annalen der Physik und Chemie, 1830, iv. p. 477—515) has written an article on "Fresh discoveries of blood-like appearances in Egypt, Arabia, and Siberia, together with a survey and critical examination of those already known." By microscopical observation he found that a blood-red tinge was given to water by cryptogami (fungi) and infusoria, and that this was the case on the banks of the Nile, on the shores of the Red Sea, and in a Siberian river. If now we accept such a condition as this, as the natural basis of *our* miracle, and imagine it heightened to the utmost possible degree and rendered universal, as the nature and design of the miracle required, we can easily explain the whole of the phenomena described in the text. The conditions which preceded the development of the microscopical algae, fungi or infusoria, may have been previously produced in the water of the Nile by the providence of God in a perfectly natural way, and may therefore have already existed in the water, which had been taken from the Nile and placed in the filters. The decay and chemical decomposition of these may have rendered the water putrid and offensive ;

and the sudden appearance of the blood-red colour may be explained on the supposition that the whole process of growth and decomposition was accelerated in an extraordinary (miraculous) way. In this explanation we also retain the *mixed* character of the miracle, which was described in § 25. 1 as essential and significant. As *Ehrenberg* discovered these blood-red phenomena, during a brief residence in Egypt, both on the banks of the Nile and elsewhere, they cannot have been unheard of or unknown to the inhabitants of Egypt at that day; but the terrible intensity and universality of the phenomenon must have convinced them, if they had *been willing* to see and believe, that the hand of God was there.

We are not informed *how long this plague lasted.* The *seven days* spoken of in ver. 25 do not refer to the continuance of the plague, but to the period which elapsed previous to the commencement of the second plague. There is no ground for the assumption that the first plague did not extend to the Israelites. Such of them as dwelt among the Egyptians were most certainly affected, but those who lived nearer to the desert would suffer less, as the distance from the Nile was such that there must have been wells and cisterns enough to supply their wants.

The *purport of this plague* is very evident, when we consider the sacredness of the Nile in the religious system of the Egyptians, the importance of the water of the river, as well as of its abundant supply of fish, and the extent to which the Egyptians depended upon these to supply their daily wants. We will mention a few of the data, however, which lead to this conclusion, taking them chiefly from the collection so diligently and carefully made by *Hengstenberg*. *Herodotus* (ii. 90) speaks of priests of the Nile; at Nilopolis there was a temple of the Nile; what the heart is to the body, says an Egyptian, the Nile is to Egypt; it is one with Osiris (Plut. de Is. et Osir. p. 363 D.), and the supreme God. On the monuments it is called the god Nile, the life-giving father of all that exists, the father of the gods, &c.—The Egyptians were and still are enthusiastic in favour of the Nile water, which is in fact almost the only drinkable water in Egypt. The Turks enjoy the water so much, that they eat salt in order that they may be able to drink all the more of it. It is a common saying with them that if Mohammed had tasted it, he would have prayed to God for immortality, that he might drink of it for ever. When the Egyptians are absent from their country, they talk of nothing so much as of the pleasure they will enjoy, when they return to drink the Nile water again, &c. On the abundance of the fish in the Nile, see *Diodorus Siculus*, Biblioth. l. i. c. 36: "In the Nile there is an incredible quantity of fish of every description. The inhabitants not only partake largely of them when fresh caught, but keep an inexhaustible supply in pickle." This is fully confirmed by modern travellers.

Hengstenberg supposes the change in the *colour* of the water to have also possessed a symbolical character. "For the Egyptians," he says, "the reddened water was to be blood, to remind them of the innocent blood which they had shed, and warn them of the future shedding of their own blood." There is an analogous passage in 2 Kings iii. 22. Moreover, *red* was regarded by the Egyptians as the colour of Typhon; and therefore was a symbol of corruption and calamity.

(2). The question "whence did the magicians obtain the water on which they tried their arts, if Moses had already turned *all* the water of Egypt into blood," has caused a great deal of unnecessary thought and gratuitous ridicule. *Hengstenberg* (p. 107) is of opinion that the word "*all*" should not be taken literally, "just as we read in chap. ix. 25, that all the trees were smitten by the hail, and yet it is said in chap. x. 5, that the locusts devoured all the trees." But the two cases are not exactly parallel. And we have seen from ver. 19, how necessary it is, that the word "all" *should* be taken literally in the case before us. *Hävernick's* explanation is equally inadmissible. In his opinion the sorcerers did not make the attempt to imitate Moses, until the plague had passed away (Einleitung i. 2, p. 417 Anm.). The question may be most easily answered, if we bear in mind that it was only the Nile water which was all changed by Moses, not the water in the wells (as ver. 24 also shows).

§ 27 (Ex. viii. 1—15).—The first plague produced no effect. The strong arm of Jehovah had therefore to be brought to bear still further upon the hard heart of Pharaoh. Fresh plagues followed one another in rapid succession, until at length his will, though not his heart, was broken. It was hardly seven days after the first plague, when the *second* ensued. Aaron stretched forth his rod over the waters of Egypt, and innumerable *frogs* came up from them, and filled all the houses and utensils of the Egyptians (1). The magicians did the same with their enchantments, and brought frogs throughout the land (3). Pharaoh seemed almost inclined to bow before the power of God He summoned Moses and Aaron, and declared his willingness to give the required permission to go and celebrate a sacrificial festival, provided the plague was removed from him and his people. This apparent change of mind was responded to by the mercy of Jehovah. Moses, who was appointed as Pharaoh's God, became his servant, as the nature of a mediator required. "Exalt thyself over me," he said, "and fix the time when the plague shall cease." But when the next day arrived, and all the frogs in the land died, with the exception of those that were in the water, Pharaoh hardened his heart again, and ceased to trouble himself about his promise (3).

(1). The Nile and the neighbouring marshes in the low grounds of Egypt are, as a rule, extraordinarily full of *frogs*; but snakes and storks generally prevent their becoming a plague. The scriptural account itself implies that they usually abounded in the Nile (ver. 9 and 11). Throughout the whole course of the plagues, there was a constant increase in the annoyance caused, or the injury inflicted, and thus the second was more troublesome than the first. In this instance, the increase in the grievance is not to be seen in the fact, that the plague was either more dangerous, or more injurious, than the previous one, but that it was more disgusting and repulsive. What rendered the plague so intolerable was, that the Egyptians could not move a foot, without treading upon one of these disgusting animals, which filled their sitting-rooms and bed-chambers, and even swarmed into their ovens and other utensils. Early writers (*e.g. Pliny* h. n. 8. 43, *Justin* 15. 2; *Aelian* anim. 17. 41) furnish accounts of similar occurrences elsewhere, and relate that whole tribes have been compelled to emigrate, from their inability to subdue the plague. There may have been a connexion between this plague and the previous one, as *Hävernick* and others suppose. It is possible, that the corrupt state of the Nile water may have favoured the extraordinary increase in the number of these animals, and thus the germ of the second plague may have been included in the first; but this would make no difference in the facts of the case, and the plague would still retain its miraculous character, whatever natural elements might be mixed up with it. This plague had also a religious significance for the Egyptians, since it was from the Nile, the source of blessings and father of the gods, that the foul abomination proceeded.

(2). The *sorcerers* demonstrated their skill by increasing the plague, instead of averting it. From the very first, they had given up all thought of counteracting the effects produced by Moses; and hence their only aim was to prove to the king, that their skill was not inferior to that of Moses. They did this in the present case, by calling up fresh swarms of frogs from the Nile by magical expedients. As the *Psylli* are able to allure the snakes from their hiding-places by means of incantations, the magicians may have possessed a similar magical power over other animals as well.

(3). The fact that Pharaoh was at first inclined to yield (and we have no reason to doubt his sincerity for a moment), is a proof that his heart was not yet thoroughly hardened, that he still possessed a certain amount of susceptibility, for impressions from the testimony of God. But his relapse, after the plague had been removed, is also a proof that the process of hardening had previously commenced, and that it was already determined and incurable. The effect produced upon that part of his inner nature, which was not yet hardened, was not sufficient to cause a reaction of adequate strength, to counteract the hardening that had already begun. On the contrary, the reaction of the latter was victorious over the former, as soon as the direct impression was weakened by the cessation of the plague (*cf.* § 21. 2).

§ 28 (Ex. viii. 16—19).—The *third plague* filled the air with immense swarms of *gnats* (1). The manner in which this plague was produced, is particularly worthy of notice. Aaron smote the dust of the ground with his rod. The second plague had issued from the fertilizing Nile; the third proceeded from the fruit-bearing soil of Egypt. The Nile represented the male or fructifying principle of the deified powers of nature; the fruitful soil of the country the female or receptive principle. Instead of the fructifying blessing, there came forth from the Nile only that which was loathsome and an abomination; and instead of a life-sustaining blessing, the soil produced only misery and suffering for man and beast. The magicians again attempted to do the same with their enchantments; but this time without success. They were therefore obliged to confess: "this is the finger of God."

(1). No one has any doubt now that the כִּנִּים of the third plague were GNATS or mosquitoes (σκνῖφες), as the *Septuagint* and *Vulgate* render the word, and not *lice*, the rendering given by the Rabbins, *Luther*, *Bochart* (and the English version—*Tr.*). Travellers are unanimous in pronouncing the Egyptian mosquitoes a terrible plague to both man and beast. *Laborde* says (p. 32): l'animal le plus inaperçu et cependant le plus terrible de la création. . . . . Combien de fois une seule de ces petites mouches ne m'a-t-elle pas conté une nuit entière. . . . . Un seul cousin d'Egypte suffit pour mettre au supplice. *Herodotus* (ii. 95) knew it to be a plague of the country, and describes the precautions which were taken by the Egyptians to defend themselves from its painful sting. It is also quite in accordance with natural history, that the biblical narrative speaks of them as coming from the dust, where the last generation had deposited its eggs.

(2). When the magicians acknowledge: "this is the finger of *Elohim*," a confession forced from them by the failure of their incantations, they are generally supposed to have meant by *Elohim* the God of Israel, and therefore to have admitted the supremacy of Jehovah. But neither the words themselves, nor the subsequent history, will harmonize with this interpretation (*cf.* chap. ix. 11). Had this been their meaning, they would have said *Jehovah* instead of Elohim, for it was by this name that Moses had always spoken to Pharaoh of the God of Israel. And if they had intended to describe their present obvious incapacity, as the consequence of an interposition and hindrance on the part of the God of Israel, we should certainly expect to find the expression, the *arm* of God, rather than the *finger* of God. The arm would denote victorious power; the finger merely means admonitory warning and instruction. For these reasons we are rather inclined to the conclusion that the Charthummim, when using this expression, did not go beyond the limits of the religious system of the Egyptians, and that by Elohim they meant, not the God of Israel, but their own deities combined in one. The refusal of their deities to assist them at the third plague, they did not regard as a proof of the weakness of those deities; but rather as a sign that the gods of Egypt themselves acknowledged the justice of Israel's demands, and for that reason alone refused to continue the contest with the God of Moses. The position of the Charthummim was so painful and humiliating, that they would have been glad to put an end to the whole affair as quickly as possible, and eagerly embraced the opportunity of representing their deities as no longer willing that they should proceed any further. The expression employed by them, undoubtedly implied a wish and suggestion on their part, that Pharaoh should accede to the request of the people; but it by no means involved a confession of the God of Israel, or of their own conversion to him. The Charthummim still retained the office and dignity which they possessed before (chap. ix. 11). But as Pharaoh did not attend to what they had announced to him as the will of the gods, he had no longer any right to call them in, to exercise their magical arts in the subsequent plagues.

§ 29 (Ex. viii. 20—32).—The *fourth plague* brought *flies* and other insects throughout the land, and into the houses of the Egyptians (1); whilst the whole of Goshen, and all the houses of the Israelites, continued perfectly free (2). This time, Pharaoh offered to allow the people to sacrifice in the land itself, but not to leave the country. But when Moses refused to accede to this, and insisted upon his original demand, Pharaoh promised all, provided he would intercede for him and remove the plague. But the king only broke his word again.

(1). The *fourth plague* is described as a plague of עָרֹב. This word is rendered by the *Septuagint*, κυνόμυια or DOG-FLIES (*tabanus caecutiens L.)*; by *Aquila*, παμμυῖα; and similarly by the *Vulgate*, *omne genus muscarum*; by *Luther*, "all kinds of insects." *Gesenius* (in his thesaurus) supports the *Septuagint* rendering, and traces the word to the verb ערב, with the meaning *dulcis, suavis fuit* ("a dulcedinis notione fortasse ductus est *sugendi* significatus). We prefer to derive it from the same verb, with another of its meanings, viz., *miscuit*, and therefore would render it MEDLEY = flying insects of different kinds (*Geschmeiss*). At the same time we willingly admit, that the *Septuagint* so far displays an acquaintance with the language and the circumstances, that the dog-fly is the most important of the insects of Egypt, and causes the most annoyance to both man and beast. We may learn from a passage, which *Hengstenberg* has copied from *Sonnini's* travels (iii. 226), how troublesome the flies are in Egypt at ordinary times: "The most numerous and troublesome insects in Egypt are the flies (muscae domesticae L.). Men and beasts are cruelly tormented by them. You can form no conception of their fury, when they want to settle upon any part of your body. You may drive them away, but they settle again immediately, and their obstinacy wearies out the most patient man. They are particularly fond of fixing on the corners of the eyes, or the edges of the eyelids, sensitive parts to which they are attracted by a little moisture." *Philo* (de vita Mos. T. ii. p. 101, ed. Mang.) speaks quite as strongly of the boldness of the dog-fly and the pain which it causes. For other explanations see *Bochart*, hieroz. T. iii. p. 30, ed. *Rosenmüller*. *Jonathan*, for example, gives mixta turba ferarum; *Saadias*, mistura ferarum; *Jarchi*, omnes species malarum bestiarum et serpentum et scorpionum inter se permixtas; *Oedmann* suggests the Blatta orientalis; *Laborde*, a species of worm, of which there are at times a fearful number in Egypt, whilst the havoc caused by them is indescribable. In confirmation of this he appeals to a passage of *Makrizi*, which is also cited by *Hengstenberg*.

(2). Although the second and third plagues were of such a

nature, that we can imagine it quite as easy that the land of Goshen should be free from them as from the fourth; yet the narrative admits of no other conclusion, than that it was not till the fourth, that the distinction was made. The solemn manner in which it is announced in ver. 22 as something peculiar, and the stress laid upon it in the words "to the end thou mayest know, that I am Jehovah in the midst of the land;—and I will put a division between my people and thy people; to-morrow shall this sign be," preclude the supposition that this had been the case before.

(3). The ground assigned by Moses, for refusing to accede to Pharaoh's proposal that they should sacrifice in the land, was this: "we shall sacrifice the abomination of the Egyptians, and they will stone us, when they see it." *Hengstenberg* (p. 114) justly opposes the current notion, that the reason why opposition was anticipated from the Egyptians, was that the Israelites would probably sacrifice animals, which the Egyptians considered sacred. The term "abomination" was not applicable to the sacred animals; and the animals which the Israelites offered in sacrifice were the same as those which the Egyptians themselves employed. In *Hengstenberg's* opinion the ground of offence was that the Israelites omitted to make a strict examination of the animals sacrificed, whilst the Egyptians attended to this with the most scrupulous care. But we agree with *Baumgarten*, in thinking such an explanation too restricted. For it must not be overlooked, that the words of Moses in ver. 27, "We will sacrifice to Jehovah our God, as he *shall command us*," show that he himself, with his consciousness of the regenerating and initiatory importance of this sacrifice (§ 20. 4), did not rightly know what its exact nature would be; many things therefore might be done, which would excite the abhorrence of the Egyptians and exasperate their minds. The fears, which Moses entertained, that if the Israelites should offer their sacrifices in the land (Goshen) itself, they might be surprised in the midst of their festival, were justified by the fact, that after the expulsion of the Hyksos (§ 43 sqq.) the eastern frontier of Egypt was strongly defended by a military force (§ 14. 6. ? 4).

§ 30. (Ex. ix. 1—12).—The *fifth plague* was a *murrain*, which destroyed all the cattle that were in the fields at the time. Israel was entirely exempt from this plague also.—As Pharaoh was equally unaffected by this plague, the *sixth plague* followed. Moses and Aaron took *ashes* in their hand from a hearth, and, casting them into the air, produced grievous *boils* and *blisters* (2) on men and beasts throughout all the land of Egypt (3). Even the magicians, whom we may suppose to have been always present at the interviews between Pharaoh and Moses, were so fearfully affected with this disease that they were unable to stand before Moses on account of the boils.

(1). There can be no doubt that the ceremony, with which the sixth plague was introduced, was selected for a particular purpose, and therefore had a symbolical meaning. To ascertain what this symbolical meaning was, it is necessary that we should first of all determine the signification of the words פִּיחַ and כִּבְשָׁן. The lexicographers, since *Michaelis* (suppl. p. 1212), are unanimously of opinion that the latter denotes a furnace, used for burning lime or melting metals, in distinction from תַּנּוּר, the ordinary oven or stove. But the etymology is so uncertain, that this meaning is by no means to be accepted as indisputable on the mere authority of a *Kimchi*. It seems to us far more suitable to trace the word to the Arabic قبس (meaning *candere, urere, incensum esse)*, than to כבש = *pedibus calcavit, subegit, oppressit* (sc. *a metallis et mineribus* domandis), which is the derivation preferred by *Gesenius*. But in that case the etymology shows no essential difference between כבשן and תנור. Nor is any such difference admitted by the *Septuagint* and the *Vulgate* (*κάμινος, caminus)*. *Κάμινος* is applied to the oven and stove, quite as much as to the smelting furnace; to the forge, as well as to the domestic hearth. פִּיחַ, (LXX: *αἰθάλη*, *Vulg. Cinis)*, is generally rendered *soot*. Etymologically it means that which can be blown away, cinder-dust, *favilla*. But the ordinary usage has not retained the original distinction, between the ashes which have been burned to powder, and the larger cinders *(cinis*, אֵפֶר). —Starting from the meanings *soot* and *furnace*, *M. Baumgarten* (p. 448), has given an explanation of the ceremony, which is certainly very ingenious, and with such premises possibly plausible: "As the water of the Nile in the first and second plagues, and the dust of the earth in the third, were not elements accidentally fixed upon, the soot of the furnace must have had some connexion with the plague to which it gave rise. In the furnaces in which metal was prepared, there was concentrated a great part of the energy, put forth in connexion with those great buildings, on which the Egyptians rested their fame. In the soot of these furnaces there was seen the baser, dirtier side of this boasted splendour; and when by the hand of Moses the soot brought out blisters on the skin of the Egyptians, this was the judgment of God upon their pride, as well as upon their magnificent buildings at which the Israelites had been compelled to labour." In addition, however, to the uncertainty of the etymological basis on which this explanation rests, it may still farther be objected that the buildings and monuments of the Egyptians were not of metal, but of stone; and though metal was used in connexion with them, it does not appear to have been sufficiently important, for such special reference to be made to it in the ceremony before us.—*Hävernick*, (p. 182), on the other hand, points to the custom, which, according to *Plutarch* (de Is. et Osir, p. 318 ed. Hutt.), prevailed among the Egyptians in the very earliest times, of scattering the ashes of the sacrifices, especially the human sacrifices, as a ceremony of purification. It is true, *Herodotus* (2. 45) denies that human sacrifices ever occurred among the Egyptians; but this only proves that they were not offered in his day. We give the preference to *Hävernick's* explanation, but would give it a less limited application. We take כבשן in its most general signification of *fire-place.* If now we might further imagine, that the fire-place referred to here was one set apart for burning the sacrificial animals, for the purpose of obtaining their purifying ashes, and that Moses could or durst take the ashes from such a fire-place, the great significance of the ceremony would be placed before us in the clearest light. The ashes, which were intended to purify, produced uncleanness; and thus it was symbolically declared that the religious purification, promised by the sacrificial worship of the Egyptians, was nothing but defilement. But even supposing that there is no foundation for any of these conjectures, the fact will still remain, that ashes in general (on account of their being used in lie), were a means of purification (*cf.* Num. xix.), and therefore that the means of purification here became the cause of defilement.

(2). Amidst the great variety of inflammatory eruptions on the skin, we do not possess the necessary data for a more particular diagnosis of the disease produced by the *sixth plague*. So much, however, is certain, that the climate of Egypt predisposes to such diseases in a most extraordinary manner.

(3). Though it is not expressly stated, that Israel was exempt from this plague, the narrative evidently implies it. And in general, there is no reason to doubt that from the fourth plague, when the distinction was not only first made between the Israelites and the Egyptians, but was so emphatically pointed out as most significant, this distinction formed an element in all the miracles, which were afterwards performed.

§ 31. (Ex. ix. 13—x. 29).—The *seventh plague—thunder, lightning, and hail*—was announced with increased solemnity (1). If any of the Egyptians had been sufficiently impressed by what had already occurred, to pay attention to the word of Jehovah and fear it; sufficient time was given them, after the announcement of the plague, to gather their servants and cattle into their houses before it commenced, and thus save them from destruction. But whatever men or cattle remained in the field, were smitten by the fearful storm of hail. Moreover the spring crop, which was nearly ripe, was entirely destroyed. In the land of Goshen alone there was no hail.—Pharaoh again promised every thing; but, as soon as the plague ceased at the intercession of Moses, he refused to perform any thing. The *eighth plague* had therefore to be announced, the devastation of the country by *locusts* (2). The people of Egypt, who were suffering severely, began now to cry to their king, to let Israel go; lest Egypt should be entirely destroyed through his obstinacy. And Pharaoh himself was sufficiently alarmed, to know that the words of Moses were not empty threats. He seemed even likely to anticipate the threatened plague by submission. But he had no sooner given the permission required, from his fear of further judgments from God, than his hardened heart was again steeled against it, and he refused to allow any but the men to depart, whilst he retained their wives and children and all their cattle, as a guarantee for their return. The messengers of Jehovah could not consent to this; and the plague immediately commenced. A continuous east wind brought such a dense swarm of locusts into the land, that the sun was obscured; and when they settled, the whole country was covered. The devastation, which they caused, was so great, that not a leaf remained upon the trees, nor a blade of grass in the fields. The pride of Pharaoh seemed broken now; he confessed that he had sinned against Jehovah, and sued for mercy. But the west wind, which carried off the locusts, took away his hypocritical repentance also. The locusts perished in the Red Sea; they were the precursors of Pharaoh with his horses and riders. Every one of the plagues had hitherto been announced to the king beforehand. This rendered it impossible, on the one hand, that he should regard them as anything but divine judgments; and on the other hand it gave him the opportunity of escaping the evil, by changing his mind. But henceforth this double precaution ceased. Without any announcement, however, the king knew whence the plague had come; and his hardness had increased to such an extent, that the rest could only be regarded as judgments from which he could not escape. Thus the *ninth plague* broke upon him without preparation, viz. such dense *darkness* (3) for three days, both out of doors and in the houses, that they could not see one another. But in the houses of the Israelites, it remained perfectly light and clear. Pharaoh capitulated again. He said he would allow the men to go with their wives and children, but the cattle and sheep must be left behind. Moses rejected these conditions, and the two parted from each other in great excitement and anger.

(1). A *thunderstorm* accompanied by *hail* is by no means a rare occurrence in Egypt, at least in the Delta; though in the usual way it is almost unknown in the more elevated districts. *Cf. Laborde* p. 42. It is important to observe the emphatic and elaborate manner, in which this plague was announced. There must certainly have been some other reason for this, than the fact that the present plague fell much more heavily, than any that had preceded it. To our mind, a stronger reason is to be found in the increasing hardness of Pharaoh's heart. It is in connexion with the previous plague (chap. ix. 12), that this hardening is first spoken of as an effect produced in Pharaoh by God himself (except in the objective announcement which was made to Moses at the first). This evidently implies, that a turning-point had been reached; and it also explains the reason, that the king was now for the first time made aware of the manner, in which his hardness and hostility to the will of God were to be made to subserve the glory of His name. Although Jehovah might have displayed his supremacy over the gods of Egypt, by the plagues that had already been inflicted; Pharaoh could still proudly boast, that with all his power and with all his efforts, Jehovah had not conquered him. Hitherto it had not been the will of Pharaoh, but that of Jehovah, which eventually succumbed. This miserable pride and defiance on his part were now put before him in their proper light; and he was made to learn, that with all his proud self-will, he was only serving the purpose and plans of God: "for this have I raised thee up, to show in thee my power, and that my name be declared in all the earth" (ver. 16); *cf.* chap. xi. 9, "Pharaoh shall not hearken to you, that my wonders may be multiplied in the land of Egypt."

The warning advice (ver. 19), to collect the men and cattle out of the fields, and shelter them in their houses from the threatened hail, was intended for the benefit of as many of Pharaoh's servants and subjects, as had learned to fear the word of Jehovah; and they profited by it (ver. 20). But it was also intended for Pharaoh, to whom it was first addressed. And, though it is not expressly stated, we may gather with certainty from the general tenor of the narrative, that he paid no attention to the warning,—another proof that the most marked provisions of mercy only increase the hardness of the hardened man. From his past experience, the king could not possibly doubt that the threatened punishment would be inflicted; but his proud and defiant spirit would not let him reap the benefit, which the warning put within his reach.

(2). There has been an incalculable amount of writing on the natural history of the *locusts* in general, and the scriptural references to them in particular. *Laborde* mentions the titles of a hundred and seventy-five different works, which he says that he consulted and used, in his complete and careful investigation of the subject (p. 44 sqq.); and yet the catalogue is far from being complete. The fact that the direction taken by a swarm of locusts is dependent upon the wind, has been confirmed by the observations of travellers a thousand times. And the thorough devastation which they are here said to have caused, as well as their eventual destruction in the sea, have been frequently witnessed. According to the biblical narrative, they were brought by the east wind, רוּחַ קָדִים. Even the *Septuagint* stumbled at this, and rendered the words ἄνεμος νότος (*Vulg.* ventus urens). This rendering has been adopted by *Bochart*, who is of opinion that קדים must here mean the *south wind*, as the east wind could only have brought the locusts from Arabia, whereas the south wind would bring them from Ethiopia, where they are much more numerous. *Hasselquist* endeavours to prove that the locusts always take the same direction, viz. from south to north. *Eichhorn* (p. 26) thinks that, as the locusts are invariably driven by a blind impulse from south to north, and never turn towards the east or west, the swarms must always have come to Egypt from Ethiopia, and never from Arabia. And *Bohlen* (Gen. p. 56) makes use of this, as a proof that our author was not acquainted with the natural history of Egypt. But *Credner* (Joel p. 286) has brought many witnesses to prove that locusts follow every wind, and (p. 288) has also shown that they not only cross over narrow straits, such as those of Gibraltar, &c., but that when their flight is favoured by the wind, they will pass over seas as broad as the Mediterranean itself. But when the wind does not favour their flight—when, for example, it rises to a tempest, or suddenly drops—the whole swarm will fall immediately into the sea. *Niebuhr* (Beschreib. p. 169) also attests the fact that the wind sometimes carries swarms of locusts across the Arabian Gulf, even at its broadest part. *Cf. Hengstenberg*, Egypt and Moses, p. 11 sqq., and *Laborde*, p. 50 sqq.

קדים is never used in the Scriptures to denote the *south wind*, but always means the *east wind.* It is the more important that we should maintain this firmly, since it is probable that in the present instance, there was some significance in the direction in which they came. They came from the same quarter as the Israelites, and they appeared as their champions and allies. But if this explanation should be given up as too far-fetched, we think that *Baumgarten* (p. 454) is certainly right in laying stress upon the fact, that they were not produced in Egypt itself, but came from a distant, foreign land, as a proof that "the power of Jehovah reached beyond the bounds of Egypt, *i.e.*, was everywhere present."

(3). The THREE DAYS' DARKNESS is now generally traced to the Egyptian *Sirocco* or *Chamsin* (*cf. Hengstenberg*, *Hävernick*, and others). The horrors of this phenomenon are described by nearly every traveller. *Du Bois Aymé* (p. 110) says: "When the Chamsin blows, the sun is of a pale yellow colour; its light is obscured, and the darkness sometimes increases to such an extent that one might fancy it was the depth of night." According to other accounts, the inhabitants of the towns and villages shut themselves up in their houses, sometimes in the lowest rooms, or even in the cellars, whilst dwellers in the desert take refuge in their tents, or in holes which they have dug in the ground. *Robinson* (i. 288) was in the desert during one Chamsin of short duration: "The wind," he says, "changed suddenly to the south, and came upon us with violence and intense heat, until it blew a perfect tempest. The atmosphere was filled with fine particles of sand, forming a blueish haze; the sun was scarcely visible, his disk exhibiting only a dim and sickly hue, and the glow of the wind came upon our faces as from a burning oven. Often we could not see ten rods around us, and our eyes, ears, mouths, and clothes were filled with sand." *Rosenmüller*, in his commentary, cites accounts from the middle ages, according to which the Chamsin covered Egypt with such dense darkness, that every one thought the last day was at hand. *Laborde*, however, will not admit that there is any resemblance between the Chamsin and the darkness referred to here: "Ce serait comparer la détonation d'un fusil au fracas du tonnerre que d'assimiler deux extrêmes de ce genre."—In the scriptural account of this plague, there is certainly no intimation of its being in any way connected with a scorching wind of this description. Still the phenomena, which accompany the Chamsin, though very different in degree, are so similar in kind, that we are inclined to agree with those who regard the Chamsin as its natural basis. It must, however, at the same time be acknowledged, that none of the earlier plagues were raised so decidedly or to such an extent above their natural basis, through the peculiar character imparted by the miracle; and that none were so completely dissevered in some respects from that basis, as was the case here. In the present instance not only was the plague extended and intensified to a degree unheard of before, but in many respects it was entirely removed from the natural foundation, and passed over into the sphere of the pure miracle, in which no known power of nature is in any way employed. This is particularly seen in the fact, that it continued perfectly light in the houses of the Israelites, some of which immediately adjoined those of the Egyptians, whilst the Egyptians were unable to escape in any way from the darkness, by which they were surrounded. For when it is said in the biblical account, that the darkness was so great that they could not see one another, and therefore that no one could rise up from the place in which he was: the meaning undoubtedly is, that even in their houses the ordinary means of procuring artificial light were entirely useless. It may also be inferred from the express statement, to the effect that no one moved from his place during the three days' darkness, and from the nature of the interview which Pharaoh had with Moses, that the latter was not sent for till the plague was over. On the meaning of this plague *Hengstenberg* correctly observes, that the darkness which covered the Egyptians, and the light which the Israelites enjoyed, represented the wrath and the mercy of God.

THE PASSOVER.

§ 32. (Ex. xi. 1—10).—All possibility of further negotiation was now, apparently, for ever gone. For Pharaoh had threatened Moses with death, if he should dare to let him see him again, and Moses had replied with equal wrath, "so let it be, I will never come into thy presence again" (x. 28, 29, xi. 8). And yet the promise of Jehovah immediately followed: "I will bring one plague more upon Pharaoh, and upon Egypt; afterwards he will let you go hence, and not merely let you go, but will himself entreat and force you to depart." Of the previous plagues some (viz., the first and second) had come, at a signal from Moses, from the beneficent river of Egypt, others (the third and fourth) from the fertile soil of the country, and others from the pure air, which pervaded the land; all the elements, which were at work in Egypt, had been one after another turned into a curse. And when that which was peculiarly Egyptian had been all exhausted, the countries round about sent their plagues into Egypt also; locusts came from the desert of Arabia, and the Sirocco with its impenetrable darkness from the Sahara. Yet all was apparently in vain. But this had been merely introductory and preparatory to the last decisive stroke. The *tenth plague* did not rest upon any natural basis, as all the rest had done. It was not called forth by either the rod or hand of Moses, nor did it proceed from the water, the earth, or the air; but the hand of Jehovah himself was stretched forth: "at midnight will I go out into the midst of Egypt, and smite *all the first-born* in Egypt, both of man and beast (1), and *I will execute judgment against all the gods of the Egyptians* (2), I Jehovah (xii. 12)—but against the children of Israel not a dog shall move his tongue, that ye may learn *how that Jehovah doth put a difference between Egypt and Israel.*" In the *tenth plague* the idea and intention of all the plagues were embodied and fulfilled. It was thought of first (chap. iv. 22, 23), but it was necessarily the last to appear. If it had also been the first to appear, the fact would not have been so completely and universally displayed, that Jehovah was *the Lord in the midst of the land* (chap. viii. 22), the Lord over the water, the earth, and the air, over gods and men, cattle and plants, and *that there was none like him in all the earth* (ix. 14). For this purpose it was necessary, that there should be *many miracles* wrought in the land of Egypt (xi. 9); and it was also requisite, that they should have both sharply defined natural features and an unmistakeably miraculous character, in order that freedom of choice might be left for faith or unbelief. But the tenth plague bore upon the face of it a purely supernatural character, and because it was the tenth, *i.e.* the one which gave a finish and completeness to the whole, it exhibited in a clear and unequivocal manner, the design of all the plagues from the very commencement; for the last furnished the key to the entire series. And inasmuch as Pharaoh's resistance was overcome by the tenth plague, although the hardness of his heart was complete; this fact alone was sufficient to prove, that the obstinacy of his refusal had only served to glorify the name of Jehovah, and that the words of Jehovah were fulfilled: "*For this cause have I raised thee up, to show in thee my power, and that my name may be declared throughout all the earth.*" (ix. 16).

(1). On the importance of the *first-born*, *Hofmann* says (Weissagung und Erfüllung i. 122): "The first-born opens the mother's womb, and thus renders all succeeding births possible; and hence the power, which deprived all the first-born of life, was also a proof of ability to control the future history of the existing generation, and the perpetuation of its life by means of posterity. The same power, which punished the existing generation, could also have annihilated all its prospects for the future." We cannot possibly comprehend, how this acute writer can have hit upon so mistaken an explanation. The notion, with which he starts, that the first birth renders all succeeding births *possible*, is completely wrong. No doubt the predicate, "that which openeth the womb," implies a precedence on the part of the first-born over the rest. But assuredly no Israelite ever explained this as meaning, that the first-begotten alone as such possessed the power "to open the womb," and that the possibility of any subsequent births depended entirely upon him. But the rest of *Hofmann's* remarks are at variance with this fundamental thought. The first-born, on whom the plague fell, were already born, they had already opened the way for further births. How then could their death appear to threaten the prospect of other births? The real importance of the first-born may be thus explained: the first-born naturally enjoyed both precedence and pre-eminence over the rest, he was the firstling of his father's strength (Gen. xlix. 3), the first-fruit of his mother. As the first-born, he stood at the head of the others, and was destined to be the chief of whatever family might be formed by the succeeding births. As he stood at the head of the whole, he represented the entire nation of the Egyptians. Hence the power, which slew all the first-born in Egypt, was exhibited as a power, which could slay all, that were born then, and, in the slaughter of the whole of the first-born, the entire body of the people were ideally slain.

(2). The question arises in connexion with chap. xii. 12: how could the death of all the first-born, of both man and beast, be regarded as a *judgment upon all the gods of Egypt?* One might be inclined to think, that the previous signs and wonders could have been much more correctly described as a victory over *all* the gods of Egypt, and a judgment upon them, than the tenth plague, which was not nearly so closely connected with the objects which the Egyptians worshipped as gods. But the fact, that this plague was intended as a judgment upon the gods of Egypt in a more eminent degree than any of the rest, is evident from the repetition of this same view in Num. xxxiii. 4: "the Egyptians buried all their first-born, for upon their gods also Jehovah executed judgment." And here we may clearly see, in what relation the death of all the first-born stood to the gods of Egypt. The gods of Egypt, as the passage before us clearly shows, were among those who were smitten by this plague. And we agree with *J. D. Michaelis* (Anmerkungen für Ungelehrte iii. 35) in the opinion, that reference is made to the *animal-worship* of Egypt (*cf. J. C. Prichard*, Egyptian mythology). A large number of animals were regarded by the Egyptians as sacred, probably because they looked upon them as incarnations of the deity. If any of these animals were found dead, there was lamentation and mourning on every hand. It was a capital offence to slay or injure them. A few specimens of them were kept in the temples, and were objects of public worship. Such was the importance generally attached to primogeniture in the whole of the ancient world, that it is very probable that the first-born were most frequently, if not invariably, chosen for that purpose. Fancy, then, what an effect must have been produced, what alarm it must have caused, what unbounded lamentation there must have been, if all the sacred animals in the temples, and thousands of them outside the temples, were struck dead *in one night.* Such an occurrence would be truly a judgment on the gods of Egypt; and for Egyptians at least, a judgment of a more fearful character, and one more calculated to produce despair, could not possibly have occurred. But the expression contained in chap. xii. 12 must not be restricted to this. The strong emphasis laid upon the fact, that judgment was to be executed upon *all* the gods of Egypt, when taken in connexion with the announcement so constantly made, that this plague would fall upon all the first-born of *men* and cattle, leads to the conclusion that *men* were also reckoned among the gods, who were to be slain. Our thoughts are naturally directed first of all to Pharaoh; not, however, in the sense in which the princes of the earth are described as gods, but rather in that sense in which, as the vain-glorious inscriptions on the monuments prove, the Egyptian kings prided themselves upon being sons of the gods, or incarnations of the deities. This explanation derives all the more weight from the fact, that during the whole of the negotiations with Moses, Pharaoh takes an *independent* stand in opposition to Jehovah. Moreover, the circumstance, that it was not merely the first-born of the god-king Pharaoh and of the sacred animals, that were slain, but all the first-born of man and beast, from the son of Pharaoh, who sat upon his throne, to the son of the slave-woman, that stood behind the mill, from the Apis, that was kept in the temple, and worshipped as a god, to the most common and unclean of the beasts, was the most humiliating part of the whole to the gods of Egypt, for it was a practical declaration of the absolute equality of both of them. In contrast with the great significance of the announcement, when thus explained, we notice the interpretation given by the Jewish expositors, who institute a comparison between this plague and the miracle wrought on the image of Dagon in the temple at Ashdod (1 Sam. v.): an interpretation which must be rejected as without foundation, and thoroughly indefensible. Thus, for example, *Jonathan* paraphrases the passage as follows: "In omnia idola Aegyptiorum edam quatuor judicia: idola fusa colliquescent, lapidea concidentur, testacea confringentur, lignea in cinerem redigentur, ut cognoscant Aegyptii me esse Dominum."

§ 33. But certain important preparations were required, before this last decisive blow could be struck. As one of the leading features of this plague it is stated in chap. xi. 7: "*Ye shall learn how that Jehovah doth put a difference between Egypt and Israel.*" The separation of Israel was the fundamental idea of the ancient covenant, the basis of its history. What, then, were the conditions and pre-requisites of this separation? They are to be found in the first stages of the history of Israel: on the one hand the call of Abraham, the creation of his seed from an unfruitful body, and the appointment of this seed to bring salvation to the world; and on the other hand, the self-surrender and self-dedication of Israel to the purposes of Jehovah in faith and obedience to his will and guidance. But nearly four hundred years had passed since then, and during that time the natural side of Israel's character, that in which every other nation perfectly resembled it, had been almost exclusively developed and in active operation. In consequence of this, the other side of its character, by which it was distinguished from other nations, had retreated so far into the background, and in the process of development had been so completely left behind, that it was necessary to renew both the election and the covenant. Moreover, Israel in the meantime had entered upon a new stage; it had passed from the family to the nation, and the covenant made with the family had to be transferred to the nation into which it had grown. The covenant with the fathers was, no doubt, in existence still; for Jehovah still continued to be the God of Abraham, of Isaac, and of Jacob (ii. 24), and Israel still bore the sign of this covenant in his flesh (xii. 48):

but the covenant itself had been in abeyance for four hundred years ; it had not made the least advance during all that period ; and in the sphere of life and motion, stagnation is equivalent to retrogression. The covenant, therefore, required to be resuscitated, enlivened, and set forth, and also to be confirmed and transferred to the nation, which now occupied the place of the family.—Moses had already been informed that this was to take place on Horeb, the Mount of God, (iii. 12) ; and for this very purpose Pharaoh was to be compelled to let the people go into the desert, that they might celebrate a festival to Jehovah. Hence, it was in Horeb, first of all, that the renewed and perfect seal of separation from the nations, the stamp, which henceforth distinguished it from all others, was impressed upon Israel. But the obduracy of Pharaoh, the hostility of his people to the nation, which Jehovah had begotten from the seed of Abraham to be his first-born son (chap. iv. 22), and the consequent necessity for executing judgment upon Pharaoh and his nation, had already shown the nature of that distinction which God was about to make between Egypt and Israel (chap. xi. 7). There had been a marked difference ever since the fourth plague (chap. viii. 22) ; but now at the tenth and concluding plague, it was to be practically demonstrated in a manner unparalleled before. The earlier plagues were chiefly intended to alarm and call to repentance ; the tenth, on the other hand, was a pure act of judgment (chap. xii. 12). The fact that Israel was Abraham's seed, was sufficient protection from the former ; but this no longer sufficed to defend them from the latter. Jehovah was now preparing to pass in judicial majesty through the land of Egypt. But judgment requires stern and impartial justice, fettered by no considerations, and admitting of no exceptions ; and it is right that judgment should begin at God's own house (1 Pet. iv. 17). If, then, there was something ungodly in Israel itself ; if the seal of its election and separation was obliterated ; if its sanctification was imperfect and faulty ; if its natural character was stronger than that imparted by grace: the judicial majesty of God could not pass over Israel, although it was Abraham's seed, but his judgment would surely fall upon the Israelites as well as the Egyptians. Nevertheless Israel was to be saved. It was necessary, however, that before the judicial wrath of Jehovah burst forth, the Israelites should be prepared by grace, or they would be unable to escape the judgment; their sins must be expiated, all ground for the wrath of God must be removed, and their fellowship with God must be renewed and fortified. This was accomplished by the institution of the *Passover*. The feast of the passover was a precursor of *that* festival, which the nation was about to celebrate in the desert in honour of its God; the paschal sacrifice was an anticipation of the sacrifice about to be offered on the Mount of God in Horeb, a preliminary demonstration of its power and its effects, a guarantee for the future.

§ 34 (Ex. xii. 1—28).—The period fixed for the last plague that was to fall upon the Egyptians, and for the celebration of the *passover* (1) by the Israelites, was the *fourteenth day of the month of green ears ;* but as early as the *tenth*, the father of every household was to select a lamb without blemish, and to keep it till the fourteenth day of the month (2), when it was to be slain *between the evenings* (3). The *lintel* of the door and the two posts were then to be marked with its blood ; in order that, when Jehovah passed through the land of Egypt, to slay all the first-born, he might pass over the houses of the Israelites, and not suffer the destroyer to enter them (4). The lamb was then to be *roasted* without *breaking a bone*, and to be eaten *with bitter herbs.* Whatever might *remain* was to be burned. The bread eaten at this meal was to be *unleavened.* Moreover they were to eat it, like persons *in a hurry to depart*, with a *staff* in their hand, with their *loins girded*, and with their *shoes on their feet* (5). To commemorate the important design and grand results of this festival, their descendants were ordered to repeat it every year, and to keep it as a seven days' feast, neither eating leavened bread nor suffering any to be found in their houses, for seven days after they had partaken of the Paschal lamb. Foreigners, and servants who were not of Israelitish descent, were prohibited from taking part in the Paschal meal, unless they had been previously incorporated in the community by circumcision. The heads of families were required to instruct their children at an early age, as to the meaning of this solemn ceremony. Moreover this month was to be henceforth regarded as the first month of the year, because it was the period of Israel's redemption, and formed a fresh commencement to Israel's history. When Moses made this announcement to the people, they bowed and worshipped, and did as Jehovah had commanded.

N.B.— It does not form part of our plan, to enter at present into a full examination of all the directions contained in the law for the observance of the passover, or of every typical and symbolical meaning which can be discovered in that institution. As we propose discussing the Mosaic legislation (including the rites and ceremonies of worship), not according to its gradual promulgation, extension, and completion, but in its systematic form as an organized whole, the only features to which we shall now refer, are those which are necessary for the elucidation of this portion of the history of Israel. See the elaborate treatise of *Bochart de agno Paschali (Hieroz.* i. 628—703, *Rosenmüller's* edition.)

(1). The solemn festival which immediately preceded the Exodus, is described as פֶּסַח לַיְהוָֹה (xii. 11), זֶבַח־פֶּסַח לַיְהוָֹה (xii. 27), and זֶבַח חַג הַפֶּסַח (xxxiv. 25). In chap. xii. 27 the derivation of the word פֶּסַח (Aramæan פַּסְחָא ; LXX. *πάσχα*;

Vulg. *Phase, transitus*) is thus explained, Jehovah "passed over (פָּסַח) the houses of the children of Israel, when he smote the Egyptians." פסח means *to step* or *leap over anything.* This leads to the notion of *sparing, exempting;* for he who steps over a thing, instead of treading upon it and crushing it, spares and exempts it. Hence, *Onkelos* has not hesitated to substitute חַיִם (misericordia) for פֶּסַח.

The phrase זבח־פסח ליהוה necessarily leads us, to regard the slaying of the Paschal lamb as a *sacrificial act,* and the eating of it as a *sacrificial meal.* And the fact, that the Paschal meal was considered (on scriptural authority, 1 Cor. v. 7) to be a type of the Lord's supper, led the theologians of the Catholic Church to seize with avidity upon the sacrificial dignity of the Paschal lamb, as a confirmation of their unscriptural theory of the repetition of the sacrifice of Christ whenever the Lord's Supper is celebrated, since it was not merely the first meal, but the first sacrifice, which was repeated at every subsequent celebration of the passover. Now, instead of contenting themselves with the reply, that the necessity for a repetition of the Paschal sacrifice, whenever the passover was celebrated, arose from the typical character, *i.e.*, from the insufficiency of the Old Testament sacrifices, and that a repetition of the sacrifice of Christ is inadmissible on account of its absolute and perpetual validity (Heb. vii. 27, ix. 28), the earlier Protestant theologians (*Chemnitz, Gerhard, Calovius, Dorschnus, Varenius, Quenstädt, Carpzov,* and others), in order that they might take away every possible foundation from the catholic theory, denied *in toto* the sacrificial worth of the passover, and would only allow that it was a *sacramentum* not a *sacrificium.* There were, however, several of the earlier theologians (*e.g., Hacspan, Dannhauer, Bochart, Vitringa,* and others), who were impartial enough to admit the opposite. Among the more modern Protestant theologians, *Hofmann* is the only one, so far as we are aware, who has reproduced the denial of its sacrificial character (Weissagung und Erfüllung i. 123, and Schriftbeweis ii. 1, p. 177 seq.). Even *M. Baumgarten* differs from him in this respect (i. 1, p. 467).

So much must undoubtedly be admitted, that the name זֶבַח is not sufficient of itself to prove that the passover possessed a sacrificial character; but this is incontrovertibly proved by the apposition לַיהוָה: a *slaying for Jehovah* cannot possibly be anything but a *sacrifice.* This is quite as convincingly demonstrated, by what is related of the blood of the slaughtered lamb. For, if the door-posts of the Israelites had to be sprinkled with this blood, in order that the judicial wrath of God might not smite them with the Egyptians; and if Jehovah spared their houses solely because they were marked with the blood: the only inference that can be drawn is that the blood was regarded as possessing an expiatory virtue, by which their sins were covered and atoned for, though otherwise they would have exposed them to the wrath of God. It cannot be disputed, however, that the blood of a *sacrifice* alone possessed this expiatory virtue. Nor can it be denied that, on subsequent occasions, when the passover was celebrated as a commemorative festival and a renewal of the deliverance of Israel, it possessed a sacrificial character. In Num. ix. 7 the Paschal lamb is expressly called a *sacrifice* (קָרְבָּן); it was slain in a holy place (Deut. xvi. 5 sqq.); its blood was sprinkled on the altar; and the fat was burnt upon the altar (2 Chr. xxx. 16, 17, xxxv. 11, 12). It is always referred to as a sacrifice in Jewish tradition; *Philo* and *Josephus* call it θῦμα and θυσία, and the apostle Paul uses the verb θύειν with reference to it (1 Cor. v. 7).

The advocates of the opposite view appeal, with some show of reason, to the fact that all the usual characteristics of a sacrifice, particularly the imposition of hands, the sprinkling of the blood upon the altar, and the burning of certain portions of flesh at the altar, were omitted from the first passover, whilst several directions were given, to which there is not the slightest analogy in any of the true sacrifices. But the *latter* circumstance merely proves, that the paschal sacrifice was not subordinate to the other kinds of sacrifice, but co-ordinate with them, and formed an independent and peculiar class. The paschal sacrifice was just as much a distinct kind of *Shelamim,* as the sacrifice offered on the great day of atonement was a distinct kind of sin-offering. The *former*, again, might be sufficiently explained from the fact that the Mosaic law of sacrifice, on which the argument is founded, was not yet promulgated; but it must also be remembered that the condition of the Israelites in Egypt did not allow of the full and practical development of the sacrificial character of the passover *(vid. e.g* chap. viii. 26). As soon, however, as the impediments were removed, and the law of sacrifice was issued, the sacrifice of the passover was assimilated to the general character common to the rest of the sacrifices, so far, that is, as its distinctive and peculiar character would allow. Thus, for example, Moses commanded that the Paschal lamb should be slain at the sanctuary (Deut. xvi. 2, 5, 6, *cf.* Ex. xxiii. 17). Again, we discover from 2 Chr. xxx. 16, xxxv. 11, that the blood of the Paschal lamb was sprinkled upon the altar, and from 2 Chr. xxxv. 12, that certain portions of the paschal sacrifice were placed (as עֹלָה) upon the altar and burned. In this it resembled other sacrifices (especially the *Shelamim*, to which it was most nearly allied).[1] Since, then, we find that in these two respects the passover was assimilated to the general idea of sacrifice, we may safely assume that the third essential characteristic of that idea, viz. the imposition of hands, was also included. The imposition of hands is treated in the law of sacrifice as something so essential that it durst not be omitted in the case of any sacrifice; but for that very reason it was so much a matter of course, that it was necessarily presupposed even where it was not expressly prescribed. Thus, for example, in the case of the not less peculiar sacrifice on the great day of atonement, no mention is made of the imposition of the hand, although it undoubtedly took place *(vid.* my Mos. Opfer., Mitau 1842, p. 296).

(2). The Paschal lamb (in cases of necessity a goat might be taken, ver. 5) was to be killed on the fourteenth of the month *Abib*, the earing month, which was afterwards called *Nisan*, but it was to be selected on the TENTH of the month. We look in vain to the greater number of commentators for any explanation of this singular appointment. *O. v. Gerlach* says that this

1 We are compelled by the context to interpret the word עוֹלָה in 2 Chr. xxxv. 12, not as denoting a burnt-offering, in the strict sense of the term, but as a comprehensive word, referring to those portions of the paschal lambs which were set apart to be burned; for, in the whole section, there is not, and cannot be, any reference (vers. 10—19) to actual burnt-offerings (which were never slain and offered at the same time as the Paschal lambs). *Cf. Chr. B. Michaelis,* Annott. in hagiogr. iii. 990: "Ver. 12, וַיָּסִירוּ הָעֹלָה: *Deinde amoverunt holocaustum, i.e.* hoc loco: eas partes paschalium victimarum, quae adolebantur et igne comburebantur, ut erant adeps, eaedemque prosiciae, quae sacrorum salutarium erant. Lev. iii. 9, 10, 11."

precept had reference to Egypt alone (? ?), where the coming judgments and the hurry of their departure left no time for a later choice (! !). As if this selection was so tedious an affair, and occupied so much time, that it would have been impossible to find an opportunity during four whole days! *M. Baumgarten* contents himself with rejecting *Hofmann's* interpretation as inadmissible, without attempting to suggest a better. *Hofmann* (Weissagung und Erfüllung i. 123) says, that the lamb had to be selected as many days before it was slain, as there had been דורות (generations) since Israel was brought to Egypt to grow into a nation. For four days the people were to be reminded of the approaching deliverance, by the sight of the lamb which had been selected. *Baumgarten* is of opinion, that this explanation is overthrown by the fact that, according to the Hebrew mode of reckoning, from the 10th to the 14th, would not be four days but five. But this objection is founded upon a misapprehension. It is certainly true that, according to the Jewish mode of reckoning, Christ is said to have lain in the grave for three days, but this was an inaccurate expression, borrowed from the current phraseology; it would have been more exact to say that he lay in the grave *upon three days*, or that he rose *on the third day*. If the selection took place on the tenth of Nisan, at about the same time of the day as that on which it was slain, on the 14th, the interval would be according to *every* mode of reckoning not *five* days, but *four*. But if the time at which it was slain ("between evenings") is to be regarded as denoting the beginning of the 15th, it might undoubtedly be said that it was killed on the fifth day after the selection was made. But even the latter would square with *Hofmann's* explanation; in fact, on any other supposition, the harmony between the symbol and the thing signified would not be complete, for at the time of the Exodus Israel had actually entered upon the fifth דור (century) of its sojourn in Egypt. We feel no hesitation, therefore, in adopting *Hofmann's* interpretation; at the same time we cannot omit to mention that *Hofmann* himself is not consistent, as this explanation of Ex. xii. 3 is clearly at variance with that which he has given of Gen. xv. 9 (i. 98). I must also retract the opinion which I have previously expressed (at vol. i. § 56. 5) with reference to Gen. xv. 9, and adopt *Baumgarten's* exposition which I have quoted there.

(3). The expression בֵּין הָעַרְבַּיִם, *i.e. between the two evenings*, has been explained in various ways. The *Caraites* and *Samaritans* suppose it to refer to the period between the disappearance of the sun below the horizon and the time when it is quite dark, *i.e.* from six o'clock till about half-past seven. Thus the *first* evening begins with the disappearance of the sun, the *second* with the cessation of day-light. *Aben-Ezra* gives the same explanation. The *Pharisees* in the days of *Josephus* (bell. jud. 6. 9. § 3) and the *Talmudists* supposed the first evening to be the afternoon from the time when the sun began to go down, the second commencing when it actually set; *Ben-haarbayim* (between the evenings) would therefore be from three o'clock till six. *Jarchi* and *Kimchi* interpret the expression as referring to the hour immediately before sunset and that immediately after, that is from five till seven. *Hitzig* (Ostern und Pfingsten, p. 16 seq.) arrives at the same conclusion. He regards the expression as denoting the indifferent boundary line between the 14th and 15th Nisan; and as the slaying and preparation of the lamb cannot have been the work of a moment, he supposes the boundary line to have been moved backwards or forwards, as occasion required. Thus, in contradiction to his own theory, he changes a *point* of time into a *space* of time. Of these different explanations the *first* is the only admissible one, as the following passages sufficiently prove: (1) Ex. xvi. 12, 13, where "between the two evenings" and "in the evening" are used as *synonymous* terms. (2) Deut. xvi. 4, where the lamb is said to have been killed *in the evening;* for the *evening* cannot possibly begin before sunset, ("evening" is the general term; "between the evenings" the more particular definition); (3) Deut. xvi. 6, where the passover is ordered to be slain "in the evening as soon as the sun goes down;" (4) Ex. xii. 6, 8, 10, from which we learn that the lamb, which had been slain between the evenings, was eaten *the same night*, and that none of it was left till the morning, for here the time called *ben-haarbayim* is evidently reckoned as a part of the night in the more general sense of the word; and (5) the occurrence of a similar phrase in Arabic (*cf. Gesenius* thes. p. 1065).—The custom of the *Pharisees* is apparently at variance with Ex. xii. 6 (compared with Lev. xxiii. 5, "on the fourteenth day of the first month at even is the Lord's passover"), for if the lamb was not slain till *after* sunset, strictly speaking it was killed on the 15th of Nasan, and not on the 14th. But all that we learn from a comparison of this verse with Ex. xii. is that, agreeably to its natural character, the first evening (*i.e.*, the time of evening twilight), could be regarded as either the termination of one day, or the commencement of another. All that we have to determine, then, is simply the point of view from which the historian was looking. If he started from the 14th of Nisan, up to which day the lamb was to be kept apart and preserved, and on which the immediate preparations for slaying it were to be made, *ben-haarbayim* would pass for the termination of the 14th; but if he took his stand at the 15th of Nisan, the first day of the feast, the time of slaying the lamb would then appear to him to be the commencement of the 15th. Thus there is no irreconcileable discrepancy in the fact, that in Ex. xii. 18 we find a command, that unfermented bread should be eaten for seven days, from the fourteenth day of the month at even until the one and twentieth day of the month at even; whereas in Lev. xxiii. 6, we read "from the fifteenth ye are to eat unleavened bread for seven days."—See the thorough examination of this question in *J. v. Gumpach's* alttestl. Studien, Heidelberg 1852 p. 224—237.

(4). The law of sacrifice had not yet been made known; the common sanctuary was not yet erected; and the sacrifices of Israel were an abomination to the Egyptians (chap. viii. 26). Hence we cannot expect to find the Israelites observing any of the general laws of sacrifice, which were promulgated afterwards. It was necessary that the sacrificial act should be performed in private houses; the dwelling of each family served as "the tabernacle of the congregation," and the *door-posts* as the altar on which the blood was sprinkled. Jacob's *one* family had grown into a number of families; and these families were still living side by side, without being organised into the unity of a nation. No single sacrifice could be offered for the community, because Israel had no existence as a community yet; nor could the Israelites assemble to offer sacrifice at a common sanctuary, for no such sanctuary had yet been provided. If Israel was to be reconciled as a whole, that it might escape the coming judgment; it was necessary that each of the separate family-groups, into which it was divided, should offer for itself the atoning sacrifice, and protect itself from the wrath of the judge with the atoning blood of the victim. When this atoning blood had been smeared upon the lintel and door-posts, the whole house was protected and everything in it; for the entrance represented the entire house. There is something insipid in the remark that "the houses of the Israelites had to be marked with the blood, in order that the destroying angel might be able to distinguish them from the houses of the Egyptians." At the same time it is a perversion of the whole meaning, to say with *Bochart* and *Bähr* (ii. 634): Itaque hoc signum *Deo* non datur sed *Hebraeis*, ut eo confirmati de liberatione certi sint. *Baumgarten*, on the other hand, correctly observes: "the sign is properly for him, who sees it and judges accordingly; now the blood was seen by Jehovah, as he himself said, and not by the Israelites who were sitting in the houses. And it was just because the blood availed as a sign for Jehovah, that it furnished Israel with a firm ground of confidence." Israel stood in need of reconciliation, because it could not continue in its sin when judgment had begun. But God was about to spare and deliver Israel, for the sake of its faith and its future destiny, and therefore he imparted an expiatory virtue to the blood of the sacrifice which was slain by the Israelites. They were to make this their own by faith, and, as a proof that they had done so, to mark their houses with the atoning blood. To disregard this precept would have been to despise and reject the mercy of God.

Baumgarten and *v. Gerlach* assign a most remarkable reason for the fact that it was the *lintel* and not the *threshold* of the door, which was to be marked with the blood, viz., that the destroying angel came from above and not from beneath. The true reason most probably was, that a threshold is not a part of the door, but is merely the basis on which it rests. A threshold is not absolutely required in a representation of a house-door, but the two posts and the lintel are indispensable. *Bähr* is much nearer the mark, when he refers (ii. 633) to Deut. vi. 9, where the lintel of the door is mentioned as being just that part of the house, which is most certain to attract the notice of any one entering or passing by, and which was therefore the most suitable place for inscriptions.—*Gerlach* is of opinion that the Jewish notion, that the marking of the door-post was only intended to apply to the first passover in Egypt, is evidently at variance with the words of the institution at vers. 24, 25. We cannot subscribe to this opinion. We cannot separate vers. 24 and 25 from their context in vers. 26, 27. The latter show that the command contained in vers. 24, 25 (to observe it as an ordinance for ever, and even to perpetuate the observance in the promised land) referred generally to the whole feast of the passover, and was not restricted to the marking of the door-posts, as it would appear to have been if we merely connect it with ver. 23. Again, from the hurried nature of the first passover it necessarily followed, that certain important modifications were required on the subsequent organization of the community and its worship. But Moses had not yet received any commission to inform the people that such modifications would be necessary, nor was there any reason why he should do so ; much less was it requisite that they should be fully described. In part they were left over for further legislation ; in part they followed as a matter of course. There were *two* reasons for marking the door-posts on this occasion. (1) It was necessary that the blood, which was intended as an expiation, should be applied ; in order that its expiatory virtue might take effect. Now there was no altar at command for the purpose ; but on the other hand the house door (and with it the house) was regarded in the light of an altar. This reason ceased after the giving of the law, and the outer court of the tabernacle was appointed as the place where the passover should be slain and offered (Ex. xxiii. 17 ; Deut. xvi. 2, 5, seq. *cf.* 2 Chr. xxx. 16, 17; xxxv. 11, 12). (2) The door posts had to be marked, because the destroyer was about to pass through the land to slay. But this reason had no force on any subsequent occasion. Nor can it be replied to this, that the festival was afterwards observed as a feast of *commemoration*, and that the door had still to be marked with the blood of the Paschal lamb, to *remind* them of the way in which their houses had formerly been passed over by the destroyer. For the passover was not merely a feast of commemoration, it was also designed to represent and renew that redemption and sanctification, whose historical foundation it served to recal. Now, the ceremony in question had no meaning after the first festival ; for it was only then, that the destroying angel was about to pass through the land. The door was marked with the atoning blood, *in order that* the destroyer might pass over. It would therefore have been a desecration of the sacred blood, if it had been applied in the same manner, *because* the destroyer had *once* passed by; this would have been to change a most significant *means* of deliverance into a very insignificant *sign* of deliverance.

(5). The *Paschal meal* had the same design as every other sacrificial meal, viz. to represent the fellowship with God, which was to be established, as the result of the atoning sacrifice (*vid.* my Mosaisches Opfer, p. 102 sqq.) "For the death of the lamb not only averted death, but originated a new life, and this new life was in the eating of the flesh" (*Baumgarten*). *Hofmann*, who denies that there was anything sacrificial in the passover, cannot of course admit that the Paschal meal was sacrificial. The only end which, in his opinion, it was designed to answer, was to give (physical) strength for the coming journey (p. 122, 123). But is it possible that these arrangements, the *symbolical* character of which is emphatically shown in so many important details, meant nothing more than: "eat as much as you can to-night, that you may be able to sustain the fatigue of your journey to-morrow morning"? Can we imagine the lawgiver looking upon a meal, which was only intended to impart physical strength, as of such great importance, that he made its annual celebration the first and most solemn of all the national festivals? It is true that *Hofmann* does not use the word "physical;" but we do not think that we have misrepresented him, by giving such an interpretation to his words. In fact, this necessarily follows from his denial of the sacrificial meaning of the passover. For if the physical strength imparted by this meal was the symbol and source of a corresponding *spiritual* strength, we do not see how it could become possessed of this character, except as the result of the sacrificial idea attached to the lamb, which was eaten at the meal.

A number of peculiar instructions were given with reference to the preparation and enjoyment of the meal, the symbolical character of which it is impossible to deny. (1). The lamb was to be *roasted* over the fire, and not boiled (ver. 9). The usual explanation of this arrangement is, that such a mode of preparing the food was most in accordance with the *hurried nature* of the whole affair. *Hofmann* (i. 123) and *Bähr* (ii. 636) both adopt this explanation. The former also adds: "the objection offered by *Spencer* (ed. Pfaff. p. 307), that it would not have taken longer to boil the meat than to roast it, if the vessels had been all made ready beforehand, contains its own refutation in the inappropriate condition and supposition." But why so inappropriate? Had not the Israelites four whole days for making these outward and unimportant preparations? Undoubtedly the Paschal meal had a compulsory and hurried character. But it was only the meal, and not the preparations; for the latter were spread over four days. If it was not inappropriate, that the lamb itself should be selected and placed in readiness four days before it was used, why should it be so inappropriate that the vessels for boiling it (if boiling had been admissible on other grounds), should be made ready a few hours or rather *minutes* before (for this is all that would have been required)? But must not the preparations for roasting on a spit, which was an unusual process, have really occupied more time, than would have been required to prepare the boiling apparatus which was in constant use? In whatever light we regard this explanation, it appears to us to be thoroughly inadmissible.

A *second* injunction was that *not a bone of the lamb should be broken* (ver. 46). Of course this did not mean, that it was not to be cut up for the purpose of eating, but for the purpose of roasting. The lamb was to be put upon the table whole. The unity, of which the undivided lamb was a representation, was communicated in a certain sense to those who ate of it. Whilst eating of the one perfect lamb, as of a provision made by God, eaten at the table of God, by intimate associates of God, they were thereby linked together as one body, being all partakers of equal fellowship with God (consult especially 1 Cor. x. 17). For this reason, as far as possible, the *head*, the *thighs*, and the *entrails* (ver. 9), were all to be eaten. It was also strictly commanded, that whatever was left should be *burned* the next morning, and not laid by for another meal. For if any portion had been eaten at another meal, this would have destroyed the idea of unity and completeness quite as much, as if only half the lamb had been cooked. "It would then have fallen into the series of ordinary meals, and this would have detracted from the holiness of that which was eaten" *(Baumgarten)*. It was evidently on the same ground, that the further direction was given, that no part of it should be carried across the street from one house to another (ver. 46). It was an act of expediency, that whatever was left should be *burned*, but this did not destroy the idea of unity. By being committed to the fire, it was safely preserved from any profane or common use, and was given back, as it were, to God in the fire.

The *third* injunction had reference to the other accompaniments, viz., *bitter herbs* and *unleavened bread* (ver. 8). The *bitters* (מְרֹרִים), with which the lamb was to be eaten (LXX.: ἐπὶ πικρίδων, *Vulgate*, cum lactucis agrestibus ; *Luther*, mit bittern Salsen), were undoubtedly *bitter herbs*. They referred to the bitterness of the Egyptian oppression, of which it is said in chap. i. 14, that "they (the Egyptians) made their lives *bitter*" (וימררו). There are other passages also, in which bitter food and drink are figuratively employed to represent suffering and distress, Ps. lxix. 21 ; Jer. viii. 14 (*Bähr*). But firmly as we adhere to this explanation, we cannot overlook the fact, that bitter accompaniments might also be regarded as a spice, by which a stronger and more agreeable flavour was communicated to the food. The sweet flesh of the roasted lamb was to be made more savoury by the bitter vegetables, for their bitterness would be lost in the sweetness of the meat, and supply to the latter its appropriate condiment. And what the bitter spice was to the sweet meat, the recollection of their oppression in Egypt was to be to their deliverance from bondage. But the recollection of their oppression was not all that was contemplated. As the sweet and the bitter relieved each other, the one supplying what the other wanted, so were the sufferings in Egypt and the deliverance from bondage intimately and essentially connected together ; for the latter could never have taken place without the former, and it was the consciousness of this which gave to the memorial its sacred worth. The words of the apostle are applicable here, "No chastening for the present seemeth to be joyous, but grievous, nevertheless, afterward it yieldeth the peaceable fruit of righteousness to them which are exercised thereby" (Heb. xii. 11).

A similar meaning has also been attributed to the direction, to eat only *unleavened bread* at the meal. In support of this explanation, reference is made to Deut. xvi. 3, where the passover bread is called bread of affliction (לֶחֶם עֹנִי), and *Winer* (in the second edition of his Real-lexicon ii. 231), is of opinion that "the

Israelite of a later age could not be more effectually reminded of the oppression endured in Egypt, than by eating for a whole week such plain and tasteless food." To this *Bähr* justly replies (ii. 630), "that if this had been the case, the whole seven days' festival would have been a period of fasting and mortification, whereas it was really a *joyous* festival, and not one of mourning and repentance. Moreover the showbread and cakes, which, according to their symbolical meaning, were intended as food for Jehovah, were ordered to be unleavened. Was Jehovah, then, to have nothing but wretched, tasteless bread offered to him?" At the same time I am of *Hofmann's* opinion (i. 124 seq.), that *Bähr's* own explanation is inadmissible, He says that "it was called bread of affliction, because it was bread, which called to mind their sojourn in Egypt, and the suffering which they there endured; though it did so, merely because it had been eaten on the occasion of their deliverance from that suffering." But this reminds one too much of the derivation "*lucus a non lucendo*." I agree with *Hofmann*, in thinking, that the explanation of the expression in Deuteronomy is to be found in the clause which immediately follows, "for thou camest forth out of the land of Egypt in haste (בְּחִפָּזוֹן *i.e.*, ye fled from it in a hurried and anxious manner)." The departure from Egypt assumed the form of a hurried *flight* (חִפָּזוֹן), and therefore was always remembered as an עֳנִי (a tribulation, or oppression). As the Egyptians compelled the Israelites to rush out of Egypt in the greatest confusion, and allowed them no time for withdrawing quietly, or making the necessary preparations for their journey; they still applied force to the Israelites, and Israel ate its last meal in Egypt בעני, *i.e.*, under the oppression and affliction of Egypt. (In confirmation of this view, *Hofmann* very appropriately refers to Is. lii. 12 לֹא בְחִפָּזוֹן תֵּצֵאוּ). Moreover we learn from Ex. xii. 39 (34), that the reckless and irresistible impetuosity of the Egyptians, and the consequent nature of their forced flight, compelled them, altogether irrespectively of the divine command (ver. 15), and any symbolical meaning in the ordinance, to continue for several days the eating of unleavened bread; for every particle of leaven had been removed from their houses on account of the feast of the passover, and they were obliged to fly before any fresh leaven could be prepared.

The account, which is given in Ex. xii. 39 (34), has been sometimes adduced as an evidence of discrepancy in the scriptural record; for, according to vers. 8 and 15 sqq., the use of leaven had been altogether prohibited, not merely on the day of the passover, but for seven days afterwards. But it was not the writer's intention in ver. 39 to assign a reason for their eating unleavened bread, either at the original festival, or on the subsequent commemoration of it. The true explanation is this. The first feast of the passover was confined to *one* day; and on this day no leavened bread was to be eaten, for *symbolical* reasons. The following days were not feast days; but, as they were spent in travelling, they were days of hardship and toil. The commemorative festival, which lasted seven days, was not intended to celebrate the day of departure, *and* the first seven days of their journey, but the day of their departure alone. The reason why *seven days* were spent in commemorating the historical events of *one* day, is to be found in the solemn character of the festival, which was observed in honour of this one day. *Seven days*, neither more nor less, were required for a full realization of the character of the festival, a perfect exhibition of the idea which it embodied. But as the eating of the Paschal lamb was the one, indivisible basis of the whole festival, and did not admit of repetition, whilst the festival itself was to last for seven days; this could only be accomplished by continuing for seven days the other essential element of the Paschal meal, viz., the eating of unleavened bread. This was the sole reason, why unleavened bread was eaten for seven days, at the subsequent commemoration of the festival. At the first festival leavened bread might have been eaten on the second, third, and following days (for then the festival was confined to *one single day)*; but the Israelites were compelled by external circumstances to continue eating unleavened bread for some days afterwards. And this is all that ver. 39 refers to. Had the writer intended to say that it was this fact, which gave rise to the future custom of eating unleavened bread for seven days, he would assuredly have referred to it in a more pointed manner, instead of omitting to make any reference to its lasting *seven days*. But in reality he is only speaking of the *first day after the departure*, and says that, on that day, they ate unleavened bread, because the dough which they had taken with them was *not yet* leavened. By the second, third, or fourth day it must have been leavened; and we may confidently assume that the Israelites ate without hesitation what they had in their possession, viz., bread made of the leavened dough.

What, then, was the *symbolical importance of the unleavened loaves?* They are called מַצּוֹת (*Sept.* ἄζυμα, *Vulg. azymi panes*). *Hofmann* has shown, in his *Weissagung* (i. 124), that this neither means *pure*, nor yet *sweet*, but *dry* loaves. The roots מצה and מצץ convey the idea of the exclusion of moisture, hence of drying, parching. In unleavened bread the moisture of the dough is driven out by the heat. It is not really baked, but parched; for the peculiar characteristic of baking is, that the leaven or yeast produces fermentation in the dough, which is thereby expanded and lightened, and at the same time the moisture, which is retained, re-acts against the parching and compressing force of the external heat. If, then, we would deduce any symbolical meaning from the *name* of the *Mazzoth;* it must be found in the fact that the *Mazzoth* were loaves, in which there was nothing but the pure meal, without any change in its nature or flavour, without the admixture of any foreign substance (the water, for example, which is driven out by the fire), and without the impartation of any foreign taste, or the least alteration by means of fermentation. There is all the more reason for adopting this interpretation, since it fully harmonizes with the course adopted in the preparation of the lamb, where (by roasting instead of boiling) every foreign substance was excluded, any change in its nature entirely prevented, and the preparation entirely effected by the pure and simple element of fire.

In the case of the *Mazzoth* everything depended upon *the absence of leaven*, and in this there was a symbolical meaning. Whether the taste of the bread was thereby improved or injured, is not taken into consideration. Leaven is dough in the course of fermentation. But fermentation is corruption, the destruction of the natural condition, the breaking up of the natural connection between the component elements. Hence from a symbolical point of view all fermentation, being an alteration of the form given to the material by the creative hand of God, is a representation of that which is ungodly in the sphere of morals, that is of moral corruption and depravity. As the lamb, which served to impart both physical and spiritual strength, and to restore communion with God, was pure; the bread, which was eaten with it, was not allowed to contain anything impure.

With reference to the command to eat the meal in *travelling costume* (ver. 11), *Baumgarten* observes that, after the Israelites had been redeemed from the death of Egypt by the *blood* of the lamb, they derived new energy from eating the lamb that had been slain, solely in order that they might immediately take their departure from the land of destruction to the Mount of God. The number of persons who formed one company at the Paschal meal is not stated. It was most natural that each household should form a separate party. But as it was desirable, as far as possible, to take care that none of the lamb should be left; it was ordered that, where a family was small, it should unite with another (ver. 4). At a later period the Jews looked upon ten as the normal number of a single company. The supplementary command (ver. 44 sqq.), that no foreign servant, or associate, or hireling, should take part in the meal, and that no foreigner, who might be dwelling among the Israelites, should keep the passover with his family, unless they had been incorporated into the community of Israel by circumcision, had its external ground in the fact, that a large number of the common people of Egypt left their country with the Israelites (ver. 38, see § 35. 7). But it is a very instructive fact, that just at this time, when everything tended to show how Jehovah distinguished between Israel and Egypt (chap. xi. 7), it was made a fundamental law that non-Israelites might enter without the least difficulty into religious and national fellowship with the Israelites, and thus participate in all the blessings of the house of Israel. We have here a proof that, even when the distinction was most marked between the heathen and the chosen people, the fundamental idea of the Old Testament history was never lost sight of, that in Abraham's seed all the nations of the earth should be blessed.

THE EXODUS FROM EGYPT.

§ 35 (Ex. xii. 29—xiii. 16).—While the children of Israel were eating the passover in travelling costume, the *tenth plague* (1) fell upon the Egyptians. At midnight the *destroying angel* (2) slew all the first-born of Egypt, both of men and cattle; and there was not a house to be found, in which there was not one dead. The terror of God came upon all the Egyptians. The same night, Pharaoh sent for Moses and Aaron, gave them permission to *depart* (3), and intreated their intercession on his behalf. The people of Egypt also urged the Israelites to depart as quickly as possible, for they said "we are *all* dead men." The Israelites then did what Jehovah had previously commanded them to do: they asked the Egyptians for *articles of gold and silver* (trinkets and jewels) and for *clothes* (festal clothing). And Jehovah caused his people to find favour in the eyes of the Egyptians, so that they gave without hesitation whatever was desired (4). The instructions to repeat the Paschal meal every year were coupled with a command, to *sanctify all the first-born* of men and cattle to the Lord (5). Thus they departed in festal costume, as an army of Jehovah (6); for the Egyptians themselves had clothed them with festal apparel and costly ornaments. The bones of Joseph were also taken by Moses, according to the promise which had been made to him on oath by the fathers of the people (§ 4), and for the fulfilment of which the people as a body were responsible. A large number of the Egyptians of the lower classes of society, who had endured the same oppression as the Israelites, from the proud spirit of caste which prevailed in Egypt, attached themselves to the latter, and served henceforth as hewers of wood and drawers of water (7). Four hundred and thirty years had been spent in Egypt by the descendants of Jacob (§ 14. 1). There were now among them 600,000 men capable of bearing arms (§ 14. 3). Raemses was the place from which the procession started; Succoth their first resting-place (§ 37).

(1). *Hengstenberg* (Egypt and the Books of Moses p. 125) pronounces the *tenth plague*, viz., *the death of all the first-born both of men and cattle*, to have been the result of a *pestilence*, a thing of frequent occurrence in Egypt. But in this instance, where the natural side of the event completely disappeared, he goes so far, in his anxiety to introduce a natural element into the miracle, that we must decidedly decline accompanying him. With greater moderation *Hävernick* says (p. 182) : " The last plague is the one, which brings us most decidedly into the sphere of the purely miraculous." The word pestilence, however, is so indefinite and general a term, that it conveys but little information after all. If by pestilence we are to understand any disease which carries men off in a sudden and unsparing way, we can offer no objection to the application of the word to the tenth plague ; for if the hand of Jehovah smote a large number of the Egyptians in one night with sudden death, the stroke itself must undoubtedly have resembled a mortal disease. But if the word be used in a more restricted sense, as denoting a disease that causes sudden death and overspreads whole districts by *contagion*, we protest with all our might against the designation. It was not by contagion, striking here and there like the electric fluid without previous warning, that so many victims were struck down by this plague, nor was it by any physical predisposition to a disease produced by some mysterious pestilential vapour, that those who fell were predestinated to die ; but the hand of Jehovah, or of the destroyer whom he sent, was the immediate cause, and not only the number of the victims, but the particular individuals, were determined beforehand by a *rule*, which had not the slightest connexion with the laws of contagion. We cannot but be surprised, that *Hengstenberg* should ever have gone so far, as to assert that " the expression '*all* the first-born' is not to be taken literally, any more than the other statement that 'there was *no house* in which there was not one dead,' which could not be strictly correct, since there were not first-born in every house." In his opinion we cannot infer from this that there were none of the first-born left alive, or that none but first-born were killed. Again, with regard to the *exemption* of the Israelites, he says that natural analogies may be adduced, for at the present day the Bedouins have very little predisposition to pestilence, and seldom suffer in the same way from its devastations. Even if we admit that the expression "*all* the cattle" in chap. ix. 6 (like chap. ix. 25) is not to be taken too literally, it is very different with chap. xii. 29. In the former case the reference is to the destruction of cattle by a general murrain, without any particular description of the individuals smitten, and therefore the historian might naturally express himself in general terms. But here, where he is speaking of particular, well-defined individuals, such a mode of expression would have been altogether out of place. The scriptural account says, "Jehovah smote *all* the first-born," and proceeds to give the greater emphasis to the word "all" by adding: "From the first-born of Pharaoh that sat on the throne, to the first-born of the maid-servant that was behind the mill, and unto the first-born of the captive that was in the dungeon" (xi. 5, xii. 29). This was the last, decisive plague. Whereas all the previous plagues had been so arranged, that it was still possible for unbelief to resort to the subterfuge, that they had been the result of nature and chance alone; this last was of such a kind, that even hardness and unbelief could not refuse to admit the interposition of the personal, living, supreme, and Almighty God. But the design of Jehovah would have been entirely frustrated by such exceptions as *Hengstenberg* refers to. There is no force in his assertion, that there cannot have been first-born in *every* house. For if, here and there, a couple may possibly have been found, where the husband was not himself *a first-born, without children*, and living in a house *by themselves*, the cases must have been extremely rare, in which these three circumstances were all combined, and therefore the writer cannot be blamed for saying "*every* house." It was not the design of the plague, that corpses should be found in every house without exception; but it was intended that *all the first-born* without exception should be slain, and if anything be pressed it must be *this*. Moreover the reference is to the male first-born on the mother's side (as chap. xiii. 2 clearly shows, *cf.* No. 5 below), so that there would sometimes be several first-born in the same family.

Again *Hengstenberg* is equally wrong, when he speaks of the supposed pestilence as being connected with the Chamsin (*i.e.*, the three days' darkness). He brings forward the evidence of travellers to the effect that, "when the Chamsin *lasts*, pestilence prevails to a fearful extent, and those who are affected die very quickly," and states immediately afterwards that "for this reason the Arabs, as soon as the Chamsin ceases, congratulate each other on having survived." But according to the biblical account, the *pestilence* did not go before the "*Chamsin*," or even accompany it; on the contrary, it did not occur for several days, perhaps some weeks afterwards, and therefore long after the Egyptians had congratulated each other on having survived the dangerous period.

(2). The infliction of the tenth plague is sometimes ascribed to Jehovah himself; at other times to a *destroyer* הַמַּשְׁחִית sent by Him and distinct from Him (chap. xii. 23). There are some, it is true, who regard מַשְׁחִית as an abstract term, meaning destruction; but ver. 13, which is adduced in support of this, does not say: "the plague will not be among you for destruction," but "there shall be no plague among you for the destroyer" (*i.e.*, no plague to inflict, no occasion for bringing a plague). So far as we have hitherto traced the operations of God in Israel and on behalf of Israel, there is everything to lead to the conclusion, that the destroyer, who was sent by Jehovah, and in whom and through whom Jehovah personally appeared and worked, was no other than *the* angel, whom we have already met with in the patriarchal history as the representative of Jehovah (vol. i. § 50. 2); before whom Moses drew off his shoes and covered his face, when he appeared to him in the burning bush (chap. iii. 2, 5, 6); and who manifested himself to Moses in the inn, when Jehovah appeared to slay him (chap. iv. 24). So far as the judgment was one of wrath and brought destruction upon the sinner, the judge was also a destroyer. But as we read in Ps. lxxviii. 49, of *an army* of angels of evil (מִשְׁלַחַת מַלְאֲכֵי רָעִים), who were actively engaged in the Egyptian plagues (for רעים, like חַיִּים, *life*, is an abstract noun meaning *evil, wickedness;* and angels, that work evil, are not therefore *wicked* angels), the question may be asked whether משחית does not indicate a plurality of angels engaged in the plagues. משחית, as *Hofmann* correctly observes (Schriftbeweis i. 310), denotes an instrument of destruction, of which there may be either one or many; and even in the latter case the many may be conceived of and described as one, on account of the unity of the principle which sets them in motion. Thus, for example, in 1 Sam. xiii. 7, an entire division of the army, which set out to devastate the land, is called המשחית, and, on the other hand, in 2 Sam. xxiv. 16, the angel of the Lord, which smote Jerusalem with pestilence during the reign of David, is called הַמַּלְאָךְ הַמַּשְׁחִית. But when we observe, that in the passage quoted from the Psalms, the work of the "angels of evil" is not restricted to the slaying of the first-born, but applies to the whole of the Egyptian plagues, and also that in the book of Exodus such emphasis is laid upon the destroyer's *passing over (i.e.,* from one door to another); it must be acknowledged, that the passage in the Psalms *does not compel us* to suppose that there was a plurality of destroying angels employed in connexion with the *tenth* plague, and that it is *much more natural* to understand the description, contained in the twelfth chapter of Exodus, as relating to a single destroyer.

(3). Different answers have been given to the question, whether Pharaoh gave the Israelites a *conditional* or an *unconditional permission* to depart, that is, whether they left with or without any obligation to return. It appears to me that the permission *must* be regarded as unconditional. It is true that at first Moses merely requested, that they might be allowed to go for a three days' journey into the desert, and thus the prospect of return was still left open (§ 20. 4); and even at the fourth plague he still presented his request in this limited form, but he never did so afterwards. It is in the nature of a war, however, that the conqueror raises the conditions of peace with every victory that he gains, whilst the vanquished are obliged to give up all their claims. The latter is just what Pharaoh did, as the scriptural record expressly declares. And we may justly assume that the former was the case with Moses, especially as a regard to Pharaoh led him to commence with a request of so limited a character, that it could not possibly have sufficed even if it had been granted, and therefore in the event of Pharaoh's compliance he would have *been obliged* to extend it. How much easier, then, was it to do this, when Pharaoh was obstinate and repeatedly broke his word! All the promises, that Moses had previously made, were annulled by Pharaoh's continued breach of faith; and when at length, after the ninth plague, the king turned the messenger of Jehovah out of doors, threatening him with death if he should ever venture to appear in his presence again, and Moses departed "with burning anger" saying: "As thou hast said, I will see thy face again no more"

(x. 29; xi. 8), all negotiations upon the former basis were for ever broken off. Henceforth *Moses* would never again request *permission* to go into the desert, but Pharaoh and his people would beg and entreat of Israel to depart from Egypt, as an act of *kindness* and *mercy*. If, then, the departure of the Israelites was regarded *at the time* as an act of kindness to Egypt, we may be sure that the Egyptians not only did not demand, but did not even desire that the Israelites should return; for they would surely fear, or rather foresee with certainty that, if they did, the former evils would sooner or later be endured again. Pharaoh's subsequent change of mind, which led him to pursue the Israelites with an army, for the purpose of bringing them back by force, is no proof to the contrary; for it was the result of his own obduracy and the hardening of his heart by God, and merely led to the full execution of judgment upon himself. The fact that Pharaoh had not the remotest idea that the Israelites would return, but on the contrary regarded it as certain that they would not, is clearly proved by his astonishment when he heard that they had not gone straight into Asia, but were still within the borders of Egypt on this side of the Red Sea (§ 22. 5). "They have missed their way," he said, "the (Egyptian) desert hath shut them in." The whole affair is so described, that we cannot possibly infer from Pharaoh's pursuit, that he had merely given the Israelites permission to take a three days' journey, and did not intend them to depart altogether. For his reason for pursuing them was not that they had passed the Egyptian frontier (in opposition to the supposed permission), but *that they were still within the limits of the Egyptian territory*. They had not acted, therefore, in violation of Pharaoh's permission, but merely contrary to his expectation. If the Israelites had taken the direct road to Canaan, or even if they had gone by the regular route to Sinai round the head of the gulf, the opposite view might possibly have been sustained. But as they had not left the soil of Egypt, and therefore cannot in any case have gone beyond the permission, which they are *supposed* to have received, we cannot see why Pharaoh should have hurried to enforce their return. On the contrary, as this was the actual state of the case, it would have been more natural for Pharaoh to conclude that the Israelites seriously intended to return after they had offered their sacrifices; and if so, why should he make use of force? Pharaoh's proceedings are incomprehensible, on any other hypothesis than that the departure of the Israelites was generally regarded as a formal, and actual exodus. And is not this *expressly* stated in Pharaoh's words: "why have we done this, that we have let Israel go from serving us?" He says *himself* that he has *let Israel go*, that they should serve him no longer. Have we not here, then, a proof of *unconditional* permission from Pharaoh's own mouth? What need we any further witness? The king fancied that the infatuated resolution of Israel, which was to him so incomprehensible, was a proof that the people had been forsaken by their God and deprived of their reason;—therefore he altered his mind and determined to pursue them.

In reply to this, however, we are referred to Pharaoh's words in Ex. xii. 31, where Moses and Aaron repeat their demand for the last time, and he answers: "Rise up and get you forth from among my people, both ye and the children of Israel; and go, serve Jehovah, *as ye have said*, also take your flocks and your herds, *as ye have said*, and begone and bless me also." Pharaoh is supposed to have expressly stated on this occasion, that he merely based his permission to depart upon their prior demand, and therefore restricted it to the limits of that demand. But so long as the maxim holds good, that the doubtful must be explained from the certain, and the obscure from that which is clear, we must interpret Ex. xii. 31 from Ex. xiv. 5, 6, and not the latter from the former. For no one will venture to assert, that Pharaoh's words in Ex. xii. 31 can only have been intended as a conditional permission to depart. He does not say a single word about making their return a necessary condition, as Ex. viii. 28 and x. 10 would lead us to expect him to do. Moreover, on the supposition that his permission was entirely unconditional, the words "serve Jehovah, *as* ye have said," are perfectly intelligible, and by no means unimportant: for if we take them in connexion with Ex. viii. 25, 26, and x. 24—26, we have good ground to suppose that Pharaoh's meaning was, that he would place no obstacle whatever in the way of their offering the intended sacrifice. Moreover his reason for mentioning this sacrifice particularly may be gathered from the expression which follows: "begone, and *bless me also!*" That is to say, he requested that the sacrifice might be made available for his own welfare and that of his people, that Israel would show its gratitude by interceding for him and entreating the mercy of Jehovah on his behalf. And it must be admitted that the words in which the permission was couched, "get you forth from among my people," look much more like an unconditional than a conditional release.

(4). On the supposed "*borrowing*" and "*purloining*" *of the gold and silver* vessels consult especially *Hengstenberg's* Genuineness of the Pentateuch (vol. ii. p. 417—432, English transl.). The passages in which it is referred to are Ex. iii. 20—22; xi. 1—3; xii. 35, 36. In the first passage, the spoiling of the Egyptians is mentioned as a divine promise; in the second, as a divine command; and in the third, as an act performed by the Israelites. All three passages have this in common, that the Israelites *asked* (*Qu.* borrowed? שָׁאַל) of the Egyptians vessels (and clothes), and that God gave the people of Israel *favour* in the eyes of the Egyptians. In the third the *gift* of the Egyptians is described as a הִשְׁאִיל (*Qu.* lending or presenting?); in the first and third the *taking* of the goods on the part of the Israelites is spoken of as a *spoiling* (*Qu.* purloining? נִצֵּל); and, lastly, in the first passage the design of the request is said to be "that ye may *not go out empty*."

Most commentators explain the word הִשְׁאִיל as meaning *to lend*. But if this be the meaning, as the Israelites were not going to return to Egypt, and knew that they were not, their borrowing must be regarded as an act of fraud, a theft in fact; and, what is still worse, God himself appears as the instigator of the robbery. Various attempts have been made to get rid of this difficulty.

In the first place, it has been said that God, the Creator of all things, is the actual owner of all created objects, and has an unconditional right to dispose of them as he will, and that, accordingly, he may justly transfer them from one steward to another. This explanation is given by *Abenezra*, *Augustine*, *Luther*, *Calvin*, *Pfeiffer*, *Calov*, *Buddeus*, &c. *Augustine* finds a more particular motive, however, for the divine decision in the fact, that the Egyptians had perverted to ungodly purposes the gold which had been entrusted to them by God, having applied it to idolatrous worship, whereas the Israelites would use it for building the tabernacle; and in this *K. v. Raumer* agrees with him. There is undoubtedly a certain amount of truth at the foundation of this explanation, and truth which is applicable to the circumstances before us; but there is by no means a sufficient amount of truth to remove the difficulty in question. If God in the administration of his government of the world, by any movement whatever, makes one man the possessor of what has previously been in the hands of another, there is nothing in this to perplex or surprise; and in such a case, the absolute right of God to the possession and disposal of the property would be most justly maintained. But it is to us most offensive and repulsive, for any one to attempt to persuade us, that an act of fraudulent borrowing was the means of transfer approved and commanded by God. We must, therefore, pronounce this attempted defence unsatisfactory, except where it is accompanied, as it sometimes is, by one of the arguments to be adduced presently as a solution of the difficulty; and even then everything depends upon whether the right selection is made.

Secondly, The purloining of the jewels has been represented in the light of a reprisal, by which the Israelites, according to the will and command of God, repaid themselves for the long continued tributary service, which they had been unjustly compelled to render. This explanation is adopted by *Philo*, *Clemens of Alexandria*, *Tertullian*, *Irenaeus*, *Theodoret*, *Grotius*, &c. As we shall presently show, we admit that this solution also contains a certain amount of truth; but it is quite as far, as the former one, from being sufficient to remove the difficulty. *Hengstenberg*, indeed, argues that there was nothing of the nature of a reprisal in the matter, since the Israelites did not stand in the same relation to the Egyptians as one independent power to another, but in the relation of subjects to the king of Egypt, and also because it was private property which they took away, and not that of the Egyptian state. With regard to the first, however, I cannot admit that there was anything like the pure relation of a subject to a monarch in the present instance. The Israelites had come to Egypt as a free and independent people, and had merely placed themselves under the protection of the king, without submitting to become his subjects. As the king, however, bestowed certain lands upon them, they certainly became his vassals, and on that relation certain rights of supremacy were probably founded. But those rights had respect to the lands, not to the individuals, and as soon as the Israelites restored the former to Pharaoh, he ceased to have any claim upon them as his subjects, and had no right to keep them in the land by force and against their will. And when he attempted this, the Israelites, who had ceased to be vassals by declaring their intention to depart, were fully justified in meeting force by force, in defence of their right to withdraw. With regard to the other objection, that the Israelites had not rendered service to the Egyptians generally, but to the king and the state, and therefore should only have indemnified themselves from the property of the state, not from private property; the difficulty is removed when we consider that the crime of oppressing Israel belonged to Egypt as a whole, and therefore the duty of rendering compensation belonged to it also as a whole. But the idea of a reprisal is entirely out of the question here; for the Israelites asked for the treasures of the Egyptians, and did not take them by force, whilst the Egyptians gave them without any compulsion from without. And if their giving were merely a lending, there would be still less room for speaking of reprisals.

Thirdly, a very peculiar escape from the difficulty has been suggested by *Justi* (über die den Aegyptern abgenommenen Geräthe, Frankfurt 1771), and in all that is essential he is followed by *Augusti* (theol. Blätt. i. 516 seq.). They are both of them of opinion, that the Israelites left their immoveable possessions in exchange, or as security for what they borrowed; and hence the failure to restore the latter cannot possibly be regarded in the light of a theft. But we read nothing about security; moreover this explanation is shown to be incorrect, by the words "they purloined (stole) it from the Egyptians."

The following mode of defence touches more closely the actual difficulty, which arises from the words *lend* and *purloin*.

It has been maintained,[1] *fourthly*, that the conduct of the Israelites was undoubtedly opposed to the universal law of nature, but that God, as the supreme lawgiver, has the right in particular cases to suspend the law of nature, or dispense with it altogether. We think it superfluous to enter into any further discussion of the inadmissibility of this assertion.

Fifthly, it has been further supposed that the Israelites borrowed

[1] According to *Hase* (Hutterus redivivus, ed. 2 p. 58), by *Eschenmayer*, in his Religionsphilosophie; but I cannot find the passage.

the vessels with the intention of restoring them ; but as God, the supreme possessor, afterwards directed them to retain them as their own, they were justified in keeping them. *Pfeiffer*, for example, says: "posito mutuo potuit id ex post-facto, intercedente assignatione juris, transire in proprietatem possidentis." Here, too, we consider a refutation uncalled for.

Sixthly, It has been said that the Egyptians lent the things, with the expectation of receiving them back again, and the Israelites borrowed with the intention of restoring; but the Egyptian king, by his breach of faith and his malicious attack, set the Israelites free from every obligation to return, and rendered it impossible that they should restore what they had borrowed. This is the opinion expressed by *J. D. Michaelis, H. Ewald, Hofmann*, &c. But it is a sufficient reply to this, that the Israelites foresaw, with all the assurance of faith in the promises of God, that they would *not* return to Egypt (Ex. iii. 16 seq.). Nor is the flaw mended by *Hofmann's* remark, that Moses undoubtedly knew that the Lord had promised to bring the people to Canaan, and therefore took Joseph's bones away, but that he did not know how this would be accomplished. That is to say, Moses knew very well that the people would not return, and yet took away the vessels, on the express or tacit condition that he should bring them back! Was not this fraud?

Seventhly, *H. Ewald* (Geschichte ii. 52) regards the explanation just referred to, as evidently in accordance with the meaning of the narrative. But in his opinion the scriptural narrative does not give the *Saga* in its original form, but in a manner so distorted and changed, that nothing less than *Ewald's* keen and prophetic glance could possibly have discovered and restored it to its original shape, which was as follows: "Israel took away *from the Egyptians the true religion* (? ! !), took away from them the proper sacrificial utensils, and therewith the true sanctuary, and even the sacrifices themselves? This *must evidently* (? !) be the meaning of the Saga. In every such period, when the fate and religion of two nations are about to be decided, the first question is, which of the two contending nations will take from the other what is really good (? !), and which will give it up (? !) ; for in the course of the conflict something higher and better will be sure to be evolved, and one of the two contending parties will eventually suffer it to be taken by the other (? !).

In this instance Israel, as the conqueror, *justly prided itself* (? ! !) upon having taken *the true sacrifice* away from the Egyptians. We find an analogy in the narrative of the robbery of Laban's household gods by Rachel, and the Grecian legend of the golden fleece." We certainly do not think that *Ewald's* true explanation will share the fate of the true religion and the true sacrifices of Egypt.

Not much more successful is (*eightly*) a perfectly new explanation given by *Schröring* (in the luth. Zeitschrift, 1850, p. 284 sqq.), who supposes that, according to the original historical form of the Saga, the Israelites were victorious in a conflict with the Egyptians, and carried off the *palladia* of the kingdom.

Ninthly and lastly, we come to the view adopted by *Hengstenberg*. It had been previously held by *Harenberg*, *Lilienthal*, *Rosenmüller*, *Tholuck*, *Winer* (Lex. Hebr.) and others, but none of them have defended it so vigorously, or carried it out so thoroughly as he. According to his explanation, the rendering of the words in question by "lend" and "purloin" can be shown to be false; and incontrovertible evidence, founded upon the circumstances of the case and supported by philological considerations, can be adduced to prove that the former actually denotes a *gift*, forced from the Egyptians by moral power, through the influence of God upon their hearts, and that the latter was an act of *spoliation* or *plunder*, the explanation of which is contained in the former. As we believe that this is the only correct explanation, we shall endeavour to sustain and, in some respects, carry out more fully the arguments of *Hengstenberg*.

The first question is, how are the words שָׁאַל and הִשְׁאִיל in Ex. xii. 35, 36 to be understood? Do they mean *postulavit* and *dedit* (to *ask* and to *give*), as *Hengstenberg* supposes, or *mutuum petiit* and *mutuum dedit* (to *borrow* and to *lend*), as nearly every other commentator renders them? In any case the two words stand in such a relation to each other, that הִשְׁאִיל expresses the granting of the request contained in שָׁאַל. If the *Septuagint* translation could always be depended upon, there would be no difficulty at all. The words are there rendered: ᾔτησαν παρὰ τῶν Αἰγυπτίων . . . καὶ ἔχρησαν αὐτοῖς. In the *Vulgate* we find *petierunt* . . . *commodaverunt*, and in *Luther's* translation, "they *asked* . . . they *lent*."

Hengstenberg, however, not merely disputes the correctness of the rendering "to lend" in this connexion, but denies that it ever has that meaning; at least he argues as if it was *a priori* wrong to introduce such a notion here, because it is inappropriate on philological grounds. But in this we differ from him. For how did such a rendering find its way into the *Septuagint* and *Vulgate* if that be true? Undoubtedly שׁאל (except in one passage) always means simply to *ask*, *request*, *entreat*, without the additional idea associated with *mutuum petiit*. Moreover there is only one other passage, in which the Hiphil of שׁאל occurs, viz. 1 Sam. i. 28; and there it cannot possibly mean to *lend*, but to give unconditionally, to *present:* for Hannah does not give the son whom she had asked of the Lord, merely as a loan to the Lord. She brings him as a gift to Him, who had presented him to her; for she gives him up to the tabernacle, with an express renunciation of all her rights and claims upon him, by setting him apart as a Nazarite for life. But there is, on the other hand, *one* passage in existence, viz. 2 Kings vi. 5, where שׁאל has undoubtedly the meaning to borrow. When the pupil of the prophets, who drops his axe into the water, mourns the more bitterly over the loss, because the axe was שָׁאוּל; this can only mean that it belonged to another, and was therefore *borrowed*. But if שׁאל; may mean to borrow, even when standing alone, there can be no doubt that השׁאיל *may* also be used in the sense of "lending."

The real state of the case is as follows: שׁאל means primarily and originally to beg, ask, desire; and השׁאיל to grant the request, to give what is asked. But there is nothing in the words themselves, to show what is the nature of the request and the gift, whether conditional or unconditional; whether what is asked for, is required and given for a permanent possession, or merely for temporary use. The context and the circumstances of the case must determine, in every instance, whether the words have any such subordinate idea at all, and if so, which of the two it may be. If we ask, which is the more probable and usual, and which therefore is to be tried first in our exposition, we must certainly decide in favour of the former. For it will be granted, we imagine, by every one, that the idea of requesting and giving conditionally is farther removed from the radical signification of the words, than that of giving and asking without conditions. And this supposition is confirmed by the usage of the language. שׁאל

only occurs once in the sense of borrowing; and השאיל, which we only meet with in one other passage, has there the indisputable meaning, "to give, or present." Our proper course is therefore plain, first of all to try whether this meaning will suit the connexion; and if we find it inappropriate, *i.e.*, if the context and the circumstances of the case compel us to give the preference to the meaning "*to lend*," then and then only are we bound, or even at liberty, to make use of the latter signification.

It appears to us, however, that both the circumstances and the context are decidedly in favour of the first meaning. The most important point to be decided in connexion with the enquiry is evidently this: did the Egyptians expect the Israelites to return, or were they led to expect it by any promise on their part either tacit or expressed? If this question were answered in the affirmative, we should not *even then be compelled* to conclude that the things were lent and not given; for the Egyptians might expect the return of the Israelites, and yet *present* to them valuable articles of plate and clothes. But if the state of the case was such that the Egyptians did not, and could not expect them to return; it follows as a matter of course that the *only* idea, entertained on either side, must have been that the things were given, and not that they were lent. That this was the actual state of the case, we have shown above in the third note to this section.

Moreover, the statement which is repeated in all the three passages, that the readiness of the Egyptians to give was the result of the operation of God upon their hearts ("God gave the people favour in the eyes of the Egyptians"), suggests the idea of *giving*, much more than of *lending*. And when it is said in Ex. iii., "ask of the Egyptians and I will give you favour in their sight, that *you may not go out empty*;" it is very evident, in the first place (especially if we compare the promise in Gen. xv. 14, which is reiterated here), that what they asked for was to become their own property; secondly, that they were to obtain possession of it, not through the impossibility of returning, but through the influence of God upon the hearts of the Egyptians; and thirdly, that the asking was not meant in the sense of purloining.

But there is still one difficulty left, and in the opinion of our opponents, it is the most important of all, viz., the words וַיְנַצְּלוּ

אֶת־מִצְרָיִם (*Angl.*, "and they spoiled the Egyptians:" *Tr.*), in chap. xii. 36. But it is just from this expression, that our explanation derives the greatest support; and to *Hengstenberg* belongs the credit, of having been the first to point this out in an emphatic manner. In the *Septuagint* the words are correctly rendered καὶ ἐσκύλευσαν τοὺς Αἰγυπτίους, and so also in the *Vulgate* "et spoliaverunt Aegyptios;" but *Luther's* rendering is decidedly false, "sie *entwendeten* es den Aegyptern" (they purloined it from the Egyptians). נִצֵּל never means *to purloin, to steal;* it never denotes the removal or appropriation of any thing craftily and secretly, by fraud and treachery, but always means to *plunder, spoil*, or take away by force. Now this does not harmonize at all with the idea of *borrowing;* for borrowing, with the intention of not returning, is an act of treachery, not of force. If we enquire how *Luther* came to lose sight of the true meaning; the most probable conjecture is, that he was misled by the erroneous idea, that השׁאיל meant to lend. He felt, no doubt, that *spoiling* did not harmonize with *borrowing* and *lending*, in fact that one excluded the other; and instead of making use of the proper explanation of נצל, to correct his false rendering of השׁאיל, he allowed himself to be misled by his regard for the *Septuagint* and *Vulgate*, and therefore altered the true rendering of the former, to suit the false or at least doubtful interpretation of the latter.

But if the things were given, was it suitable to speak of this as robbery and spoliation? At first sight it does appear somewhat strange, that we should answer in the affirmative. But on closer investigation it will not be thought so. The author intends to lay stress upon the fact, that constraint was put upon the Egyptians; that they were plundered of their possessions in consequence of the contest; and that Israel had marched away, "laden, as it were, with the booty of their powerful foes, as a sign of the victory, which they in their weakness had gained through the omnipotence of God." It was the fulfilment of the promises contained in Gen. xv. 14, and Ex. iii. 19 sqq.: "Afterward shall they come out *with great substance*," and "I know that the king of Egypt will not let you go except by a strong hand, and I will stretch out my hand and smite Egypt with all my wonders, and I will give this people favour in the sight of the Egyptians, and it shall come to pass that when ye go, *ye shall not go away empty.*" "The author," says *Hengstenberg*, "represents the gifts of the Egyptians as spoil which God had awarded to his army (xii. 41), and thus calls attention to the fact, that the bestowment of these gifts, which outwardly appeared to be the effect of the liberality of the Egyptians, proceeded from another giver; that the act of the Egyptians, which was performed without external compulsion, was the result of a divine constraining influence within, which they were utterly unable to withstand." Without this divine constraint, the Egyptians would rather have done all the injury they could to the despised and hated shepherds, from whom they had received such harm. "At the same time the expression is chosen with reference to the conduct of the Egyptians, for which they had now to make some compensation to God and his people. They had plundered Israel (by the tributary service which they had unjustly compelled it to render); and now Israel carried away the plunder of Egypt." That which happened to Egypt here, however, was a type of every similar conflict between Israel and heathenism, for Egypt was the first-fruit and representative of the whole heathen world in its relation to the kingdom of God. Hence it is said in Zech. xiv. 14, with evident reference to this event, that "the wealth of all the heathen round about shall be gathered together, *gold* and *silver*, and *apparel*, in great abundance." In the present instance, therefore, on the occasion of the first and fundamental conflict with heathenism and victory over it, it would have been at variance with the divine rule of propriety, if Jehovah had allowed his people to depart without compensation for the privations and injuries, to which they had been unjustly exposed in Egypt. It would have been but a partial victory, and, therefore, (according to the primary intention of this portion of history), no victory at all, if Israel had not thus obtained complete satisfaction.

Thus, and thus only, can the passages quoted from the book of Exodus be freed from every difficulty; thus only can the terms of the announcement, and the demands of all the circumstances, be fully satisfied; thus only can the transaction itself be brought into harmony with the whole of the miracles and events which occurred in Egypt; and only thus can we explain the fact, that the author regarded it as so important and significant, that he describes it three times with all the details, and in every form in which it could possibly be presented (as a promise, a precept, and a fulfilment) ; and not only so, but that it was also included in the promise made to Abraham 650 years before (Gen. xv. 14). With this explanation, it appears to us as the climax of all the signs and wonders, or rather of all the events associated with the deliverance of Israel from the yoke of the Egyptians. Every explanation, in which the notion of a loan is retained, comes into conflict with the idea of the holiness of God ; whilst all the elements of truth, which lie at the root of the unsuccessful attempts referred to above, are not only retained, but receive ample justice in this explanation of ours.

Before leaving this subject, there is one more question to which we must direct our attention. It has generally been overlooked altogether, and where this has not been the case, it has been incorrectly answered. I refer to the question, what kind of gifts the Israelites asked for, what determined their choice, and to what purpose they intended to apply them ?

Ewald is of opinion, that the *vessels and clothes*, for which they asked, were indisputably intended for the sacrificial festival, which the Israelites were about to celebrate, and for which Moses had requested Pharaoh to let the people go. They must, therefore, have been *sacrificial utensils* and *priestly clothes*, which had already been set apart by the Egyptians for the same religious object. But how does this square with the injunction contained in Ex. iii. 22, that " every woman shall request of her neighbour *(fem.)*, and of *her* that sojourneth in her house," these clothes and other articles ? Does *Ewald* suppose, that all the Egyptian *women*, or at any rate *all* those, who lived with or among the Israelites, were in possession of sacrificial articles and priestly clothes ? Any one who is versed in Egyptian antiquities, must know very well that the system of caste was too strictly maintained in ancient Egypt, for such a thing to be credible. *Justin*, it is true, says of Moses, " sacra Aegyptiorum furto abstulit" (hist. 36. 2. 13 ;) but he certainly does not represent the Israelites as stealing the *Sacra Aegyptiorum* from the wives of the Egyptian Fellahs. His notion undoubtedly was, that Moses broke into the Egyptian temples at the head of an Israelitish army and plundered them. And if any one imagines that *Justin's* account is more correct, than that of the book of Exodus, we will not deprive him of his pleasure by any untimely discussion, or refutation of his views.

Hofmann also supposes, that the articles of gold and silver were sought (borrowed) by the Israelites, for the approaching festival in the desert. But he imagines them to have been, not bowls and other sacrificial vessels in use among the Egyptians, as *Ewald* fancies, but articles of furniture and table utensils. It can be easily proved, however, that this explanation is erroneous and inadmissible. For, like the other, it is opposed to the command in Ex. iii. 22, that *every* woman should request of her neighbour, and of her that sojourned in her house, the articles referred to. Can it for a moment be supposed that cooking utensils, furniture, and table services of gold and silver were possessed by *all* the Egyptians, even of the poorer and lower orders (for it is these and not the rich and noble that we must chiefly think of as dwelling in the midst of the Israelites and lodging in their houses)? But apart from this, the Egyptians and Israelites were both of them too scrupulous and particular, with regard to the cleanness and uncleanness of their vessels, for the latter to use either sacred or profane vessels belonging to the unholy Egyptians at their most holy festival; or, *vice versa*, for the former to *lend* them to the Israelites for their abominations (Ex. viii. 26). Moreover, the supporters of this explanation are at a loss to know how to dispose of the *clothes*, which are referred to in just the same terms as the rest of the things. However the issue proves that it is incorrect. No doubt the Israelites were altogether ignorant, at the time when they asked for the golden vessels and the clothes, of the *manner* in which the festival was to be celebrated and the sacrifices to be offered in the desert (Ex. x. 26); and hence it is not *impossible* that they may have thought, that every Israelitish woman (Ex. iii.) and every Israelitish man (Ex. xi.) would require one or more of these vessels of gold or silver. But this is certainly not very probable. Besides, even if neither Moses nor the Israelites knew, what kind of festival was to be celebrated and what would be required (Ex. x. 26), *God* certainly knew all this, and it was *He* who commanded the Israelites to ask for the articles of gold and silver. Now, if we look at the result, viz., at the manner in which the sacrificial festival was actually celebrated in the desert (Ex. xxiv.), we find not the least hint of their having used the articles, which they had borrowed from the Egyptians; in fact it is evident, that only one sacrificial basin was employed (xxiv. 6). We must therefore conclude, that at all events it could not have been the intention of God, that the articles should be employed in their sacrificial worship. If, however, this was not the purpose for which God commanded the Israelites to ask for the vessels of gold and silver, He must have had some other object in view. And in our opinion, the scriptural narrative states clearly enough what His purpose was. But we must postpone the discussion of it to a later period.

First of all, we must endeavour to ascertain clearly, what we are to understand by the כְּלֵי כֶסֶף וּכְלֵי זָהָב וּשְׂמָלֹת in this passage. There is no difficulty with regard to the particular words. כְּלִי plur. כֵּלִים, derived from כלה, means, according to *Gesenius* (thesaurus): quidquid factum, confectum, paratum. It is a "vocabulum late patens" to which it would be difficult to find a parallel; for it is used in the Old Testament, with reference to articles of every possible kind. House-furniture, tools, sacrificial utensils, vessels of all kinds and for all purposes, armour, clothes, ornaments, jewels, and other things, are all called by this name. But when the כלים are said to have been of gold and silver; of course the range, which is covered by the word, is considerably limited. It must then be understood as referring to such articles only, as were generally or frequently made of the precious metals. Now as a rule the articles of gold and silver would be merely ornaments (rings, bracelets, chains, &c.), things used in the temple (sacrificial vessels, &c.), and certain articles of furniture or table-service to be found in the houses of the rich and noble. As the precious metals were most frequently employed in the manufacture of the articles first-named, it is most natural to suppose that כלי כסף וכלי זהב were jewels and ornaments; there being nothing in the context or the words themselves, on which to found a more minute description. Hence, starting from this abstract standpoint, we have good reason to suppose that the primary reference is not to articles of furniture or table-service, but to ornaments, trinkets, and jewels. In this sense the expression occurs in Gen. xxiv. 53 without further details: "And Eliezer produced כלי כסף וכלי זהב וּבְגָדִים, and gave them to Rebekah." But although we have no particular description, can we reasonably suppose the presents to have consisted of anything but *trinkets* and *ornamental articles of female attire?* כלי זהב is used in the same sense in Ex. xxxv. 22, clasps and ear-(nose)-rings, finger-(signet)-rings, and necklaces being mentioned by way of example; and also in Num. xxxi. 50, where, in addition to the ornaments just named, foot-chains and bracelets are specified.

These passages show very clearly, that in common parlance the phrase "articles of gold and silver" was generally understood as referring to ornaments. And if we look more closely at the verses before us, we shall soon arrive at the conviction, that nothing else can be intended there. The very fact, that the articles of gold and silver are mentioned in connexion with *clothes*, leads to the conclusion that the two are to be placed in the same category; that is to say, that they were both of them ornamental, since the *clothes* were evidently *festal* dresses. This conclusion is still more decidedly forced upon us, when we consider that in Ex. iii. none but *women* are mentioned either as *givers* or *receivers*. For it is evident that only such things are referred to, as are generally to be found in the possession of *women*, and such as *women* care most about. We have not to think, then, of either furniture or table service, but of ornaments alone. And thus the difficulty, to which *Hofmann's* explanation was exposed, is entirely removed. We certainly cannot imagine that every Egyptian family was in possession of gold and silver plate, much less of sacrificial basons, dishes, &c., of gold and silver; but without wandering beyond the range of probability, we may assume, that an Egyptian woman, though otherwise poor, might possess a ring, bracelet, clasp, or some other ornament of gold or silver.

We find, indeed, in Ex. xi., *men* as well as women mentioned as givers and receivers; but no one, we trust, will think that this is at variance with our explanation. For *clothes* are spoken of, as well as gold and silver כלים; and they were certainly required by men, as much as by women. Moreover the men sometimes wore either gold or silver ornaments. Judah, for example, had a signet ring (Gen. xxxviii. 18), and Joseph a ring and gold chain (Gen. xli. 42).

A further proof that this is the only correct explanation, is to be found in the words וְשַׂמְתֶּם עַל בְּנֵיכֶם וְעַל בְּנֹתֵיכֶם (Ex. iii. 22).

Luther follows the *Septuagint* and *Vulgate*, and renders these words, "ye shall lay them upon your sons and daughters," *i.e.*, that they may carry them. This is a mistake, which necessarily arose from the false interpretation of כלים (*Gefässe*, *σκέυη*, *vasa*), and which a correct interpretation at once removes. Even on the supposition that the כלים were vessels (basons, kettles, dishes, &c.), there would be something startling in such a rendering. Had the Israelites, we might ask, whose chief pursuit was the rearing of cattle, no asses and other beasts to carry their baggage? Was it necessary that their sons and daughters should supply the place of beasts and burden? If we refer to Ex. xii. 38, we shall find that the Israelites went out with a very large quantity of cattle of every kind. Moreover the verb שׂים with the preposition עַל can *only* mean to *put on*, when used of *clothing*, *armour*, *jewellery*, &c. (*cf.* Gen. xxxvii. 34, xli. 42; Lev. viii. 8, seq.; Ruth iii. 2; Ezek xvi. 14, and many other passages), and this must be the meaning here.

But we have still one more way open, of fully satisfying ourselves as to the meaning of the words "articles of gold and articles of silver," in the passage under review. As every woman was to ask, and receive from her female neighbour, and every man from his neighbour also; there must have been a great abundance of the articles referred to, in the possession of the Israelites. Let us see now, whether we do not find some further reference to them in the course of the history. First of all we meet with them in Ex. xxxii., where the men and women, sons and daughters of Israel, take their golden ear-rings out of their ears, that Aaron may make them into *Elohim* to go before them.

Again, although the quantity of golden ornaments, required to make the golden calf, must have been far from inconsiderable, and this calf was subsequently destroyed (Ex. xxxii. 20); we read shortly afterwards (Ex. xxxv. 21, seq.), that the whole community, both men and women, brought "clasps and ear-rings (nose-rings?), signet rings, necklaces, and all sorts of trinkets of gold," as a free-will offering towards the erection and furnishing of the tabernacle. Let any one think for a moment what a mass of gold must have been used in connexion with the tabernacle, when the beams were all plated with gold, and the articles of furniture were either made of solid gold or at least covered with it, and he will be obliged to admit not only that the quantity of gold in the possession of the Israelites was extraordinary, but that if we were not acquainted with the circumstance narrated in Ex. xii. 35, 36, it would be incredible and inconceivable.

We shall, perhaps, be reminded, however, that according to Num. iv., there were gold and silver dishes, bowls, cups, cans, lamps, snuffers, extinguishers, and oil vessels in the tabernacle, and told that these were probably the " articles" which the Israelites had received from the Egyptians (*cf. v. Raumer*, Der Zug der Israel. p. 3. 4. Anm.). But this was evidently not the case, for we learn from Ex. xxv. 29, 38, that these dishes, bowls, cans, cups, snuffers, and extinguishers were made in the desert; and in ver. 3 sqq. we read, that they were made from the freewill offerings of ornaments and jewellery, that were brought to Moses by both men and women. We may, perhaps, be also referred to Num. vii., where it is said that every one of the *twelve* princes of Israel brought as his offering, at the dedication of the altar, a silver dish weighing a hundred and thirty shekels, a silver bowl of seventy shekels, and a golden spoon of ten shekels, in which his meat offering was placed; and these again will probably be pointed out as Egyptian vessels. But let it be observed that every woman, and every man, requested and received the articles of gold and silver; whilst here it was only the twelve princes of Israel, who brought such offerings as these. Moreover, they were not brought till the tabernacle was finished; and therefore the twelve dishes, and bowls, and spoons, had most probably been made in the desert for that purpose, as well as the things already named in Ex. xxv. 29, 38, and Num. iv. 7, 9. The gifts of the Egyptians may possibly have been employed; but if any objection be felt to this, it must be borne in mind, that the offerings were made by the *princes* of Israel, and as they were the richest and most eminent among the people, they may very well have possessed both gold and silver, and, for aught I know, silver dishes and bowls, as well as golden spoons, among their private property. Still, as the dishes, bowls, and spoons, offered by the twelve princes, were all of exactly the same weight, we are forced to the conclusion, that they had been prepared expressly for the purpose.

The character and drift of the whole narrative are brought out more clearly by this explanation; and on the other hand it serves to confirm the opinion, that the articles were not obtained by borrowing and purloining, but were spoils which came to the Israelites in the shape of presents, though they were forced from the Egyptians by moral constraint.

After the severe oppression, under which Israel had groaned so long, the resources of a large portion of the nation must necessarily have been considerably reduced, through the loss of the property which they once possessed. Under such circumstances, unless some provision had been made, the departure of the Israelites would have been upon the whole but a very miserable one; and the last impression left by the people of God, on their exodus from Egypt, could only have been that of a wretched and contemptible horde of beggars and of slaves. This would undoubtedly have been opposed to the divine rule of propriety; for the reproach of the people was the reproach of Jehovah, just as in other cases the glory of the people was Jehovah's glory. It was not to be with great difficulty, and with hardly a sound skin, that the Israelites were to depart; but as a victorious and triumphant people, laden with the treasures of Egypt, in *festal attire*, and adorned with jewels and costly ornaments, and with necklaces and bracelets of gold. They were going to the celebration of a festival, the greatest and most glorious that ever occurred in their history; such a festival demanded festal attire, and this was to be furnished by their bitter and obdurate foes, without (and this was the climax of their triumph) the least external compulsion, and yet without resistance or refusal, on the simple request of the Israelites alone. To such an extent had the pride and intolerance of the Egyptians been broken; so completely were the tables turned, that Egypt now entreated as a favour the very departure, which it had hitherto so obstinately opposed, and it was no longer the Egyptians but the Israelites who prescribed the conditions of their departure, whilst the former assented at once to every condition, however humiliating it might be.

(5). The Old Testament divides the *First-born* into two classes, the first-born of the father, and the first-born of the mother. The former alone possessed the civil rights of primogeniture, namely, the headship in the family, and the double inheritance, which secured to them the title of *primogeniti haereditatis* (*cf.* Deut. xxi. 15—17). The latter, who were called, in distinction from the others, "every first-born that openeth every womb," had no civil pre-eminence; unless they were also the first-born of the father. In the case before us, the first-born of the mothers are intended; and as they were to be sanctified to Jehovah, they were designated as *primogeniti sanctitudinis.* (*cf. Selden*, de success. in bona defunctt. c. 7. p. 26 sqq. and *Iken* diss. ii. p. 37).—The question arises here, what are we to understand by the *sanctification of the first-born?* That they were not to be set apart to the priesthood is proved most conclusively by Ex. xiii. 2, 13, where the first-born of men are ordered to be sanctified, in exactly the same sense as the first-born of beasts. It was not as *sacerdotes*, but as *sacrificia* to the Lord, that the first-born were to be set apart. "Sanctify unto me all the first-born both of man and of beast, for they are *mine*," are the terms of the command in ver. 2 (*cf. Vitringa*, observv. ss. ii. 2, p. 272 sqq.). When Jehovah passed through the land and smote all the first-born of the Egyptians, he had passed over all the houses of the Israelites that were marked with the atoning blood of the Paschal lamb, and spared the first-born in them; but notwithstanding this, he had the same claim to the first-born of the Israelites, as to those of the Egyptians. This claim of Jehovah to the possession of all the first-born was founded upon the fact, that He was the Lord and Creator of all things, and that as every created object owed its life to Him, to Him should its life be entirely devoted. The earliest birth is here regarded as the representative of all the births; so that the dedication of the whole family was involved in that of the first-born. The difference between the first-born of Israel and the first-born of Egypt was this: the Egyptians refused to render to Jehovah that which was due, and continued most obstinately to resist his will; Israel, on the other hand, did *not* draw back from the dedication required, and covered their previous omissions by the atoning blood of the sacrificial lamb. Now the law of the kingdom of God is, that every thing which will not *voluntarily* consecrate, itself to the Lord, for the purpose of receiving life and blessedness through this self-dedication, is *compulsorily* dedicated in such a manner as to receive judgment and condemnation. The slaughter of the first-born of the Egyptians is therefore to be regarded as of the nature of *a ban* (חֵרֶם), an involuntary, compulsory, dedication. But Israel's self-dedication to Jehovah had hitherto been insufficient, and hence the necessity for the expiatory sacrifice to cover the defects. The necessary complement of reconciliation is sanctification. By virtue of the atoning Paschal blood the first-born of Israel had been spared; but if they were to continue to be thus spared, the sanctification of the first-born must follow. And as the first-born of Egypt represented the entire nation, and in their fate the whole people were subjected to a compulsory dedication; so was the voluntary dedication of the *whole* nation of Israel set forth in the sanctification of the Israelitish first-born. It is true that the sparing of the first-born, like the redemption from Egypt, did not occur more than once in history; but future generations reaped the benefit of both events; and therefore in the particular generation which was spared and delivered, every succeeding generation was spared and redeemed at the same time (and it was for the purpose of keeping this in mind that the annual commemoration of the passover was enjoined). Hence it was not sufficient that the first-born of that first generation should be consecrated to the Lord, in order that the protection and deliverance afforded should be subjectively completed; but it was required that the first-born of every succeeding generation should be also sanctified to the Lord, as having been also spared and redeemed. Therefore the command was issued, that this first sanctification of the first-born should be repeated in the case of all the first-born in every age.

The words "*for they are mine*," (chap. xiii. 2) show, in the most general terms, in what the consecration of the first-born consisted. The first-born was the Lord's; it was not *sui juris*, but the property of Jehovah, Jehovah's *mancipium.* Knowing then, as we do, from the next stages in the development of their history, that Jehovah had determined to fix his abode in the midst of the Israelites, and that his dwelling-place was to be the sanctuary of Israel, the tabernacle of assembly, where they were to meet with their God and serve him; we naturally expect that the consecration of the first-born, that is, their dedication to Jehovah, should take place either in or at this sanctuary, and this expectation was fully realized in the subsequent course of their history (*vid.* Vol. iii. § 20. 3). But the sanctuary was not yet erected; therefore, the sanctification required here cannot have been anything more than a provisional *separation* for that purpose, not the actual realization of it. But Israel was already to be made to understand, that after that solemn night of protection and deliverance, the first-born of its families and the first-born of its cattle were no longer *its own*, but belonged to *God*. It was no longer at liberty to dispose of them according to its own pleasure; but must wait submissively, till God in his own time should determine what they were to do. So much, however, was already made known (chap. xiii. 13), that only clean animals, *i.e.*, such as were fit for sacrifice, were to be actually and irredeemably set apart as sacrifices to the Lord; whilst all the rest of the cattle were either to be slain, or redeemed by a clean beast, and the first-born children were also to be redeemed. But it was not declared till a later period, how this was to be done (Num. iii. 8; viii. 17; xviii. 14—18). At the same time, they were already made perfectly conscious of the meaning of the whole transaction (vers. 14, 15): "When thy son asketh thee in time to come, saying, what is this? thou shalt say unto him: By strength of hand Jehovah brought us out of Egypt, from the house of bondage, and it came to pass, when Pharaoh would hardly let us go, that the Lord slew all the first-born of Egypt, ... therefore I sacrifice to Jehovah all that openeth the matrix, being males; but all the first-born of my children I redeem."

An expression occurs in Ex. xiii. 16, with reference to the sanctification of the first-born, which is similar to that which has already been used in ver. 9 respecting the yearly celebration of the passover: "It shall be for a token upon thine hand, and for a memorial-band (וּלְזִכָּרוֹן; ver. 16, for a frontlet, וּלְטוֹטָפֹת) between thine eyes. The pharisaic custom of later times was founded upon these passages; just as the practice of wearing תְּפִלִּין or *φυλακτήρια* (Matt. xxiii. 5), *i.e.*, strips of parchment with passages of Scripture written upon them, which were tied to the forehead and the hand at the time of prayer, was based upon Deut. vi. 8 and xi. 18; whilst others interpret the passages as symbolical only. That the latter is the only admissible explanation of the two passages in the Book of Exodus, must be apparent to every one; but whether the same may be said of the passages in Deuteronomy, is a question that we must reserve for a later occasion.

(6). According to Ex. xiii. 18 the children of Israel departed from Egypt חֲמֻשִׁים. The *Septuagint* rendering is πέμπτη δὲ γενεᾷ ἀνέβησαν. *Clericus* explains it in the same manner, with special reference to Gen. xv. 16, and Ex. vi. 16 sqq. (Jacob, Levi, Kohath, Amram, Moses). *Fuller* adheres firmly to the derivation of the word from חָמֵשׁ *five* (Miscell. ss. 5. 2). He renders it by πεμπτάδες, and supposes it to mean that they were drawn up in five columns. But neither of these renderings corresponds to the sense, in which the word is used in other places (Josh. i. 14; vi. 12; Judg. vii. 11). In Num. xxii. 30, 32, and Deut. iii. 18, the men who are called חמשים in Josh. i. 14, and vi. 12, are described as חֲלוּצִים (= *accincti, expediti ad iter s. ad proelium). The Vulgate* translates it *armati; Aquila*, ἐνωπλισμένοι; *Symmachus*, καθωπλισμένοι. A more suitable rendering of the passages cited would be "equipped for battle, in battle array," which certainly includes the notion of being armed. The etymology is doubtful. *Gesenius* refers to the cognate roots חמץ = *acer fuit*, חמס = *violenter egit, oppressit*, and to the Arabic حَمِسَ = *acer, strenuus fuit in proelio.* It has been objected to our explanation, that the Israelites went away unarmed. But this is nowhere stated; and the panic, which seized them afterwards (chap. xiv. 10 sqq.), does not prove that they were not armed. On the other hand, we read shortly afterwards of their fighting a regular battle at Rephidim with the Amalekites (xvii. 10 sqq.). There could have been no reason whatever for dividing the people into five companies. The *Septuagint* rendering has still less to commend it; were it only because there is no ground for the assumption, that Moses was the fifth in order of descent from Jacob (vol. i. § 6. 1). But the rendering "equipped for battle" or "in battle array" furnishes a good, appropriate, and very significant meaning. This was a necessary part of the triumphant and jubilant attitude, in which Israel was to depart from Egypt.

(7). The Egyptians, who attached themselves to the Israelites on their departure, are called עֵרֶב (from ערב *to mix)* in chap. xii. 38, and in Num. xi. 4 אֲסַפְסֻף (from אסף to collect). *Luther* renders both words *Pöbelvolk* (a mob); the *Septuagint*, ἐπίμικτος; the *Vulgate, vulgus promiscuum*. The Hebrew expressions describe them as a people that had flocked together (the formation and meaning of the words correspond to the German *Mischmasch)*, and lead to the conclusion that they formed the lowest stratum of Egyptian society, like the Pariahs in India, and did not belong to any of the recognised castes *(cf.* § 45. 4). Even among the Israelites they occupied a very subordinate position; for there can be no doubt that they were the hewers of wood and drawers of water mentioned in Deut. xxix. 10, 11. At the same time we perceive from this passage, that in spite of their subordinate position, and their performance of the lowest kinds of service, they were regarded as an integral part of the Israelitish community.

PASSAGE THROUGH THE RED SEA, AND DESTRUCTION OF PHARAOH.

§ 36. (Ex. xiii. 17—xv. 21 ; Num. xxxiii. 3—8).—The nearest route to Canaan, the ultimate destination of the children of Israel (chap. iii. 17), would have been in a north-easterly direction, along the coast of the Mediterranean; and by this route their pilgrimage would not have lasted more than a very few days. But Jehovah had his own good reasons (1) for not leading them straight to Canaan, but causing them to take a circuitous route across the desert of Sinai (2). The regular road from Egypt to Sinai goes round the northern point of the Heroopolitan Gulf (the Red Sea), and then follows a south-easterly direction along its eastern shore. In this direction the Israelitish procession started, under the guidance of Moses. The point from which they set out was *Raemses*, the chief city of the land of Goshen. The main body, which started from this city, was no doubt joined on the road by detachments from the more distant provinces. Their first place of encampment was *Succoth*, the second *Etham*, "at the end of the desert." But instead of going completely round the northern extremity of the Red Sea, so as to get as quickly as possible beyond the borders of the Egyptian territory, and out of the reach of Egyptian weapons ; as soon as they reached this point, they received orders from Jehovah to turn round and continue their march upon the western side of the sea. Thus they still remained on Egyptian soil, and took a route, which apparently exposed them to inevitable destruction, if Pharaoh should make up his mind to pursue them. For they were completely shut in by the sea on the one hand, and by high mountains and narrow defiles on the other, without any method of escape which human sagacity could possibly discover. In such a position no prudence, or skill, or power, that any human leader, even though he were a Moses, might possess, could be of the least avail. But it was the will of God; and God never demands more than he gives. When He required that Israel should take this route, He had also provided the means of escape. In his own person he undertook the direction of their march, and that in an outward and visible form, and by a phenomenon of so magnificent a character, that every individual in the immense procession could see it, and that all might be convinced that they were under the guidance of God. Jehovah went before them, by day in a *pillar of cloud*, that he might lead them by the right way, and by night in a *pillar of fire*, to enlighten the darkness of the night. This pillar of cloud never left the people during the day, nor the pillar of fire during the night (3). Tidings were quickly brought to Pharaoh from Etham of the unexpected, and, as it seemed, inconceivably infatuated change which the Israelites had made in their course. And Pharaoh said, " they have lost their way in the land; the desert has shut them in." The old pride of Egypt, which the last plague had broken down, lifted up its head once more. " Why have we done this, they said, to let Israel go from serving us?" Pharaoh collected an army with the greatest possible speed, and pursued the Israelites, overtaking them when they were encamped within sight of the sea, between Pihahiroth, Migdol, and Baalzephon. Shut in between mountains, the sea, and Pharaoh's cavalry, and neither prepared nor able to fight; enveloped, moreover, in the darkness of night, and without the least human prospect of victory, deliverance, or flight; the people now began to despair. "Were there no graves in Egypt," they cried out to Moses, "that thou shouldest lead us away to die in the wilderness?" Nor did Moses see any human way of escape. But he expected deliverance from Jehovah, and from Jehovah it came. "Fear not," said he to the desponding people, "stand firm, and see the salvation which Jehovah will effect for you to-day. Jehovah will fight for you, and ye shall be still." It was now to be clearly shown, that the ways of God, though they may appear to be foolish by the side of the wisdom of men, ensure the result in the safest, quickest, and most glorious way. "Forward!" sounded the command of the leader of Israel, "straight through the midst of the deep sea," through which the omnipotence of Jehovah was about to open a pathway on dry ground. The angel of God, who went before the army of Israel in the pillar of cloud and fire, passed over their heads and placed himself as a rampart between the Egyptians and the Israelites. To the former he appeared as a dark cloud, deepening still further the darkness of the night; to the latter as a brilliant light, illuminating the nocturnal gloom. Moses did as Jehovah commanded him; he raised his staff and stretched his hand over the sea. Jehovah then caused an *east wind* to blow, which continued the whole night, until it had laid bare the bottom of the sea, and divided the waters asunder. The children of Israel passed through the midst of the sea on dry ground, and the waters were *as walls* unto them on the right hand and on the left. The foe, bewildered, driven forward by the vehement determination to prevent a second escape of those whom they had regarded as so sure a prey, and unable, from the darkness that surrounded them, to discover the extent of the danger to which their attempt exposed them, rushed on with thoughtless haste in pursuit of the fugitives. As soon as the morning began to dawn, Israel had reached the opposite shore, and the Egyptians found themselves in the midst of the sea. Then Jehovah looked out from the pillar of cloud and fire, upon the army of the Egyptians; terror came upon them; wild confusion and thoughtless uproar impeded their march, and they shouted to turn back and fly. But Moses had already stretched out his hand over the sea again; and the waters, which had hitherto been standing like walls on either side, began to give way at the western end. The Egyptians rushed back and met the torrent; and Pharaoh, with all his horses, his chariots, and horsemen (5), was swallowed up by the sea. When morning came, the bodies of the Egyptians were washed up by the current upon the shore. Then Israel saw that it was the hand of Jehovah, which had been lifted up and had saved them; and they feared the Lord, and believed on him and on his servant Moses. The strong emotions of gratitude which filled the heart of Moses, burst forth in a lofty song of praise to their exalted deliverer. The anthem was sung by Moses and the chorus of men, whilst Miriam, the prophetess, Aaron's sister, at the head of a chorus of women, accompanied the choral-anthem of the men with timbrels, dances, and songs (6). This was the farewell to the first passover, which ended as it had begun, with deliverance and salvation.

(1). With regard to *the circuitous route by Sinai*, it is said in chap. xiii. 17: "God led them not by the road through the land of the Philistines (along the shore of the Mediterranean), which was the nearest way; for God said lest peradventure the people repent when they see war, and return to Egypt." It is necessary, first of all, to do away with a misapprehension, which has crept in here, to the effect that reference is already made in this passage to the forty years' sojourn in the desert, and that it was necessary that a new generation should grow up, before the conquest of Canaan could be thought of. The removal of the difficulty spoken of here could have been effected in much less than 40 years. The sojourn at Sinai, which lasted for an entire year, and the wonderful works of God with which they became acquainted by that time, must have been amply sufficient for this. It was not by the natural facilities afforded by their new unfettered life in the desert, so much as by *faith*, that the cowardice of Israel was to be overcome, and courage infused into their minds; and the events of the first year surely supplied all that was needed to strengthen faith. The sentence of rejection, which condemned the Israelites to wander in the desert for forty years, was simply a punishment for the want of that faith, which could and should have been matured by the works of God. The despair of the Israelites, when Pharaoh pursued them with his chariots and horsemen (xiv. 10 sqq,), soon showed how necessary such a precaution had been. The Philistines were a thoroughly warlike and powerful nation. Moreover, they were not destined to be exterminated, as the true Canaanites were; nor was their land, for the present at least, to be taken possession of by Israel.

Many readers may, perhaps, have been surprised to find only the negative and subordinate reason for their circuitous route mentioned here, viz. the impossibility of avoiding a collision with the Philistines; whilst no reference is made to the more important and positive motive, namely, the necessity for the conclusion of the covenant and giving of the law, previous to their entrance into the promised land. But we find a solution of this difficulty in the fact, that the immediate circumstances brought the negative reason more prominently before the author's mind;—the conflict with Pharaoh, which he was just about to describe, keeping his attention rivetted for the time to this particular point of view.

(2). We must not confound the reason for their *turning round*, and remaining within the Egyptian territory (chap. xiv. 3), with the reason for their circuitous route across the desert of Sinai. For the former the following causes may be assigned: (1) it was no circuitous route, to turn round the mountains as they did towards the Red Sea; for the road to the sea was the *straight* road from Egypt to Sinai. Such a leader as Jehovah, who knew how to make for his army a dry pathway through the depths of the sea, did not need to keep the circuitous caravan-road which goes round the sea. That which Pharaoh and every one else regarded as the wrong road (xiv. 3), was under the leading of God (and so it is called in Ps. cvii. 7) a "right way."—(2). Whilst the road *through* the sea was the most direct, and therefore outwardly the most expedient; there were other internal grounds, which concurred to render it the most desirable. The first of these had reference to Pharaoh. We find it expressed in chap. xiv. 4: "I will harden his heart, that he shall follow after them; and I will be honoured upon Pharaoh and upon all his host, that the Egyptians may know that I am the Lord." And there was also a reference to the *nations round about*, with whom the history of Israel was immediately to come in contact. The importance of the event, considered in this light, is hinted at, and foretold in Moses' song of praise (chap. xv. 14 sqq.):

"The people hear it, they are afraid,
Terror seizes the inhabitants of *Philistia*.
Amazed are the princes of *Edom*,
And trembling takes hold of the mighty in *Moab*,
All the inhabitants of *Canaan* despair.
Dread and fear fall upon them,
By the might of thine arm they are stiff as a stone,
Till thy people are through, O Jehovah,
Till the people, which thou hast purchased, are through."

And, lastly, there was a reference to the *Israelites*. Their faith was greatly strengthened thereby; it led them to place greater confidence in Jehovah, and to trust more in Moses, the servant of Jehovah (xiv. 31); and they also learned, and proclaimed in their song of praise, that Jehovah was "the true man of war" (xv. 3), "a king to all eternity" (xv. 18).

(2). On the *pillar of cloud* and *pillar of fire*, which went before the children of Israel when they departed from Etham, and accompanied them during the whole period of their journeying through the desert, consult *Camp. Vitringa*, Observ. ss. v. c. 14—17. Even this miracle has not been left, without some attempts to explain it as a natural phenomenon. In the opinion of *Herm. v. d. Hardt* (Ephemerid. philol. discurs. 6, p. 86, and 210 sqq., and also Ephemerid. illustr. p. 93 sqq.) the pillar of cloud and fire was the sacred fire of the Israelites, which had been preserved by the patriarchs both before and after the flood, from the time of the first sacrifice, which was consumed by fire from heaven. This fire, he says, was brought by Jacob into

Egypt; and when the Israelites departed from Egypt, it was carried by Aaron in front of the army, as a symbol of the presence of God. A similar custom is said by *Curtius* to have been observed by the Persian armies (iii. 3—9): "*Ordo agminis fuit talis. Ignis, quem ipsi sacrum et aeternum vocant, argenteis altaribus praeferebatur.*" *Toland's* opinion (in his *Tetradynami* Disc. 1) has met with more approval than *Hardt's*. He believed it to have been nothing but the regular caravan-fire, which was carried before the people at the top of a long pole, by the guide who was appointed for that purpose, as a signal of the route they were to take. That this custom was adopted in ancient times, not merely by the large trading caravans, but also by the armies of the East, especially when they were travelling along unknown and difficult routes, has been sufficiently attested by witnesses of earlier times. *Curtius* says with reference to the march of Alexander the Great (v. 2—7): "Tuba, cum castro movere vellet, signum dabat: cujus sonitus plerumque, tumultuantium fremitu exoriente, haud satis exaudiebatur. Ergo perticam, quae undique conspici posset, supra praetorium statuit, ex qua signum eminebat pariter omnibus conspicuum. *Observabatur ignis noctu, fumus interdiu.*" To the present day the same custom is adopted by trading caravans, according to the testimony of *Harmar* (Beobachtungen i. 438 seq.), the author of the Description de l'Egypte (viii. 128), and others. Even in the cities of the East, when there are any evening processions, iron fire-baskets, with pine-wood burning in them, are carried in front, on the top of a long pole (*Russegger* Reis. ii. 1, p. 38).—*Toland* still takes the trouble, to endeavour to bring everything contained in the scriptural narrative respecting the pillar of cloud and fire, into harmony with this custom. But the modern supporters of the same view (*e.g. Winer* ii. 696, *Ewald* ii. 164 sqq., and many others) for the most part admit without hesitation, that the author intended to relate a miracle, and treat his account as a mythical embellishment of the simple custom we have just described. *Köster* is the only one who admits the historical character of the biblical narrative, regards the whole transaction as suited to the circumstances and worthy of God, and yet adheres to this natural explanation. He says: "We have undoubtedly to understand by this pillar of cloud and fire the ordinary caravan-fire, used upon their march by the armies of the east. . . . . The cloud was a symbol of the presence of God, and in it Israel saw Jehovah face to face (Num. xiv. 14). It derived its worth entirely from the belief of the Israelites, that Jehovah was visibly among them in the cloud and in the fire; and therefore what we have here is not a mythical embellishment of a simple fact, but the simple fact itself, exalted by faith and a religious idea. Hence we are told with the greatest candour, that the guidance of Jehovah did not render careful reflection on the part of Moses himself, or the good advice of others, at all superfluous (Num. x. 31)." We should not make the slightest objection to this explanation of *Köster*, if there were any possibility of showing that it was in harmony with the text. But it must be apparent to every one that this is not the case. The only course open to us, in fact, is either to admit the historical character of the miracle, however incomprehensible it may be, with all the startling phenomena and effects attributed to it in the scriptural narrative, or else to pronounce everything that is supernatural in the account a mythical embellishment of later times. From the standpoint which we have adopted, in relation to the sacred history and its original records, it follows as a matter of course, that we adhere to the first of these.

Still we cannot but acknowledge, that in the pillar of cloud and of fire, in which Jehovah himself accompanied and conducted his people, there was some reference to the ordinary caravan-fire, which served as a guide as well as a signal of encampment and departure to the caravans and armies of the East. For, in the design and form of the two phenomena, we can trace exactly the same features; the difference being, that the one was a merely natural arrangement, which answered its purpose but very imperfectly, and was exceedingly insignificant in its character, whilst the other was a supernatural phenomenon, beyond all comparison more splendid and magnificent in its form, which not only served as a signal of encampment and departure, and led the way in an incomparably superior manner, but was also made to answer far greater and more glorious ends.

The following is our idea of the connexion to which we have alluded: As the armies and caravans of other nations required that a caravan-fire should be carried before them, whose ascending smoke by day and brilliant light by night could be seen by the whole procession; so did Israel stand in need of some such visible signal in its journey through the desert. But, whilst a caravan-fire carried at the head might suffice for a trading caravan of some hundreds or even thousands of persons, and for an army of some thousands or even tens of thousands of soldiers; no contrivance upon so small a scale could possibly have sufficed for two millions of men, with wives and children, besides a large quantity of cattle and a considerable amount of baggage. Even on their journey, a mere fire carried at the head of the procession would have been almost invisible to those behind, and it would have been entirely useless on the breaking up of the camp after a period of rest. For, whenever they rested for any considerable time, the different parties were obliged to scatter themselves far and wide for the purpose of seeking suitable spots on which to pitch their tents, and fertile oases in which to feed their flocks. Now, if Israel had had no other means of accomplishing all this, than such as are possessed by nomad tribes in general, it would have had to contend with just as many difficulties, hindrances, and dangers, as they are inevitably exposed to. But Israel was not to migrate like any other tribe. Jehovah had delivered them from Egyptian bondage, and led them out by His powerful arm; Jehovah had determined what their destination should be, and He himself would lead them thither. As He had already removed every obstacle in Egypt by signs and wonders; so would He remove them in the desert also. And as He afterwards caused water to flow from the rock to satisfy the thirsty people, and gave them quails and manna in rich abundance when they hungered; so, *instead* of the miserable caravan-fire, which would have been of very little use to such a procession, He gave them a more glorious and totally different signal to guide them through the desert, namely, a *pillar of cloud and fire*, which did not ascend from the earth, but came down from heaven. In this pillar He himself dwelt; it rose and fell according to circumstances; it sometimes spread itself out, and at other times was closely condensed; at one time it went before the procession; at another it hung suspended over it, and again it settled behind it, spreading impenetrable darkness on the one side, and lighting up the darkness of the night on the other. The ordinary caravan-fire bore the same relation to the pillar of cloud and fire, as the miserable tricks of the Egyptian Charthummim to the magnificent, all-embracing miracles, which Moses wrought in Egypt; or rather the difference is even greater and more apparent in the case of the pillar, than in that of the miracles in Egypt.

That we may form at the outset as clear a conception as possible of this wonderful phenomenon, we will collect together all the most important particulars in our possession, respecting its appearance and effects; but the fuller discussion of each of these we shall reserve, till we reach the period of its historical manifestation.

When we read in Ex. xiii. 21 that Jehovah went before Israel, by day in a pillar of cloud, and by night in a pillar of fire; we might be led to suppose, that there were two different pillars, the one appearing by day and the other taking its place in the night. But it is soon apparent that this was not the case. For in Ex. xiv. 19 and Num. ix. 21 only *one* pillar is mentioned, and in Ex. xiv. 24 it is called a pillar of cloud and fire (עַמּוּד אֵשׁ וְעָנָן). The cloud was undoubtedly the vehicle of the fire, and entirely lighted up by the fire, which caused it to shine with brilliant splendour on the dark background of the night; whilst it looked like a mere light cloud, in contrast with the brightness of the sunshine. We have no reason to depart from the literal meaning of אֵשׁ and עָנָן, and therefore regard the latter as an actual cloud, formed of the same material as every other cloud, and the former as actual fire, produced, it may be, so far as the natural cause is concerned, from electricity. The ordinary form, which it assumed, was that of a *pillar*, which moved forward at the head of the Israelites, and showed the way to the hindmost ranks of the whole procession (Ex. xiii. 21; Deut. i. 33). Still we must not imagine that this form was fixed and unchangeable. When the cloud descended between the army of the Egyptians and that of the Israelites, showing a bright light to the latter and making the darkness more intense to the former, "so that the one came not near the other the whole night," its form was no doubt that of a wall rather than a pillar. When the people were to encamp for the purpose of resting, the cloud descended; and when they were to set out again, it was taken up (Ex. xl. 35, sqq., Num. ix. 16, sqq.). So long as they rested, it remained suspended above the camp, at a later period above the tabernacle (Num. ix. 16). According to Ps. cv. 39,

Jehovah spread the cloud over Israel as a *protecting* (overshadowing) covering (*cf.* Wisdom x. 17, xviii. 3, xix. 7; 1 Cor. x. 1, 2; Is. iv. 5, 6),—a poetical description, which may, however, be literally in harmony with Num. x. 34, ("the cloud of Jehovah was *over* them by day, when they went out of the camp.")

The fire in the pillar was a symbol of the holiness of God, which moved before the sanctified people, both as a covering and a defence. It was the same fire which Moses had seen in the bramble (chap. iii. 2), and the same which afterwards came down upon Sinai, with thunder and lightning, and enveloped in a thick cloud (chap. xix. 16). Moses had covered his face before it, being afraid to look on God (chap. iii. 6). And so would Israel now have been obliged to cover its face, if the fire, which represented the presence of God, had appeared without a screen. But in that case it could not have fulfilled its purpose, of going before the Israelites to light their way. God, therefore, condescended to the weakness of his people; and from the very first He caused the fire to shine upon them through an appropriate medium. Hence, as the fire was a symbol of the *holiness* of God, the cloud in which it was enveloped was a symbol of his *mercy*. Nor was it merely a symbol, unattended by the thing which it signified. In and with the symbol was Jehovah himself, with his holiness and mercy (xiii. 21, xiv. 24), or the angel (xiv. 19) who represented him in the Old Testament, and foreshadowed his future incarnation (*cf.* Vol. i. § 50, 2). Thus in the pillar of cloud and fire there dwelt the *holiness of the Lord* (כְּבוֹד יְהוָֹה *cf.* Ex. xvi. 10, xl. 34; Num. xvi. 42). This relation between the symbol and that which it represented, was afterwards designated in Jewish theology the *Shechinah*. From it proceeded all the commands of Jehovah; from it he terrified the enemies of Israel (xiv. 24); from it he threatened the murmuring people (Ex. xvi. 10; Num. xvi. 42); and from it went forth the devouring flame, which slew the disobedient and rebellious (Lev. x. 2, Num. xvi. 35).

In the paragraph above, we have already expressed our opinion that the pillar of cloud and fire first appeared as the leader of the procession, when its course was altered at Etham. If we read attentively chap. xiii. 17—22, it is scarcely possible to come to any other conclusion. For if the pillar accompanied the procession when it set out from Raemses; we should naturally expect to find a statement to that effect, in the same passage in which an account is given of their setting out "*equipped,*" and taking Joseph's bones along with them, (viz. in vers. 18, 19). But there is no such statement; on the contrary they are said to have journeyed from Raemses to Succoth, and from Succoth to Etham; and then, for the first time, it is recorded that (henceforth) the pillar went before them. The internal grounds, on which we may explain the fact, that this miraculous guidance was neither granted nor required before, have been pointed out in the paragraph.

Stickel (Studien und Kritiken 1850, p. 390) is of our opinion, that the pillar of cloud and fire made its first appearance at Etham. But he sees in it nothing but an ordinary caravan-fire. In order that we may do justice to this rationalistic opinion, we will quote the subtle arguments adduced by the critic. "A signal of smoke, which this caravan-fire was evidently intended to be, would have failed of its object, so long as the procession was moving along the fertile and well known road, *when columns of smoke were rising on every side.* But in Etham, at the end of the desert, where it entered upon the open southern steppe, such a precaution was not only judicious but necessary." On chap. xiv. 19, 20, he says: "The arrangement of Moses with reference to the position of the pillar of cloud (namely, that the caravan-fire should no longer be carried *before*, but *behind* the procession) fills us with great admiration of his inventive mind, by which all the necessities of the moment were fully grasped. When the passage through the sea was about to commence, he had the signal fire, which had hitherto been carried at the head, transferred to the rear, and placed at the north or north-west of the camp, between the Egyptians and the Israelites. The result of this arrangement was, that the east wind, which was blowing, necessarily caused a *dense cloud of smoke to* pass between the two, whereas the same cloud would have blown in the faces of the Israelites, if the fire had been carried before them through the sea. By this means, and because the pillar ceased to move forward, the departure of the Israelites was concealed from the Egyptians; whilst at the same time the light of the fire, which shone towards the east upon the surface of the sea, enabled those who were passing through, to distinguish between the water and the dry land. But when at length the pillar of fire left the spot, the Egyptians saw in the vacant ground the proof that their designs were frustrated ; and their rage and eagerness to plunge forward into the dangerous road through the sea were truly human, and can easily be understood."

Stickel's treatise, from which we have taken these passages, has filled us with sincere admiration of the diligence and learning, the grasp and acuteness of mind displayed by the author; for it is truly a perfect master-piece of searching and thorough investigation, in a difficult and untrodden domain. But all the greater is our astonishment, that, in spite of the powers of mind which he has displayed in connexion with this subject, he should so strikingly have failed in his attempt to reduce the pretended mythical elements in the account before us to their historical foundation; and that he could first of all bring *himself*, and then require *others*, to look upon such an expedient, as he here imputes to Moses, as something extremely clever, adapted to its end, and worthy of the highest admiration. Truly it requires such a faith as "is not given to every man," a faith in comparison with which it is a very small matter to believe in the miracle which was wrought by God, if we are to believe in all that *Stickel* tells us, with regard to the wondrous effects produced by a caravan-fire. Just imagine a procession, composed of two million human beings, with an immense quantity of cattle, and behind them an insignificant caravan-fire, so insignificant that the smoke which ascended from it could not possibly have been distinguished from that which rises from the first good chimney you may meet with,—could any one but a critic, who takes fright at a miracle, possibly believe that the light from such a fire shone over the heads of two millions of men, and (at a low estimate) two million head of cattle, and then "lighted up the surface of the sea, so that those who were passing through were able to distinguish the water from the dry ground?" Who, again, can possibly believe that a caravan-fire produced "a dark cloud of smoke," of such a volume that it stood like a wall between Pharaoh's chariots and the army of Israel; a pillar of smoke, we repeat, which cannot have been larger than that which ordinarily rises from a single chimney?

Criticism like this can only expect *its own disciples*, to have faith enough to believe such things as these.

(4). According to the biblical account, the dividing of the sea was entirely effected by "*a strong east wind.*" Nothing is said about its being favoured by the *ebbing of the tide.* Still it is not improbable, that both these powers of nature may have been associated, in the accomplishment of this stupendous miracle. At all events the narrative makes no mention of the ebb, which may have contributed to some extent. This was but a subordinate auxiliary and every day occurrence, and therefore it lays the whole stress upon the *east wind*, as the instrument and messenger of the miraculous power of Jehovah. It was not the ebb, but the east wind, which rendered the opening of a dry pathway through the sea an extraordinary, unheard of, and miraculous event. It may also have been the case, that at the spot at which the passage took place, the bottom of the sea was raised by sand-banks, and was therefore higher at this point than in any other; but if the scriptural account is to be relied upon as true, this can only be regarded as a thoroughly subordinate and auxiliary feature, which it did not come within the province of the author to mention, seeing that the miracle was a miracle still. Moreover, if the returning waters entombed the whole of Pharaoh's army, without the least possibility of escape, the place cannot have been what is ordinarily termed a ford. And lastly, the passage may certainly have taken place at one of the narrowest parts of the gulf. But if Pharaoh's 600 war-chariots, with a proportionate number of horsemen, were in the midst of the sea when the waves returned, and though they turned back in the greatest haste, were unable to reach the Egyptian shore, the breadth cannot have been so very inconsiderable.

There is a diversity of opinion as to the direction of the *wind*, by which the miracle was effected. In the *Septuagint* we have again the rendering νότος (*cf.* § 31. 2). Modern commentators maintain that the expression רוּחַ קָדִים was currently employed to denote any "*strong*" wind, so that a west wind or a north wind would have been called by the same name. Hence it has for the most part been assumed, that in the present instance it was a north wind, this being regarded as the most suitable wind to dry up a ford at the northern end of the gulf. But there is no foundation for such an opinion. The words, as they stand, can only refer to a wind from the east. But as the author did not select his terms with mathematical precision, according to the points of the compass; there does not appear to be any objection to the supposition, that the wind blew from the north-east or south-east. The latter is the more probable of the two. For רוח קדים literally means a wind which blew from קֶדֶם, and in biblical phraseology קֶדֶם is generally suggestive of Arabia. Such a wind would drive the water away from the point in question, towards the northern end of the gulf, which to all appearance formerly extended much farther northwards than it does now (*cf.* § 39. 1). But under any circumstances so much at least must be firmly maintained, that it was not an ordinary wind, but one which was made to blow with unwonted violence by the omnipotence of God, and which therefore sufficed to produce phenomena, such as no other wind, however strong, could possibly effect. But inasmuch as the writer himself gives prominence to the fact that it was a *strong* wind, and that it blew *the whole night*, an expositor is justified in laying stress upon the power of nature, which served as the medium, as well as upon the peculiarly miraculous power. And in doing this it is important to remember that a very small force, if it be regular, uninterrupted, and long-continued, can produce stupendous and almost incredible results. A suspension bridge, for example, which scarcely moves beneath the tread of persons walking irregularly, is thrown into the most dangerous oscillations by a regiment of soldiers keeping step.

When it is stated that the water stood firm, like *walls on the right hand and on the left*, the figurative character of the expression must not be so far overlooked, as that we should think ourselves obliged to assume that the water really formed a *perpendicular* wall on both sides. But we must also not refuse to admit, that the meaning is, that the water was *forced back* on *both sides*, and kept back by the uninterrupted blowing of the wind; and yet was as surely prevented from flowing together again, as if there had been walls erected between.

In conclusion, we may mention the passage, which *Eusebius* (praep. ev. 9, 27) quotes from *Artapanus*, according to which the inhabitants of Memphis maintained that Moses, who knew the ground most thoroughly, took advantage of the ebb, to lead the people through the bed of the sea while it was dry, after they had borrowed many costly vessels and clothes from the Egyptians. This can hardly be supposed to be an ancient Egyptian tradition, but must be one of modern date, originating in the Grecian period, and is nothing more than the biblical account interpreted to suit the interests of Egypt, by those who were acquainted with the *Septuagint* translation. We can hardly attribute any greater importance to the statement made by *Diodorus Siculus* (iii. 39), to the effect that among the Ichthyophagi, the inhabitants of the district in question, there was a legend current, that the bottom of the gulf had once been entirely exposed by an extraordinary ebb (*μεταπεσούσης τῆς θαλάττης εἰς τἀναντία μέρη*) ; but that as soon as the bottom of the deep was visible, the flood suddenly set in, and the sea returned to its former condition.

(5). The strength of Pharaoh's army consisted chiefly in his *war-chariots*. *Hengstenberg* (Egypt and Moses, p. 126 sqq.) has shown, how strongly this account is supported by information derived from the monuments, respecting the customs of ancient Egypt. The chariots on the Egyptian monuments are drawn by two horses, and generally hold *one* driver and *one* warrior. Frequently, however, three men may be seen in one chariot (this is the rule on the Assyrian monuments), and in that case the third is an armour-bearer. This custom may serve to explain the use of the Hebrew word שָׁלִישִׁם to denote an armour-bearer (Ex. xiv. 7, *cf. Gesenii thesaurus* p. 1429). We may be surprised to find that the number of chariots, with which Pharaoh pursued the Israelites, was so small, viz., six hundred *picked* chariots (ver. 7), when *Diodorus* states that Sesostris possessed 27,000 war chariots. But we must not overlook the accompanying clause וְכֹל רֶכֶב מִצְרַיִם, which *Luther* has very correctly rendered : *and whatever other chariots there were in Egypt.* Pharaoh hastily gathered together all the available chariots that could be procured, and did not wait till the entire force could be brought from the most distant military stations. The six hundred " picked " chariots probably belonged to his body guard. In addition to the charioteers, he was also attended by a proportionate number of horsemen (פָּרָשִׁים *cf.* xiv. 9, 23; xv. 1). It is true that *Hengstenberg* takes a great deal of trouble to prove that there is no reference to horsemen in the text, by forcing upon פרשים the meaning: chariot-soldiers. He agrees with *Champollion* in doubting whether the military force of the ancient Egyptians included any cavalry at all, seeing that there is no representation of any on the monuments. But *Wilkinson* has shown, that the command of cavalry is spoken of in the hieroglyphics as a very distinguished post, and that *Diodorus* describes the army of Sesostris as consisting of 600,000 infantry, 24,000 cavalry, and 27,000 war-chariots. It is true that, afterwards (viz. in chap. xiv. 9, 23, and xv. 1), only horses, chariots, and riders are named; but the riders were actually riders (פרשים), not chariot soldiers (שלישים). The latter, as a matter of course, are included in the term "chariots," especially when the word is accompanied with the express statement contained in ver. 7, that *all* the chariots were manned with chariot-soldiers (שלישים). *Hengstenberg's* anxiety lest "an objection, by no means inconsiderable, should be raised against the credibility of the narrative," in consequence of our regarding the riders as *horsemen* and nothing else, is in our opinion entirely uncalled for. Nothing is said concerning any infantry, and there is sufficient reason for this in the fact, that the object of the king was to pursue and overtake the Israelites as quickly as possible. *Josephus*, indeed, on his own authority, adds 50,000 cavalry and 200,000 infantry to the six hundred chariots mentioned in the Bible (Ant. ii. 15, 3), and the Jewish tragedian, *Ezekiel* (in *Eusebius*, praep. evang. 9. 29) makes Pharaoh set out with an army of a million men.

(6). The *anthem*, which Moses here composed in the name of the whole nation, was, as it were, the *nuptial song* of Israel (*cf.* Jer. ii. 2). Jehovah had rescued his chosen bride from the hands of her oppressors, and was about to lead her to the marriage altar at Sinai. With her deliverance from bondage still fresh in her memory, and looking forward with a longing heart to her approaching marriage (ver. 17), she uttered her feelings of joy in a song of praise. There is not much weight in the objections, which have been made by certain critics, to the authenticity of this song. The weakest of all, and utterly unworthy of refutation, is *De Wette's* remark (Krit. d. Isr. Gesch. p. 216) that the anthem is too long for an *impromptu*. Others have found evident marks of a later age in vers. 14—17. But though we have there a description of the fear and amazement into which the Philistines, the Moabites, the Edomites, and the Canaanites, have been thrown by the tidings of this miracle, it is not even necessary to appeal to the prophetic character of Moses in order to account for this. Without the gift of prophecy, it was possible to foresee with certainty that these nations would be alarmed, when they heard the report of the mighty acts of the God of Israel; for it was but natural that they should be anxious, lest Israel's approaching march might disturb them in the possession of their land, and that they should also feel that they could not hope to do much to resist the power of such a God, who had broken the pride of the haughty Egypt; and therefore Israel could assume it as a fact. There is more weight in the argument founded upon ver. 17:

> "Bring them in, and plant them in the mountain of thine inheritance,
> In the place, which thou hast made to dwell in, O Jehovah,
> In the sanctuary, O Adonai, which thy hands have established."

But that Israel's hopes should stretch beyond the desert, after so glorious a deliverance from Egypt; that they should look forward with certainty to the possession of the land, which had been promised them (Gen. xv. 16; Ex. iii. 17); that they should feel that a sanctuary would be there required, and that a settled spot must already have been selected for it by Jehovah—all this was so natural, that no reasonable critic can possibly take offence at it. Nor is there anything to object to in the fact that the song assumed it as indisputably certain, that Jehovah had already chosen a high and stately mountain in the promised land, as the place where his sanctuary should be erected. For a mountain was the most natural and appropriate spot for the offering of sacrifice and prayer. Abraham had to take a three days' journey, that he might offer the most important sacrifice of his life upon a *mountain* (Gen. xxii.); Moses had been told that the sacrifice connected with the conclusion of the covenant was to be offered upon a *mountain* (Ex. iii. 12); what else, then, could Moses and Israel expect, than that in the promised land the place of worship would still be a *mountain?*

Miriam, who appears as the leader of the daughters of Israel on this occasion, is described as the sister of Aaron and a prophetess. There is nothing accidental or unmeaning in either of these notices. At the very outset the position is indicated, which she afterwards occupied in the community of Israel. She is called the sister of Aaron, and not the sister of Moses, because her position was co-ordinate with that of Aaron, but subordinate to that of Moses. Although Aaron was the brother of Moses, yet in his official position he was only the mouth, the prophet of Moses, and Moses was Aaron's God (§ 20. 8). And in the same way, although Miriam had been the saviour and protector of Moses in his youth (Ex. ii. 4 sqq.), she was placed in a position of subordination to the brother she had saved; for Jehovah had chosen him to be the mediator of his covenant, and placed him at the head of Israel. Hence, she entirely mistook her position, when at a later period (Num. xii.), she took upon herself to command and rebuke him.

(7). According to Jewish tradition, the passage through the sea and the song of Moses belong to the seventh day after the celebration of the passover in Egypt. We have no decisive evidence to the contrary; at the same time it cannot be positively established from the original narrative. In chap. xii. 39, however, it is clearly intimated that the first days of the journey fell within the limits of the feast of the passover. This feast was a feast of deliverance from Egypt, and the deliverance was not complete till they had passed through the Red Sea. There is nothing improbable in the supposition, that the appointment of seven days, for the subsequent commemoration of this deliverance (xii. 19), had a historical foundation in addition to the sacred character of the number seven. This will be apparent at once, if we consider how frequently the ideal element contained in prophecy and revelation corresponds in a most striking manner to the accidental, historical element, observable in the particular events connected with the development of the sacred history. At the same time no peculiar importance is attached to the latter; and therefore they are only important, as they produce in the thoughtful observer of the movements of God in history, a salutary consciousness of the perfect symmetry and harmony, which exist even in the most trivial and casual occurrences.

This Jewish tradition would necessarily fall to the ground at once, if the opinion held by most commentators were correct, that only *three days* intervened between the departure of the Israelites on the night of the passover, and their encampment by the sea at Baal-Zephon, since only three places of encampment are named (Succoth, Etham, and Baal-Zephon). But it has already been repeatedly shown (*Raumer* Beitr. p. 2 sqq., *Lengerke,* Kenaan i. 432), that the word מַסַּע did not denote a day's journey, but a station or place of encampment, where the tents were set up and every preparation was made for a longer period of rest than usual. The day's journeys are called יָמִים. We may see from Num. xxxiii. 8, how great the difference was between a day's journey and a station. According to this passage, the Israelites, after passing through the Red Sea, went three days' journeys (יָמִים) through the desert of Etham, and then encamped in Marah. Here, then, there was evidently a journey of *three* days between two stations. So also do we read in Num. x. 33: " And they deparated from the Mount of the Lord three days' journey, and the ark of the covenant of the Lord went before them in the three days' journey, to show them a resting-place" (מְנוּחָה not מַסַּע a place of encampment). And even supposing (though we do not grant it), that the place of encampment and place of rest were identical, it is in itself a very improbable thing, that the Israelites only spent three days in their journey from Raemses to their place of encampment at Baal-Zephon by the Red Sea. Even if they only *travelled* three days, it would certainly be necessary to assume, as *Tischendorf* does (de Israel. transitu, p. 23), that there were periods of rest of longer duration, *i.e.*, actual days of rest between the three marching days. Just fancy two million men, with large herds of cattle, and all the baggage of emigrants, with their wives, children, and old men, obliged to start in the most hurried way (chap. xii. 33)! What confusion, what difficulties would inevitably impede them during the first days of their journey! An ordinary caravan may travel fifteen or twenty miles a day; but such a procession would hardly be able to do the half of this. Let it be remembered, too, that fresh parties were constantly joining them, and that this must have caused some disturbance and delay. (We cannot imagine it possible that two millions of Israelites, whose residences were scattered over the whole of the land of Goshen, should all have met together in Raemses, many of them merely to retrace their steps: moreover, if we consider that they were ordered to eat the passover at the early part of the night in *their own* houses, and not to leave their houses till the morning (chap. xii. 22), we shall see that it must have been actually impossible for them all to meet in Raemses on the next morning, many of them from the most distant parts of Goshen. Raemses was the capital of the province. There, no doubt, Moses and Aaron were residing. The procession started thence; and after the main body had set out, smaller parties came from all directions as speedily as possible, and joined it at the point of the road nearest to their own dwellings). The following considerations also serve to show, that the Israelites must necessarily have spent more than three days, on their march from Raemses to their encampment by the sea. It is true that the site of Raemses is not precisely known. But it is certain that it must have stood somewhere in the immediate neighbourhood of the king's palace; sufficiently near, at all events, for a communication to pass from one city to the other in a very few hours. Now, whether we suppose the palace to have been in Heliopolis, Bubastis, or Zoan (and we have certainly only these three points to choose from, *cf.* § 41. 2), the shortest route from Raemses to the sea, taking into account the circuitous way by which the Israelites went (chap. xiv. 2), would be so long that it would be necessary to travel seventeen or twenty miles a day in order to accomplish the whole in three days. Others may believe it if they please; but I cannot believe that such a procession as we have described, could keep up a journey of seventeen or twenty miles a day for three days running. Again, we find from chap. xiv. 5, that information was brought to the king that the Israelites had turned round at Etham, and entered the Egyptian desert on the west of the Red Sea. This message must have been sent to the king from Etham itself, and of course it was not sent till after the Israelites had changed their course in the manner described. Now the Israelites had already occupied *two* days at least in going in a straight course from Raemses to Etham; and the king's palace was certainly farther from Etham than Raemses was. Hence the messenger, who was sent from Etham, may be safely supposed to have taken *one* day in reaching the king; and Pharaoh's chariots and army (even if, to please our opponents, we assume the possibility of that which is certainly impossible) must also have required at least *one* day (!) to travel by a forced march from Heliopolis, Bubastis, or Zoan, to the neighbourhood of Baal-Zephon on the sea-coast. Thus, even granting the correctness of our opponents' premises, at least four days must have intervened between the departure from Raemses and the passage through the sea. This will show how little foundation there is for the assertion, that " the longest space of time allowed by the biblical narrative for these events is three days."

GEOGRAPHICAL INTRODUCTION TO THE EXODUS.

LITERATURE: *J. Clericus*, Diss. de Maris Idumaei trajectione, ad calcem comment. in Exod.—*S. Deyling*, Observv. ss. P. iii. p. 45 sqq. and P. v. p. 31 sqq.—*A. Calmet*, Biblical researches, with notes by *Mosheim*, ii. 56 sqq.—*Du Bois Aymé*, Description de l'Egypte, T. viii. sur le séjour des Hébreux en Egypte. —*K. v. Raumer*, der Zug. der Israeliten aus Aegypten nach Kanaan. Leipzig, 1837; Beiträge zur bibl. Geogr. Lpz. 1843, p. 1—5; and Palästina Ed. 3 Lpz. 1850, p. 437—442.—*J. V. Kutscheit*, H. Prof. Dr Lepsius u. d. Sinai. Berlin, 1846.—*Const. Tischendorf*, de Israelitarum per mare rubrum transitu Lps. 1847; and Reise in d. Orient, i. 174 sqq.—*J. G. Stickel*, der Israeliten Auszug aus Aeg. bis zum rothen Meere, in the Studien und Kritiken 1850, ii. p. 328—398.—*Robinson*, Palestine i. 74 sqq.—*Hengstenberg*, Egypt and the Books of Moses, p. 55—60 transl.—*L. de. Laborde*, commentaire géographique sur l'Exode et les Nombres. Paris and Leipzig 1841, p. 75 sqq.—*C. v. Lengerke*, Kenaan i. 430 sqq.—*Ewald*, Geschichte ii. 52 sqq.

§ 37. The district, which forms the subject of our present remarks, is bounded on the *south* by the so-called Valley of Error; on the *west* and *north-west* by the Nile and its Tanitic or Bubastic arm; on the *north* by Lake Menzaleh; and on the *east* by a line drawn from the southern point of Lake Menzaleh, and through the Birket-Temseh (the so-called Crocodile Lake), the Bitter Lakes and the Heroopolitan gulf. We will commence with a minute description of the *southern boundary*. The *Valley of Error* (Wady et-Tih) runs due east from the village of *Besatin* on the Nile to the Red Sea, and terminates in the broad plain of Baideah on the coast. The name Wady et-Tih is frequently restricted to the western end of the valley; further east it is then called the *Wady er-Ramlijeh* and still nearer to the sea the *Wady et-Tawârik*. The whole valley, from the Nile to the sea, is shut in on both sides, viz., on the north and south, by high mountain-ranges. Of these the northern range deserves a closer investigation. It rises from the valley of the Nile, not far from Cairo, stretches in a straight line towards the east, and terminates at the Isthmus of Suez in the promontory of (Ras-) Atâkah. But near the centre the range is entirely broken. About twenty-three miles from Besatin, not far from the *fountain of Gandelhi* (the only drinkable water in the whole Wady), another valley branches off from the Wady et-Tih. This valley runs in a north-easterly direction through the northern range of mountains to the north of the gulf. The western half of the range is called *Jebel Mokattem*, the eastern *Jebel Atâkah*. At the present day caravans sometimes travel from Cairo to Suez through the Wady et-Tih, but they naturally turn into the north-eastern valley at the fountain of Gandelhi. This road is now called the *Derb el-Besatin*. In ancient times it was, no doubt, the regular road from Memphis to Klysma (or Suez) and thence into Asia.

Let us now turn to the *western boundary*, and follow it from the village of Besatin along the Nile and its two eastern arms to the Lake Menzaleh, into which the latter empty themselves. On the Nile itself we first of all arrive at the ancient *Latopolis* or Babylon (ancient Cairo); a little farther north is *Cairo*, and somewhat to the east of this, at a distance from the Nile, the old city of *On* or *Heliopolis.* Within the limits of this district the Pelusiac arm branches off from the main stream, and about thirty miles further north the *Bubastic arm.* The latter flows into Lake Menzaleh on the western side, and not far off, on the southern side, the *Tanitic arm.* Near the mouth of the latter stood ancient *Zoan* or *Tanis*, and further to the south on the Bubastic arm, near the point at which it leaves the main stream, *Bubastis* (now Pi Beset). The *Pelusiac* arm flows into the bay at the south of Lake Menzaleh. On the fertile strip of land which fringes this arm on the eastern side stand, or stood, the important cities of *Belbeis* (Raemses ?) and *Abasieh* (Pithom or Tum), both towards the south. The *northern boundary* is formed by the southern side of Lake Menzaleh.

§ 38. The *eastern boundary* causes the greatest difficulty, when we attempt to form a precise conjecture as to its condition in the time of Moses. For the moveable sand has been driven about by violent winds, and has evidently made considerable changes in the face of the country during the four thousand years, which have intervened between our days and the age of Moses, and the ground has not yet been surveyed with sufficient care, to enable us to determine with certainty of what nature these changes have been. At present the principle features are the following : The breadth of the *isthmus*, measured in a straight line from the southern point of Lake Menzaleh to the northern extremity of the Gulf of Suez is about eleven geographical miles. From the isthmus to *Ras Atâkah* (§ 37. 1) there is a road, about eighteen miles long, on the western coast of the *gulf.* To the south of this promontory the *Wady et-Tawarik* opens into the plain of *Baideah,* and to the north there is a narrow pass, which widens at Suez into a large, barren plain. At this point, too, a tongue of land runs into the gulf to such a distance, that, acccording to *Niebuhr*, it is not more than 3450 feet or two-thirds of a mile across. To the south of Suez the coast describes a sharp curve, and runs so far to the west, that at a very short distance off the sea is three or four miles broad; this breadth is maintained as far as Ras Atâkah, to the south of which it becomes considerably greater. And even to the north of Suez the sea is broader. At this point stood ancient *Klysma*, a harbour in former times. At present it is buried in the sand, but the site is still undoubtedly marked by the ruins at *Tel el-Kolzum*. To the north of el-Kolzum the gulf contracts again, and still runs northwards for a considerable distance, terminating in a narrow strip of water from 1000 to 1500 paces broad. At Suez, where the gulf is narrowest, there are sandbanks, which stretch from the eastern to the western shore, and when the ebb is strong these are to some extent exposed, whilst the water which covers the rest is so shallow, that it can easily be waded through. On the other hand, when the flood is strong, the water is as much as seven feet deep.[1] At Suez and round the northern part of the gulf there are "evident traces of a gradual filling up of this part of the Red Sea," (*Robinson* i. 71). Around the head of the inlet, there are also obvious indications, that the water once extended much farther north, and probably spread itself out over a wide tract towards the north-east. The ground bears every mark of being still occasionally overflowed, (*Robinson* i. 71).

§ 39. From the head of the gulf, running towards the north, traces are still visible of the *old canal* which was cut for the purpose of connecting the Nile with the Red Sea, and was very frequently renewed. It ran due east from the Nile through the Wady Tumilat, crossed the dam of Arbek to reach the

[1] In the year 1799, when Napoleon was returning from Ayin Mousa on the eastern shore, he attempted to cross the ford. "It was already late and grew dark; the tide rose and flowed with greater rapidity than had been expected; so that the General and his suite were exposed to the greatest danger, although they had guides well acquainted with the ground." (*Robinson* i. 85).

Bitter Lakes, and finally passed through the dam of Ajrud. The *Bitter Lakes* are formed by a depression of the soil, to a depth of forty or fifty feet below the level of the Red Sea. They were once eight or ten miles broad, but at the present day are nearly dry, there being only a few shallow pools of salt water, and occasionally patches of marshy ground Their length, reckoning from the north east, has been variously stated. According to *Seetzen*, the distance from Arbek, at the north western end of the Bitter Lakes, to Suez, is only about twenty miles; whereas *Du Bois Aymé* states that the large basin of the Bitter Lakes terminates at a point about forty miles to the north of Suez. *Stickel* (Studien und Kritiken 1850, p. 367 seq.) reconciles this discrepancy between two trustworthy travellers, by assuming that on the eastern side there was a narrow tongue, running up from the basin which is ten miles broad, and reaching much farther north than Arbek. Such a supposition is not at all at variance with the fact, "that on the western side, along which *Seetzen* travelled, the lake terminated at a point much farther south. In this case the northern border of the Bitter Lakes must have described a curve, from the south-west to the north-east." On the side towards the gulf a broad, sandy strip of land, which is only about three feet higher than the surface of the gulf, prevents the confluence of the waters. On the south-eastern slope of this strip of land stands the present *fortress of Ajrud.* The basin of the Bitter Lakes is separated from the district washed by the Nile, and from the Crocodile Lake, by a similar but much greater elevation of the soil, on the western slope of which *Arbek* is situated *(cf. Stickel* p. 366). There are many facts, which afford the strongest evidence, that the gulf of Suez once stretched as far as this dam, and therefore that the basin of the Bitter Lakes formed the most northerly part of the gulf (1). The isthmus between the Crocodile Lake and Lake Menzaleh is about fifteen miles broad.

(1). *Du Bois Aymé* enters into a thorough investigation of the *ancient limits of the Red Sea*, in his Description de l'Egypte T. xi. 371, sqq.; *cf. Rosenmüller's* Altherthumskunde iii. 263, and *Stickel* p. 369 sqq. Travellers are all agreed that the strip of desert, which fringes the northern end of the gulf, bears the most unmistakeable marks of having once formed the bottom of the sea. The neighbourhood of the Bitter Lakes has very seldom been visited by travellers. But *Du Bois Aymé*, who went through the basin several times, says (according to *Rosenmüller's* Altherthumskunde iii. 263): "This basin has the appearance of having once been covered by the sea. Strata of sea-salt are still found there, and sometimes they assume the form of caverns. In such places the earth resounded under our feet. There were also small fissures, and at a depth of four or five *metres* we found water, which tasted like sea-water. The ground is generally marshy, with pools of salt-water. In the sandy spots, after digging down twelve or fifteen *decimetres* at the most, salt-water is found beneath a stratum of clay and loam. The ground is covered with shells, and is much lower than the surface of the Red Sea, from which it is divided by a sand-bank, the height of which is seldom more than a *metre* above the water of the Arabian Gulf. Lastly, along the hills surrounding this basin we can trace a line formed by the remains of marine vegetation, exactly resembling the line, which the flood-tide leaves upon the shore; and what is very remarkable, this line is exactly of the same height as the high-water mark of the gulf."

We might, indeed, be led to suppose that the basin of the Bitter Lakes was first of all filled by the water of the Red Sea, in consequence of the dam at Ajrud having been cut through for the purpose of forming a canal from the Sea to the Nile. But *Stickel* has adduced historical testimony to disprove this opinion (p. 372 sqq.). First of all, *Strabo* states (xvii. 1, 25, 26) that in consequence of the cutting of a canal from Egypt to the Bitter Lakes, the water of the lakes, *which had previously been bitter*, was changed through the admixture of the water from the Nile. Now, this canal was cut before the one from the Bitter Lakes to the Arabian Gulf, and therefore the lakes must have been connected with the Red Sea, before they were joined by a canal. And as *Stickel* observes (p. 373), the proximity of the northern boundary of the Red Sea to the Nile (the waters of which, during the inundations, flow through the Wady Tumilat to the Arbek dam) affords the only explanation of the fact, that Sesostris undertook to connect the two by means of a canal. But in the course of centuries the sea retired, in consequence of the accumulation of sand, and hence the same operation was repeated at different periods by Necho, Darius, and Ptolemy II., the excavation being always made from the north towards the south, so that the last piece was the most southerly of all. And in our opinion this piece was nothing more than the piercing of a sand-bank, which had gradually accumulated between the basin of the Bitter Lakes and the present extremity of the gulf. While this bank, now the *Isthmus of Ajrud*, was gradually accumulating, it must have rendered it difficult for ships to pass into the deeper water towards the north, and hence the necessity for building Klysma. Moreover, when this bank at length reached the surface, it must have formed *a ford across the sea*, which was dry at the ebb-tide, but covered with water at the flood. There was thus a dry road from Africa into Arabia between two basins filled with water, similar to that which may still be seen to the south at Suez. The same views have been expressed by *Du Bois Aymé* (sur le séjour des Hébreux, in his *Description* viii. 114 sqq. *cf. Rosenmüller* iii. 264 sqq.).

If we assume that the Bitter Lakes formed part of the gulf of Suez in the time of the Ptolemies, this will throw light upon many passages in the works of ancient writers with reference to distances, sites, &c., which would be otherwise inexplicable, such, for example, as the statement of Ptolemy that the city of *Klysma* was six miles to the south of the northern extremity of the Arabian Gulf, and that *Heroopolis* (which is identical with Abu-Keishid in the Wady Tumilat, *cf.* § 40. 1) was only two geographical miles from the same point. The same may be said of *Strabo's* statement, that the road from Heroopolis to the extreme point of the gulf formed an angle with the gulf (xvi. 4, 2, 5). Moreover, unless this city was formerly much nearer to the gulf than it is now, it is impossible to explain the origin of the name Heroopolitan gulf.

§ 40. If we turn now to the interior of this tract of land, the boundaries of which we have just described, we find it divided into two halves, a northern and a southern half, by the *Wady Tumilat.* This Wady commences at Abasieh (the ancient Pithom), in the lowlands of the Nile, and stretches eastwards in a straight line as far as the downs, which divide the Bitter Lakes from the Crocodile Lake, and from the Nile when it overflows (§ 39). To the south of this Wady, which is broad, well watered, and therefore fertile and well adapted for cultivation, lies the *Egyptian desert*, which is bounded on the other sides, by the lowlands of the Nile on the west, the Valley of Error on the south, and the Red Sea and Bitter Lakes on the east. The fertile district to the west and north of this desert, reaches as far as the Bubastic arm of the Nile and Lake Menzaleh, and forms at the present time the province of *es-Sharkiyeh.* In the days of Moses, it was called the *land of Goshen.* On the western slope of the desert stands the town of Belbeis; at the north-western corner (by the entrance to the Wady Tumilat) *Abasieh*, and near the eastern end of the Wady are the ruins of *Abu Keishid.* Modern researches have shown that *Abu Keishid* is probably identical with the ancient city of *Heroopolis*, and *Abasieh* with *Pithom* (1). There is more uncertainty about the question, whether Belbeis is to be identified with any known city of antiquity, and if so, with which? (*cf.* § 41).

(1). The identity of *Heroopolis* with Abu-Keischid, and ot Abasieh with *Pithom*, was first proved by the French expedition to Egypt (*cf. Hengstenberg*, Egypt and the Books of Moses, p. 42 sqq.). *Lepsius* has taken the lead in objecting to this conclusion (Chronol. i. 345 sqq.). He endeavours to prove that Heroopolis is rather to be associated with the ruins of el-Mukfâr, which lie farther west. The question is one of no moment to us, and therefore we need not enter upon the discussion here.

§ 41. In order that we may trace the road, by which the Israelites travelled, it is necessary first of all to search for the point from which they started. This is everywhere said to have been *Raemses.* As the land of Goshen is also frequently called the *land of Raemses (cf.* § 1. 5), the name Raemses is sometimes supposed to have been used, not merely as the name of the capital, but as that of the province also. And this has led *K. v. Raumer, L. de Laborde,* and others, to conclude, that the Raemses from which the Israelites set out was the province, and not the capital. But this explanation may easily be shown to be inadmissible. If Raemses is spoken of in other places as a city (and this has never been disputed yet), then the term אֶרֶץ רַעְמְסֵס (Gen. xlvii. 11) can only be regarded as meaning the land *of* Raemses, *i.e.*, the land of which Raemses was the capital. Moreover, if the first places of encampment, Succoth and Etham (Ex. xii. 37), were towns, there can be no doubt that Raemses, the place from which they set out, was also a city, and not a province. But it is not so easy to decide where *Raemses* was situated. *Hengstenberg*, *Robinson*, and others, identify it with *Heroopolis* in the Wady Tumilat (§ 40). But the difficulties in the way of this assumption are so numerous, and of so serious a character, that we must decidedly reject it as erroneous (1). *Stickel*, on the other hand, endeavours to prove that modern *Belbeis* stands on the site of the ancient Raemses (2). And the arguments, which he has adduced in support of this opinion, are sufficiently weighty to convince us, that no other place, with which we are acquainted, has such strong claims to be regarded as the representative of that ancient city.

(1). *Hengstenberg* (Egypt and the Books of Moses, p. 49 sqq. transl.) adduces the following argument in support of the *identity* of Heroopolis and Raemses. In Gen. xlvi. 28, where the messengers whom Joseph sent to meet his father in "the land of Goshen" are spoken of, the *Septuagint* paraphrases the passage thus: καθ' Ἡρώων πόλιν εἰς γῆν Ῥαμεσσῆ, and in the same way the clause in the next verse, where Joseph himself is described as going "to Goshen" to meet his father, is rendered καθ' Ἡρώων πόλιν. Now *Hengstenberg* is undoubtedly correct in maintaining, that the clause in the *Septuagint* cannot be regarded as a mere arbitrary conceit. And if his opinion could be substantiated,—that the *Seventy* substituted the name Heroopolis, which was current in their day, for the name Raemses, which had then become antiquated,—we could certainly give them credit for sufficient acquaintance with the antiquities of their own country, to take their word for the fact, that Heroopolis and Raemses were the same. But this is not the actual state of the case. There is no ground for speaking of any such *substitution*, simply because the name Raemses does not occur in the Hebrew text. The clause, *καθ' Ἡρώων πόλιν*, is an explanatory *addition* and nothing more. The statement in the text, that Joseph sent to meet his father in *the land of Goshen* (in Gen xlvii. 11, both the *Septuagint* and the Hebrew text have "the land of Raemses"), and that he afterwards came thither himself, appeared to them to be too indefinite. They therefore thought it desirable to introduce a more precise account of the exact spot, from their own knowledge of the country. The information which we have gained since the French expedition, with reference to the site of ancient Heroopolis, fully establishes the correctness of their account. For if Joseph set out from the heart of Egypt to meet his father, who was coming from Palestine, he could hardly take any other route than that through the Wady Tumilat; nor was Jacob, who came from Canaan and crossed the isthmus between Lake Menzaleh and the Arabian gulf, likely to choose any other road than the beaten caravan-track through the Wady Tumilat. And if we attempt to fix upon any particular locality, as the precise spot at which the meeting occurred; Heroopolis, the most easterly city of Egypt by this route, has certainly the strongest claims. But when *Hengstenberg* cites Gen. xli. 45, where the *Septuagint* substitutes *Heliopolis* for the *On* of the text, as a perfectly analogous case, he forgets again that in the passage before us Heroopolis is not introduced in the place of Raemses, but as a more precise definition of what the original means by Goshen. Moreover, so far as there is any analogy between the two passages, it is decidedly *against Hengstenberg's* opinion. For if Raemses had been the antiquated, and Heroopolis the current name, the place where the *Septuagint* should have substituted the latter for the former is Ex. xii. 37, and nowhere else. *Hengstenberg* has a further argument, derived from the meaning of Heroopolis, *hero-city*, which is said to be the Greek rendering of the ancient name *Raemses* (the city being evidently so called in honour of the hero-kings who bore that name); but this argument, it appears to me, is purely visionary and without the least force.

On the other hand, positive proof can be brought that Raemses and Heroopolis are not the same. A city which stood *so close to the eastern frontier of Egypt*, as Heroopolis did, cannot possibly have been the point from which the procession set out, however probable it may be that it passed through the city or in its immediate vicinity. The point of departure, as Ex. xii.—xiv. clearly shows, must have been quite close to the palace of the king. And whether we suppose the palace to have been situated in On, Bubastis, or Zoan (see below, note 2), in either case the identity of Heroopolis and Raemses appears to us an impossibility.

The same argument may be adduced against the assumption of *Lepsius* (Chronol. i. 348 seq.), who disputes the identity of Raemses and Heroopolis (= Mukfâr), but brings forward a new, and, as he thinks, a decisive argument to prove that Raemses and Abu-Keishid are the same. On the ruins of Abu-Keishid a group was discovered, at the time of the Franco-Egyptian expedition, consisting of three figures, hewn from a block of granite, representing the two gods Ra and Tum, with the King Ramses II. between them. But this is, to our minds, by no means a conclusive argument. For the discovery of a statue bearing the name of the great King Ramses, is by no means a proof that the city in which it was found was built by Ramses, or that it must have been called by his name. Ramses the great, who led such magnificent expeditions into Asia, may very well have caused such a memorial to be set up in Abu-Keishid, as being the first important city of Egypt into which he entered on his triumphant return, whatever the name of the city may have been.

(2). The arguments adduced by *Stickel* (p. 377 sqq.), in support of *the identity* of Belbeis, and Raemses, are sufficient at least to show that such an assumption is highly probable. First of all, it is sustained by the authority of the geographer *Makrizi*, who was well informed in all matters relating to his native country, Egypt. He states that Belbeis was an ancient city, which was in existence before the land was conquered by the Moslems, and was the capital of the province, which is called in the *Pentateuch* the land of Goshen. The situation answers extremely well to all that we can gather from the *Pentateuch* with reference to it. For, standing as it does at one of the most westerly points of Goshen, its position coincides exactly to the statement, that it formed the starting-point of the Israelitish procession. As far back as the earliest period to which the reports of ancient authorities reach, the city and neighbourhood of Belbeis appear to have formed the actual starting-point of the expeditions to the East, as well as of all the traffic that was carried on between Egypt and the Arabian gulf. It stood upon one of the principal canals from the Nile, by which it was brought into connexion with the southern provinces of Egypt.

Moreover, the account contained in Ex. i. 11, that the Israelites were compelled to build for Pharaoh the store-cities of *Pithom* and *Raemses*, answers very well to the situation given above, whether we suppose the cities to have been newly built, or merely enlarged and fortified. These cities were not intended, as *Ewald* supposes (i. 479), for royal commercial cities, but for military stores and provisions, in other words they were arsenals on a large scale, erected for the purpose of providing the troops, which were stationed in the desert at the eastern extremity of the land, with provisions and munitions of war. Hence they were not actual fortresses, in which case they would have been placed further to the east, but store-houses from which the fortresses were supplied. It was therefore necessary that they should be so situated, that the road to the fortresses, and also the approach to the Nile, should be both easy and convenient. *Pithom* (*Πάτουμος*), which was identical with Abasieh, stood at the entrance to the Wady Tumilat, the high road to the east, and met these requirements in every way. The manner in which Pithom and Raemses are linked together, justifies us in supposing that the two cities, which were intended for the same purpose, were both erected in the district which lies between the valley of the Nile and the eastern frontier of the country. Abasieh stood on the road from Bubastis to the frontier; and Belbeis was also in the way from Memphis to the same boundary.

A more difficult matter, to which we must now direct our attention, is the determination of the site of the ROYAL PALACE at that time. For it necessarily follows from the history of the night, in which Israel was to prepare to depart, that it cannot have been far from Raemses. If, then, the opinion which we have ourselves expressed at § 1. 5 be correct, viz. that the palace was in *Zoan or Tanis*, which stood near to the point at which the Bubastic arm of the Nile enters Lake Menzaleh; the result of the foregoing investigation, viz. that Raemses was the same as Belbeis, or at any rate stood in the neighbourhood of Belbeis, must necessarily fall to the ground. The former assumption rests upon passages of Scripture. One of these, Num. xiii. 23, states that "Hebron was built seven years before Zoan in Egypt;" but this merely proves that, in the time of Moses, Zoan was already a comparatively old and important city in Egypt. It says nothing whatever with regard to the residence of the Egyptian kings. Nor is there a reference to this point in any other part of the *Pentateuch*. The second passage is apparently more important. In Ps. lxxviii. 12, 43, the wonders in the land of Egypt are said to have been wrought *in the field of Zoan* (שְׂדֵה־צֹעַן). But even these words cannot be said to be conclusive. The "field of Zoan" may denote the whole Delta; and it is the more probable that it does so, because the Egyptian plagues were not restricted to the immediate vicinity of Zoan, and it is expressly and repeatedly said that they extended over the *whole* of the land of Egypt. If, however, the Psalmist gives the name of Zoan to the Delta, this certainly proves that Zoan was *at that time* the most important city, and possibly the royal residence; more than this it does not prove. But if we consider, that the history before us relates to a period, when the yoke of the foreign Hyksos-dynasty had just been thrown off, and there was constant fear lest they should again attack the eastern frontier of the land, across which they had been driven (Ex. i. 10), it seems hardly credible that the Egyptian kings can have fixed their residence in the most northerly district of the country. The national dynasty, which was now in power, had come down from Upper Egypt. The source of its strength, and its chief supporters, were still to be found there, and it can hardly, therefore, have fixed its abode at so great a distance to the north; for Tanis is very far to the north of the line by which an attack from the east would be made. Any such attack would be sure to be directed against the heart of the land, which was much farther to the south, and for this the Wady Tumilat afforded a convenient and well beaten road. Hence if the king's forces had been stationed so much to the north, they could very easily have been cut off from upper Egypt. One of the cities to the south would therefore be much more suitable, either *Heliopolis*, as the Arabian geographer *Kaswini* (in Stickel, p. 383) supposes, or what would answer much better to the probable site of Raemses (= Belbeis), the ancient and celebrated *Bubastis*. All the other references contained in the *Pentateuch* apply equally well, or even better, to one of these two cities, than they do to Zoan.

§ 42. Assuming then, for the reasons just assigned, that Belbeis was most probably the starting-point of the procession, we have now to determine the direction which it took. It was to proceed to Sinai. To accomplish this it was necessary that it should go round the northern extremity of the Arabian gulf. Now, if the northern boundary of the gulf was exactly the same in the time of Moses as at the present day, the procession will probably have gone by the caravan-road direct from Belbeis to Suez (Derb el-Bân); and in this case we must look for *Succoth* about the centre of the road, and *Etham* to the north of Suez. But, as we have shown in § 39. 1 that it is almost certain, that at that time the northern end of the gulf reached much farther to the north than it does now, viz., to the downs of Arbek, which are almost in a straight line with Belbeis, we are brought to the conclusion that instead of taking a south-easterly direction, the procession travelled due east through the well watered and cultivated district of the Wady Tumilat. The site of the first station, *Succoth* (which means tents), is then easily determined. The second place of encampment is called *Etham*, and is further described as being "at the end of the desert." For this we shall have to seek upon the downs of Arbek (§ 39), between the Bitter Lakes and the Crocodile Lake (1). At this point Moses received the command to turn round, and cause the people to encamp with Pihachiroth on the north, Migdol on the west, the sea on the east, and Baal-Zephon to the south. From the words of the command so much at least may be inferred with certainty, that the procession, which had already arrived at the boundary between Egypt and Arabia, did not go round the northern extremity of the gulf, as Moses at first intended, but remained within the territory of Egypt, going southwards along the western shore of the gulf, and at length arriving at a point, where it was completely shut in, by the sea and mountains in front and on the two sides, and by Pharaoh's chariots in the rear. If we look for a spot on the western shore of the gulf, which answers to this description, we find it in the *plain of Suez.* This plain is large enough to hold two millions of men ; it is bounded on the west and south-west by the mountains of Atakah (§ 37), and these mountains approach so nearly to the sea, which is here considerably widened in consequence of a rapid curve to the west, that very few men could pass side by side along the shore.—If the procession came from the north or north-east, the *third place of encampment* cannot have been any other than the plain of Suez, and this, too, is the only point at which we can justly suppose *that the passage through the sea* occurred (2).

(1). The name Etham is explained by *Jablonsky* from the Egyptian, as meaning *sea boundary.* If it were of Semitic origin, it would necessarily be connected with אֵיתָן, *perennitas.* It would in this case denote a place watered by perennial streams, in contradistinction to the brooks of the desert, which are so quickly dried up. The question, whether the "end of the desert," where Etham was situated, is to be understood as referring to the Egyptian desert (as in chap. xiv. 3. 11), or the Arabian, both of which touched each other at Etham, may perhaps be decided in favour of the latter, when we consider that the whole strip of desert land on the eastern coast of the gulf bore the name of *desert of Etham* (Num. xxxiii. 8, *cf.* § 47. 5).

(2). The *plain of Suez* "is not far from ten miles square; extending with a gentle slope from Ajrud to the sea west of Suez, and from the hills at the base of Atâkah to the arm of the sea north of Suez" (*Robinson* i. 65). In the boundaries of the plain which are given here, we think we can discover with comparative certainty the places mentioned in Ex. xiv. 2. The words which we find there are, "speak to the children of Israel that they turn round and encamp before Pihachiroth, between Migdol and the sea, and, before Baal-Zephon, opposite to it shall ye encamp by the sea." *Pihachiroth*, we find, even by name, in *Ajrud ;* for *Pi* is merely the Egyptian article, and hence the place is also called *Hachiroth* in other passages (Num. xxxiii. 8), and there are many instances of analogous changes (*cf. Stickel*, p. 391). *Migdol*, in any case, must be looked for in the direction exactly opposite to the sea (according to Ex. xiv. 2), and therefore near Mount *Atâkah*, whether Migdol (which means a tower) was a fortress upon or by the side of the mountain, or, as *Tischendorf* supposes, the summit of the Atâkah itself. In the *Septuagint* it is rendered *Μαγδώλον*, and *Hengstenberg* (p. 59) thinks himself justified in connecting it with the fortress of *Magdolum*, which stood at a latter period twelve Roman miles to the south of Pelusium. But the supposition that, whereas the other three places mentioned as the boundaries of the encampment were all in the neighbourhood of Suez, the fourth was fifteen geographical miles to the north, is perfectly incredible, and is not rendered a whit more probable by the remark that a frontier-garrison was stationed there. Moreover, apart altogether from the distance, the Israelites would not then have been between Migdol and the sea, but the sea between Migdol and the Israelites. Baal-Zephon (the place of Typhon) cannot be more particularly described ; but according to the description contained in Ex. xiv. 2, it must be looked for at the south of the plain of Suez.

The point at which the sea was crossed was, therefore, in all probability, *near to Suez.* But there are strong reasons for doubting whether the ford of Suez was in existence then, and also whether the sea was then only 3450 feet broad at this spot, as it is now. For if that had been the case, the return of the waters which had been divided by the east wind, would hardly have been sufficient to drown Pharaoh's entire army. It is necessary, however, that the peculiar configuration of the sea at Suez should be kept in mind. If the pathway through the sea went in a south-easterly direction, and *not* due east (a supposition by no means improbable, seeing that the direction depended upon the wind which opened the way), then even at present the breadth and the depth would be quite sufficient to hold and to drown an entire army.

The only point in which *Du Bois Aymé* and *Stickel* differ from the views we have expressed, is with reference to the last place of encampment, and the spot at which the sea was crossed. They both of them fix upon the supposed ford at Ajrud (§ 39. 1), instead of Suez, as the place where the passage occurred. But there are many objections to this. First of all, the ground about Ajrud does not answer in the least to the description of the last place of encampment, which is given in the text. There is no plain sufficiently large to hold two millions of men, nor is there the steep impassable mountain wall which reached the sea, and caused the Israelites to be hemmed in on three sides. It is true that *Du Bois Aymé* says (*vid. Rosenmüller* iii. 265): "The biblical account is in perfect harmony with the position which I have assigned to the Israelitish army; for the chain of mountains, which is visible towards the south, *appears to stretch as far as the shore*." But in reading these words, we cannot escape the feeling that, in spite of the confidence with which the author speaks, he was conscious of a certain incongruity between the locality referred to, and the description contained in the Bible. Again, the order in which the boundary-points are named in Ex. xiv. 2 does not square with this view, for, according to Ex. xiv. 2, Ajrud must have been situated to the north of the place of encampment, whereas, if *Du Bois Aymé's* opinion were correct, it would have been to the south-east. He also adds (*Rosenmüller* iii. 268): "Moreover, there is so little difference between the two opinions (that which fixes upon Ajrud, and that which selects Suez, as the spot at which the Israelites crossed), that it does not matter much which of the two we choose. My opinion rests upon the situation of the castle of Ajrud, before which the Israelites encamped, and the *great probability that the sea at Suez was much deeper then*, than it is now." This we can fully comprehend, for in the opinion of the learned Frenchman, the sea must have been crossed in a perfectly natural way, without any miraculous intervention on the part of God. But the greater depth of the sea at Suez is, to our mind, one of the very reasons why we should prefer that spot, not from any love of miracles, but because we are anxious to do justice to the text. *Stickel* gives the preference to Ajrud for another reason. This keen-sighted scholar would no doubt have fixed upon Suez,—as the description contained in the text, when compared with the shape of the ground, unconditionally requires,—were it not that his foregone conclusion, that only three days can have elapsed between the departure of the Israelites from Raemses and their arrival at the opposite shore of the sea (§ 36. 7), compelled him to relinquish such on opinion. For the distance from Etham to the plain at Suez is certainly too great, for any one to bring himself easily to believe, that the Israelitish procession could traverse it in a single day. But we have already pointed out, that the journey from Etham to the point at which the sea was crossed, must have occupied a longer time, seeing that the message was sent from Etham to the palace, and the royal army marched from the palace to the sea, whilst the Israelites were travelling from Etham to the same spot. This must have required *at least* two days (§ 36. 7).

(3). We shall conclude by giving a short sketch, and, where necessary, our own criticism of *the different views which have been entertained, with regard to the crossing of the sea.* Among the earliest is one which has lately been defended with great firmness and confidence by *K. v. Raumer*, and of which *v. Lengerke* has most remarkably expressed his approval (Kenaan i. 432 sqq.). In all that is essential *J. V. Kutscheit* also adopts it. It originated with *Sicard* (*cf. Paulus* Sammll. v. 211 sqq.), who had travelled by the road in question. *Sicard*, however, places the city of Raemses, the starting-point of the procession, in the neighbourhood of the village of Besatin (§ 37), whereas *Raumer* does not regard Raemses as the name of a city, but of the land of Goshen, and supposes the procession to have been first formed in the vicinity of Heliopolis or On, from which point it went southwards to Latopolis or Babylon, and then turned towards the east into the Valley of Error, in the first instance with the intention of following the ordinary caravan road, which leads through this valley to Suez, and then going round the northern extremity of the gulf. Succoth would in this case be in the neighbourhood of the village of Besatin; and Etham, near the fountain of Gandelhi, at which point the caravan road turns towards the north-east, between the two northern ranges of mountains (§ 37). But, instead of carrying out the first intention, the procession turned away from the ordinary caravan road at the express command of God, and had to take the road through the Wady er-Ramliyeh and the Wady et-Tawarik to the plain of Baideah by the sea-shore. It was here that they were overtaken by Pharaoh and his army. With the deep sea, which is here about fifteen miles broad, before them; with Mount Atâkah on the north, and, opposite to this, Mount Kuaibe on the south; and with Pharaoh's chariots behind, they were to all human appearance utterly lost. But God caused an east wind to blow during the night, and thus opened a way through the heart of the sea. They followed this road; and the next morning, they found themselves safe on the other side, at a place which is still called by the Arabs the wells of Moses (Ayun Musa).—At first sight there is something very plausible in this view. But on closer investigation we find it beset with insuperable difficulties. Its main features are not derived from scriptural data, but from the statements of *Josephus* (Ant. ii. 15, 1), who says that the Israelites started from Raemses above the place where Babylon was afterwards built (Latopolis, Old Cairo). But no particular evidence is required to prove, that the authority of *Josephus* is of little value in questions of this kind. Moreover, his account is founded upon the *tradition*, which has given to the valley the name of Wady et-Tih (Valley of Error), and which fixed upon Ayun Musa as the spot near to which the passage took place. But with regard to the first, the name Wady et-Tih originated with *Sicard*, and for the second we must bear in mind the warning given by *Niebuhr* (Beschreib. v. Arabien p. 404), who says that the Arabs always declare the spot, at which the question is proposed to them, to be the very spot where the children of Israel went through the Sea. As decisive objections, however, the following are of especial importance: (1), *Raemses* is always the name of a city, never of a province (*cf.* § 41). (2), Justice is not done to the word שׁוּב, which always means to turn. (3), The same remark applies to the expression "*Etham, at the end of the desert;*" for, according to *Raumer's* hypothesis, Etham was not at the end, but in the middle of the desert. (4), Without the least ground for so doing, it gives us *two* Ethams, one in the Egyptian desert and the other in the desert of Arabia Petræa.

(5), It places the passage through the sea at a point where the sea is too broad, not indeed for it to be *miraculously* divided, but for the *natural* part of the event, namely their crossing over in the time stated. The breadth of the gulf at this point is fifteen miles. Now a few hours of the night had certainly gone, before the sea was sufficiently dried up by the east wind, to allow the passage to commence; and yet at the morning watch (two o'clock), they were on the opposite shore.

A second class of commentators fix upon *Suez* as the point at which the passage took place. This class includes *Niebuhr*, *Robinson*, *Hengstenberg*, *Laborde*, *Ewald*, *Tischendorf*, and many others. But whilst they agree upon this point, they differ in many respects as to the road by which Suez was reached. *Hengstenberg's* opinion is that the Israelites started from Raemses, which he supposes to have been the same as Heroopolis; that Etham was at the point, which now forms the northern extremity of the gulf; and that when the procession had reached that point it turned round, that is went back into the Egyptian territory, and proceeded along the western shore of the gulf, till it reached Suez, where it passed through the sea upon dry ground. *Robinson* gives upon the whole the same route, but leaves it an open question, whether Heroopolis was identical with Raemses; though he has not the least doubt that Raemses was situated in the Wady Tumilat, not far from the northern extremity of the Bitter Lakes. From our previous enquiry, however, it necessarily follows that this opinion is erroneous. *Ewald's* view is closely related to that of *Hengstenberg*, only much more confused *(cf. Stickel's* critique, p. 358 sqq.). *Laborde* looks upon Raemses as a name applied to the whole of the land of Goshen, and supposes the Israelites to have assembled at Succoth, whence they proceeded in a straight line to Etham, which was somewhere in the neighbourhood of Ajrud. There they received a command from God, not to travel any farther in a easterly direction, and went towards the south-east to Suez. From this point they crossed the gulf, still in a south-easterly direction, and emerged at Ayun Musa. After what we have already said, we regard it as unnecessary to criticize, that is, to refute this opinion. *Tischendorf* supposes the procession to have started from Heliopolis, whence it proceeded to a spot somewhere near the northern end of the Bitter Lakes (which in his opinion was at that time the northern boundary of the gulf). At this point it turned towards the south-east and proceeded to Suez. In several essential points his view agrees with our own.

Lastly we may mention *Thierbach's* romantic conjecture (Erfurter Osterprogramm, 1830); though we do so, merely to make the list complete. According to his view, the Israelites set out from Heliopolis (*i.e.*, Raemses). They then journeyed to Pithom, (or Etham) on the Mediterranean (the sea of reeds). From this point they proceeded through Lake Menzaleh. Here the phosphorescence of the water supplied them with light; and at the same time a cloud, which hung suspended like a pillar over the surface of the water and was strongly charged with electricity, was driven behind them by a change in the wind, and discharged its electric fluid upon the foe. Thus death and destruction fell upon the Egyptians, whilst light and safety were afforded to the Israelites. Compare *Stickel* p. 331, 332.

THE HYKSOS AND THE ISRAELITES.

Sources: *vid. Bunsen's Urkundenbuch*, an appendix to the third part of his work on Egypt; *C. Meier*, Judaica, Jena 1832; and *Stroth*, Aegyptiaca, Gotha 1782.

Literature: *Jac. Perizonii* Aegyptiarum originum investigatio. 1711. c. 19 p. 327 ss.—*Fr. Buddei*, Historia ecclesiastica V. T. I. iii. § 24, Ed. iv. p. 560 ss.—*Thorlacius* de Hycsosorum Abari. Copenh. 1794.—*J. Chr. C. Hofmann*, unter welcher Dynastie haben die Israeliten Aegypten verlassen? (in the Studien und Kritiken 1839. ii. p. 393 sqq.), and also, Aegyptische und israelitische Zeitrechnung, ein Sendschreiben an Dr Böckh. Nördl. 1847.—*E. Hengstenberg*, Manetho and the Hyksos, in his Egypt and the Books of Moses, p. 227 sqq. transl.—*E. Bertheau*, zur Gesch. d. Israel. p. 227 sqq.—*H. Ewald*, Gesch. d. Israel. i. 445 sqq.—*C. v. Lengerke*, Kenaan i. 360 sqq.—*A. Böckh*, Manetho und die Hundssternperiode, Berl. 1845.—*Chr. C. J. Bunsen*, Aegyptens Stelle in der Weltgeschichte, 3 vols. Hamburg 1843.

—*R. Lepsius*, die Chronologie der Aegypter, i. Berlin 1849, and Herzog's Real-Encyclopädie d. prot. Theol. i. 144 sqq.—*J. L. Saalschütz*, Forschungen auf dem Gebiete der hebr. ägypt. Archäologie, Königsb. 1851, iii. Die manethonischen Hyksos, p. 41 sqq.—Consult also: *J. G. Müller*, Krit. Untersuchung der Taciteischen Berichte über den Ursprung der Juden, in the Studien und Kritiken, 1843, iv. p. 893 sqq.—*Fr. Werner*, chronologische Bemerkungen über einige Gegenstände der alttestamentlichen Gegenstände (in the lutherische Zeitschrift, 1845, i. p. 29 sqq.) —*K. B. Stark*, Forschungen zur Geschichte der Alterthumskunde des hellenischen Orients: Gaza, oder die philistäische Küste, Jena 1852, p. 82 sqq.—*Fr. Delitzsch*, Commentar zur Genesis, 2. Aufl. 1853, ii. 71 sqq.; and Nachtrag p. 221 sqq.—*A. Knobel*, Die Völkertafel der Genesis, Giessen 1850, p. 208 sqq.; and Genesis p. 271 sqq.—*Raoul-Rochette*, in the Journal des Savants, 1846 and 1848, (review of *Bunsen's* work), particularly 1848, p. 354 sqq.

§ 43. The *Pentateuch* does not inform us what dynasty was in power, or what king was reigning, either when the Israelites went down to Egypt, when the oppression commenced, or at the time of their departure. We must therefore turn to the Egyptian and other profane history before we can solve these questions; and what we have now to do, is to determine from these sources, to what periods of time the events described in the *Pentateuch* respectively belong. The simplest means of obtaining the information we need, would be to compare the two chronologies; but unfortunately, both in the biblical and Egyptian histories, there is so much uncertainty, obscurity, and even confusion, in the matter of dates, that comparative chronology is a most uncertain, and therefore impracticable, method of ascertaining the points of coincidence between the two. Our knowledge of the facts, connected with the early history of Egypt, is for the most part confined to bare catalogues of dynasties, which do not of themselves afford any information, that can be brought to bear upon the history of Israel. But *Josephus* has preserved two considerable fragments from the old historical work of *Manetho*, the contents of which coincide in many respects with the history of the *Pentateuch.* The *first extract from Manetho* treats of the *Hyksos dynasty*, and contains unmistakeable traces of the relation which existed between this dynasty and the Israelites (1) ; the *second* identifies the Israelites with a number of lepers, whom the king, Amenophis, is said to have banished from Egypt (2). The same tradition, in a somewhat modified form, is found in *Chaeremon* and *Lysimachus*, and on the authority of the latter it has been repeated by *Apion*, *Diodorus Siculus*, *Tacitus*, and *Justin* (3).

(1). The *first extract from Manetho* (on the reign of the Hyksos) is found in Josephus against Apion, i. 14. *Josephus* there says: " Manetho was a man who was by birth an Egyptian; yet had he made himself master of the Greek learning, as is very evident: for he wrote the history of his own country in the Greek tongue, by translating it, as he saith himself, out of their sacred records; he also finds great fault with Herodotus for his ignorance and false relations of Egyptian affairs. Now this Manetho, in the second book of his Egyptian history, writes concerning us in the following manner (I will set down his very words, as if I were to bring the very man himself into court for a witness): 'There was a king of ours whose name was *Timaus.* Under him it came to pass, I know not how, that God was averse to us, and there came, after a surprising manner, men of ignoble birth out of the eastern parts, and had boldness enough to make an expedition into our country, and with ease subdued it by force, yet without our hazarding a battle with them. So when they had gotten those that governed us under their power, they afterwards burned down our cities and demolished the temples of the gods, and used all the inhabitants after a most barbarous manner; nay, some they slew, and led their wives and children into slavery. At length they made one of themselves king, whose name was *Salatis*; he also lived at Memphis, and made both the upper and lower regions pay tribute, and left garrisons in the places that were most proper for them. He chiefly aimed to secure the eastern parts, as foreseeing that the Assyrians, who had then the greatest power, would be desirous of that kingdom and invade them; and, as he found in the *Saitic Nomos* a city very proper for this purpose, and which lay upon the Bubastic channel, but in a certain ancient theological account was called *Avaris*, this he rebuilt, and made very strong by the walls he built about it, and by a most numerous garrison of two hundred and forty thousand armed men, whom he put into it to keep it. Thither Salatis came in summer time, partly to gather his corn, and pay his soldiers their wages (σιτομετρῶν καὶ μισθοφορίαν παρεχόμενος), and partly to exercise his troops and thereby to terrify foreigners. When this man had reigned nineteen years he died; after him reigned another, whose name was *Beon*, for forty-four years; after him reigned another, called *Apachnas*, thirty-six years and seven months; after him *Apophis* reigned sixty-one years; and then *Janias* fifty years and one month; after all these reigned *Assis* forty-two years and two months. And these six were the first rulers among them, who were all along making war with the Egyptians, and were very desirous gradually to destroy them to the very roots. This whole nation was styled *Hycsos*, *i.e.*, shepherd-kings (βασιλεῖς ποιμένες); for the first syllable Hyc, according to the sacred dialect, denotes *a king;* Sos, according to the ordinary dialect, is a shepherd; and *Hycsos* is compounded of these. But some say that these people were Arabians.' In another manuscript (ἐν δ' ἄλλῳ ἀντιγράφῳ), however, I have found that *Hyk* does not denote *kings*, but on the contrary *captive shepherds;* for *Hyc*, with the aspirate (δασυνόμενον) means in the Egyptian tongue *prisoners;* and this seems to me the more probable opinion and more in accordance with sacred history. But *Manetho* goes on: 'These people whom we have before named kings, and called shepherds also, and their descendants, kept possession of Egypt five hundred and eleven years. After this, however, the kings of Thebais and the others parts of Egypt made an insurrection against the shepherds, and a terrible and long war was waged between them. And under a king named *Alisphragmuthosis*, the shepherds were subdued by him, and were driven out of other parts of Egypt, but were shut up in a place that contained ten thousand acres of land; this place was Avaris. The shepherds built a wall round all this place, which was a large and strong wall, and this in order to keep all their possessions and their prey within a place of strength. But *Thummosis*, the son of Alis-Phragmuthosis, made an attempt to take them by force and by siege, with four hundred and eighty thousand men to lie round about them, but upon his despair of taking the place by that siege, they came to a composition with them, that they should leave Egypt and go, without any harm being done to them, whithersoever they would. After this composition was made, they went away with their whole families and effects, not fewer in number than two hundred and fifty thousand, and took their journey from Egypt, *through the wilderness to Syria.* But as they feared the Assyrian power, which had then the dominion over Asia, they settled in the country which is now called *Judea*, and there they built a city, large enough to contain so many thousand men, and called it *Hierosolyma.*' In another book of the Aegyptiaca, *Manetho* says, that the shepherds are described as captives in the sacred books. And this account of his is the truth, for feeding of sheep was the employment of our forefathers in the most ancient times, and therefore they were called *shepherds;* nor was it without reason that they were called *captives* by the Egyptians, since one of our ancestors, Joseph, called himself a captive before the king of Egypt" (*Josephus* contra Ap. i. 14, *Whiston's* transl.).

(2). The *second extract from Manetho* (on the expulsion of the lepers) is found in the same book of *Josephus*, (c. Apion i. 26 seq.), who says: "*Manetho* promised to interpret the Egyptian history out of the sacred writings, and first of all relates, that our forefathers came in many myriads into Egypt and subdued its inhabitants. But in the next age they were expelled, took Judea and there built Jerusalem and the temple. So far he follows the ancient records. But after this he takes the liberty of introducing incredible fables and legends (*μυθευόμενα καὶ λεγόμενα, λόγους ἀπιθάνους*;—according to Bk. i. c. 16: *οὐκ ἐκ τῶν παρ' 'Αιγυπτίοις γραμμάτων, ἀλλ' ὡς αὐτὸς ὡμολόγηκεν, ἐκ τῶν ἀδεσπότως μυθολογουμένων προστέθεικεν*), concerning the Jews, confounding our forefathers with a number of leprous Egyptians, who were driven out of Egypt on account of their leprosy and other diseases. For this purpose he brings in a king *Amenophis*, whose name is a fictitious one, on which account he does not venture to give the length of his reign, which he always on other occasions most scrupulously does. With this king he associates the fables referred to, and forgets that, according to his own statements, 518 years must have passed since the shepherds were expelled. For they left Egypt in the reign of *Thutmosis* (Thummosis). Now from him to *Sethos* there were 393 years; Sethos reigned fifty-nine years, and his son *Rampses* sixty-six. It is not till this point that he introduces the fabulous *Amenophis*, of whom he gives the following account: 'Amenophis desired to see the gods, as King *Horus* had formerly done. He made known this wish to a wise man, who was also named Amenophis, and was told by him that he must first of all cleanse the land entirely from lepers and unclean persons. The king then had all the unclean persons gathered together out of the whole of the land of Egypt, 80,000 in number, and sent them to work in the quarries to the east of the Nile. Among these lepers there were some learned priests. In the meantime Amenophis repented that he had advised the king to expel the lepers, fearing that the wrath of the gods might be excited thereby, and, as a revelation was made to him a short time afterwards, that the lepers would rule for thirteen years over Egypt, supported by foreigners, and he durst not make this known to the king, he killed himself, and left a written document behind him, which greatly troubled the king. After the lepers had continued for a long time to do hard work in the quarries, the king listened to their request, and gave them the city of *Avaris*, which had formerly been occupied by the shepherds, but at that time was desolate. In the ancient theological documents this city is called the city of Typhon (*Τυφώνιος*). Now, when the lepers had settled there, they chose a priest of Heliopolis, named *Osarsiph*, to be their leader, and swore that they would yield obedience to him in everything. He first of all commanded them to worship no gods, to cease to abstain from the animals which were regarded as sacred in Egypt, to slay and eat without distinction, and to hold fellowship with no man, who did not belong to them. He also gave them many other laws, which were directly opposed to the customs of Egypt. After this he had the city fortified with walls, and prepared to make war upon Amenophis. He sent messengers to the shepherds at Jerusalem, who had been expelled by Thutmosis, and urged them to join in a common attack upon Egypt. The shepherds gladly listened to his appeal, and came to Avaris with 200,000 men. King Amenophis remembered the prophecy, and lost all his spirit. He gathered together the sacred animals, hid the images of the gods, brought his son *Sethos*, who was five years old, and had been named *Ramesses* after his father Rampses, placed him under the protection of a friend, and then advanced with 300,000 men to meet the foe. From fear of the gods, however, he did not venture to attack them, but withdrew into Ethiopia, taking with him the sacred animals, and there he remained in voluntary exile for thirteen years; the king of Ethiopia being bound to him by ties of gratitude. The Solymites, in conjunction with the lepers, inflicted the greatest cruelties upon the Egyptians, who were left behind. They set fire to the cities and villages, destroyed the temples, and used the wood of the images of the gods to cook the flesh of the sacred animals. The priests were compelled to slaughter the sacred animals with their own hand, and were then driven naked from the spot. The founder of this state had formerly been a priest of Heliopolis. He was named *Osarsiph* after the god Osiris, who was worshipped there; but afterwards he was called *Moyses.* After an exile of thirteen years, Amenophis and his son Rampses returned from Ethiopia to Egypt, each at the head of a powerful army. The shepherds and lepers were speedily subdued, and driven as far as the frontier of Syria.'"—*Josephus* then proceeds to demonstrate the absurdity of this fictitious account.

(3). The account, which *Manetho* gives of the lepers, is found in the works of other authors, but with various alterations. CHAEREMON (in *Josephus* c. Apion i. 32) relates that the goddess *Isis* appeared in a dream to King *Amenophis*, and complained that her temple had been destroyed in war. By the advice of the priest, *Phritiphas*, who informs him that he will not be disturbed by the goddess any more, if he cleanses Egypt from all its lepers, he has 25,000 of them banished. Their leaders, the scribes *Moyses* and *Josepos* (whose Egyptian names were Tisithes and Peteseph) conducted them to Pelusium. There they united with 380,000 men, whom Amenophis had placed there with orders not to enter Egypt, and with these they invaded that land. Amenophis was unable to resist their attack, and fled to Ethiopia. His wife, whose time of delivery was drawing near, could not accompany him in his flight, and hid herself in a cave. There she gave birth to a son, who, when he had grown up, drove out the Jews, at that time numbering 200,000 men, chased them to Syria, and recalled his father from Ethiopia.

The same legend is given by LYSIMACHUS (*Josephus* c. Apion i. 34) in a still more romantic form: During the reign of King *Bokchoris*, the people of the Jews, having been attacked with leprosy, the itch, and other diseases, took refuge in the temple, and got their living by begging. In consequence of this, the land was visited by famine and pestilence. The oracle of Ammon ordered the temple to be purified from the unclean and wicked men, who were all to be sent into the desert, with the exception of those afflicted with leprosy and the itch. The latter were to be rolled up in lead and thrown into the sea. This was done. The others, who had been transported to the desert, then took counsel what they might do. They lighted torches and lamps as soon as the night came on, set watches, and fasted, for the purpose of propitiating the gods. The next morning a certain *Moyses* advised them to go forward in a regular procession, till they came to some inhabited country. He also commanded that in future they should do good to no one, and should destroy every temple and altar that they might happen to meet with. After many obstructions, they reached Judea, where they plundered and burned all the temples, and built a city, which they called *Hierosyla* in commemoration of their deeds. But as this name was afterwards regarded as a term of reproach, they altered it to *Hierosolyma*.

APION, in the third book of his history of Egypt (*Josephus* c. Apion ii. 2), adopts the account given by *Lysimachus*, but he also embellishes it with a "trustworthy" explanation of the manner in which the Sabbath originated. The Jews, he says, arrived at Judea after a six days' march through the desert. On the seventh day they were attacked with internal ulcers, which compelled them to rest on that day, and as this disease was known in Egypt by the name of Sabbatosis, they called that day the Sabbath.

TACITUS (Hist. 5. 2—5), in his description of the destruction of Jerusalem, refers to the origin of the Jewish people. He cites different reports, with which he was acquainted; but does not decide in favour of either of them. According to some, he says, the Jews originally came from Crete, at the time when Jupiter dethroned his father Saturn, and settled first of all on the frontier of Lybia;—this opinion rests upon the supposed derivation of the word *Judæi* from *Idæi*, the inhabitants of Mount *Ida* in Crete. Others trace the origin of the Jews to Egypt. They say that when *Isis* sat upon the throne, the number of men in Egypt was too great, and therefore a portion of them emigrated under the guidance of *Hierosolymus* and *Juda* and settled in a neighbouring country. Others again suppose the Jews to be descendants of the *Ethiopians*, who were led to emigrate by their fear and hatred of the Ethiopian king *Kepheus*. According to a fourth opinion they were *Assyrians*, who first of all took possession of part of Egypt, and then settled in the neighbouring Hebrew and Syrian lands. Others imagined that they were the *Solymi* mentioned by *Homer*. But the most general opinion of all was, that the Jews were originally *leprous Egyptians*. The account given by *Lysimachus* is then served up again, and enlarged in the following way. On their march through the desert, they were all threatened with destruction from want of water. Suddenly there appeared a number of wild asses, which, after grazing, went back up a rock that was covered by a dense wood. Moses thought that where there was wood there must also be water, and following the asses actually discovered some copious springs. After a further march of six days' duration they arrived at the Jewish land, expelled the inhabitants, and founded a city and temple. Moses then introduced a variety of customs, which were opposed to those of other nations; and among other things had an image of the animal, which had saved them from perishing with thirst in the desert, set up in the holiest place as an object of worship.—Consult the article, already noticed, by *J. G. Müller* in the Studien und Kritiken 1843.

Justin (Hist. 36. 2) traces the origin of the Jews to Damascus. The first king of that city, from whom it derived its name, was *Damaskus*. He was followed by *Azelus*, *Adores*, *Abraham*, and *Israel*. Israel had ten sons, and divided his kingdom among them. Shortly after the division, *Judah*, one of his sons, died. His share was distributed among the rest, and henceforth the whole people were called *Judæans*. The youngest of the brothers was sold by the rest to some foreign merchants, who took him to Egypt. There he learned the arts of magic, interpreted omens and dreams, predicted a famine some years before it occurred, and thus saved Egypt from perishing with hunger. He had a son named *Moses*, who not only inherited his father's learning, but was distinguished for his extraordinary beauty. When the Egyptians with the itch were banished from the land, according to the sentence of an oracle, he offered himself as their leader, and stole the sacred relics from the Egyptians. The latter pursued him, for the purpose of recovering their sacred things by force of arms; but violent tempests arose, which compelled them to return. Moses then led his followers to his native place, Damaskus, and took possession of Mount *Syna*. As he arrived there with his people, worn out after a seven days' fast, he set apart the seventh day, the Sabbath, as a regular fast-day. They avoided all intercourse with the inhabitants of the district in which they settled, from a desire to transfer to the latter the hostility which had previously existed between themselves and their fellow-countrymen in Egypt. This separation gradually became a religious law. *Aruas*, the son of Moses, combined the royal with the priestly dignity, and from his time the Jews continued to be governed by priestly kings.

§ 44. The earliest attempt, with which we are acquainted, to reconcile these accounts with the *Pentateuch* history, is that made by *Josephus*, who identifies the lepers with the Hyksos, and both with the Israelites. He pronounces the first account of *Manetho*, which was taken from early Egyptian documents, in all its essential features, historically trustworthy; and makes use of it, to establish the great antiquity and historical importance of his nation, in opposition to the insults and slanders of *Apion*. In the second account of *Manetho*, on the other hand, which is evidently at variance with the first, and according to *Manetho's* own confession, was not derived from any *written* historical source, but merely taken from a vague, unfounded, popular legend, *Josephus* will not allow that there is anything trustworthy. At the same time there is reason enough for entertaining very strong doubts, as to the sincerity of his belief in the historical character of the first account, or at least in the correctness of the explanation which he gives of that account (1). The nearest approach to the views of *Josephus*, as expressed and defended in the book against *Apion*, is to be seen in the opinion entertained by *Delitzsch*, who is inclined to regard it as a historical fact, that Egypt was subjugated and governed by the Israelites for several hundred years (2). *Perizonius*, *Buddeus*, *Thorlacius*, *Hofmann*, and *Hengstenberg* look upon the two accounts as different forms of the very same legend; at the same time they maintain that in both of them the actual facts, which are to be found in a credible form in the *Pentateuch* alone, have been distorted to favour the national interests of Egypt, and that in such a manner that all the cruelty, violence, and oppression, in a word, all the evil done by the Egyptians to the Israelites, is transferred to the latter, a transfer suggested to a certain extent by the measures of state adopted by Joseph (Gen. xlvii. 13—26) (3). *Hengstenberg* also paves the way for his line of argument with reference to the Hyksos, by describing the pretended *Manetho* as a "miserable subject," and a "windmaker by profession," of the time of the Roman emperors, who perpetrated this distortion of the *Pentateuch* history purely out of his own head (4). A closer investigation of the arguments which have been adduced, from the time of *Josephus* till that of *Hofmann*, *Hengstenberg*, and *Delitzsch*, in support of the identity of the Hyksos and the Israelites, will show that such an opinion cannot possibly be sustained (5).

(1). The object of JOSEPHUS was to rebut the insults, heaped by *Apion* upon the Jewish people, and to bring forward witnesses to the respect and esteem, which the history of that people had secured. *Biblical* writers would have had but little weight with the heathen Apion, as being witnesses in their own cause, but of so much greater worth would be the testimony of *heathen* writers, who could not possibly be charged with partiality. Again, the unfavourable manner in which heathen writers, as a whole, spoke of the Jews, must have heightened the pleasure, with which he appealed to so distinguished a historian as *Manetho*. If *Manetho's* account of the Hyksos could be applied to the Jews, it was drawn up in such a way, that *Josephus* could hardly have desired anything better suited to his purpose. For nothing was so likely to affect a man of *Apion's* cast of mind, as the proof that the Jews, whom he so despised, had ruled for half a millennium over the most powerful and magnificent, the wisest, richest, and most distinguished nation of ancient history. Now, we are fully convinced that *Josephus* himself did not believe in the identity of the Hyksos and the Jews. The proof of this opinion we find partly in the fact that, in his Antiquities, which was just the place for it, he does not seem for a moment to have thought of inserting the account of the Hyksos given by *Manetho*, in fact does not even refer to it; and partly in the care, with which (even in the book against *Apion*), he avoids going more thoroughly into the difficulties, that arise from a combination of the Hyksos legend and the *Pentateuch*-history. To any one, who is in the least acquainted with the latter, a number of questions, doubts, and difficulties, inevitably suggest themselves, both as to the admissibility of such an identification of the Jews and the Hyksos, and also as to the historical character of many of the statements contained in *Manetho's* account; and any writer, who is sincerely convinced of both of these, must be prepared to remove all such difficulties, and not pass them over in silence or explain them away. But of this we cannot find the least trace, in the whole of the elaborate argument of *Josephus*. He speaks and argues in all respects just as if the identity of the Hyksos and the Jews were a fact so fully established, that no reasonable man ever had doubted or ever could doubt its reality. The book of *Josephus* against *Apion* is not intended to subserve the interests of historical research; its object is exclusively polemical. Hence he lays hold of every argument, whether good or bad, and however sophistical, and does not hesitate to throw dust in his opponent's eyes, on all fitting occasions. Assurance and confident assertions are expected to cover the weakness of his own convictions. With an opponent, better acquainted with the *Pentateuch* than *Apion* was, he would have been unable to make such random and unsupported statements with impunity, and would hardly have dared to do so. But with an ignorant and conceited man like *Apion*, who to all appearance had obtained his knowledge of the Isr elitish history exclusively from hearsay, something might be ventured, especially as the occurrence of the march through the desert, the settlement in Judea, and the building of Jerusalem, in the account given by *Manetho*, were quite sufficient to convince an ignorant man that the identity was indisputable, especially as he knew nothing of the incongruities and contradictions which preponderate to such an extent.

(2). DELITZSCH (p. 75) says: " What if the three Hyksos dynasties consisted of three different tribes of Israel? Is it not possible that after Joseph's almost royal supremacy (according to Artapanus in *Eusebius* praep. ev. 9. 23, in the end *fully* royal supremacy) over the Egyptians, the native princes were brought into a condition of dependence, in which they were treated with warlike cruelty by the Israelitish tribes, and that the oppression of Israel did not begin, till Amosis had recovered both the Egypts after a tedious war? The four centuries, about which the *Pentateuch* is silent, because they presented no points of interest, so far as sacred history was concerned, may have been of all the greater importance in connexion with the history of the world. When we read in Ex. i. 7: " The Israelites were fruitful and increased abundantly, and multiplied and waxed exceeding mighty, and the land was filled with them,"—it follows that they had extended beyond the boundaries of the province originally assigned them. And when we find it stated immediately afterwards, that a new king arose over Egypt who was hostile to the Israelites, and endeavoured to keep them down by forcing them to work as slaves, because they had become " more and mightier" than the Egyptians, it is evident that the king, here referred to, was the first king of the native dynasty, which had been overpowered and confined to Thebais, but had now recovered its supremacy. That the prophecy contained in Gen. xv. 13 (" they shall serve them; and they shall afflict them four hundred years") was not irreconcileable, from an Old Testament point of view, with a gradual increase of the power of Israel in Egypt, is evident from such passages as Deut. xxvi. 5 and Ps. cv., where, with reference to Ex. i. 7, it is said that Jehovah made the Israelites in Egypt stronger than their enemies. Moreover, we know, that during the period of their sojourning in Egypt, war-like expeditions were undertaken by them. Thus in 1 Chr. vii. 21, the Ephraimites make a predatory incursion into Philistia, and in 1 Chr. iv. 22 we read of the descendants of Judah having dominion in Moab. Again, the fact that an Israelite, named Mered, married a daughter of Pharaoh, and that her name in the Semitic dialect was *Bitjah* (1 Chr. iv. 18, *cf.* chap. vii. 18, where a sister of Gilead the Manassite is called הַמֹּלֶכֶת Ham-molecheth) favours the assumption of an Israelitish government. The term 'ancient histories' is applied by the writer of the Chronicles to chap. iv. 22."

These are the views of *Delitzsch*. We have copied the whole argument without abbreviation, because we think it deserving of very careful examination. There is, in fact, much that is plausible in it. It might be even extended and strengthened by a reference to Gen. xxxiv. 25 sqq.; inasmuch as the crime, committed by the sons of Jacob against the Sichemites, proves that before the Israelites went down to Egypt, and even in the case of the patriarchs themselves, the disposition to indulge in the cruelties of war, and commit aggressive acts of violence, was so deeply rooted in their nature, that we might naturally expect it, when fully developed, to lead to some such result as the conquest and subjugation of Egypt. Still we must reject the hypothesis as inadmissible.

In the first place, we do not think that *Delitzsch* is successful, in his attempt to account for the fact, that we find no reference whatever in the *Pentateuch*, to an event so stupendous and unexampled a character, as the subjugation of Egypt would have been. With regard to the assertion that "the four centuries about which the *Pentateuch* is silent,—because they presented no points of interest so far as *sacred history* was concerned,—may have been of all the greater importance in connection with the *history of the world*:" we regard such an assertion as not only erroneous and calculated to mislead, but as altogether at variance with the general analogy of the sacred history. We are prepared to maintain that every thing connected with the history of Israel, which had any important bearing upon the history of the world, was *eo ipso* of importance in relation to the sacred history, either as promoting or else as impeding and disturbing its course. An event of world-wide importance, occurring within the limits of the history of Israel, could not possibly be a matter of indifference, so far as the development of the sacred history was concerned; for the peculiar characteristic of the history of Israel was just this, that from its position in relation to the history of the world, the history of salvation received its development in such a manner, that the political history of Israel became a history of salvation at the same time. If the rule laid down by *Delitzsch* be correct, there is a great deal in the historical books of the Old Testament which must be regarded as irrevelant. If the author of the *Pentateuch* thought it necessary to relate the injury done by the sons of Jacob to the Sichemites, he would certainly have felt a still greater inducement, or rather have felt compelled to record the supposed subjugation of Egypt; for if this actually occurred, it must have been of infinitely greater importance in connexion with the development of the history of salvation, and must have exerted a much deeper influence upon that history, or at least upon its passing phases, and the obstructions which impeded its course. How thoroughly would the corrupt, ungodly, and unsubdued natural disposition of Israel, which the sacred history so constantly and emphatically refers to, have been set before us in the clearest light by such an event? How would the divine Nemesis, which the sacred history no less emphatically describes on every fitting occasion, have found a distinct expression in the Egyptian bondage and all the misery which Israel had then to endure! How would this fact have furnished the future generations of Israel with a sermon, and lesson of warning and reproof, which would have sounded through all their subsequent history! And is it possible that the sacred record can have regarded as utterly unimportant, an event of such magnitude as this, and one that spoke so eloquently to Israelites of future ages? Or can we suppose that the spirit of God, the spirit of prophecy, which directed even the writing of Israel's history from a foresight of its future necessities, and which knew that the subsequent history of Israel would continue to go more and more astray into the foreign domain of purely political action and reaction, and thus eventually cause its own destruction, passed over this splendid opportunity of engraving on the very portal of Israel's history a fact so full of warning and instruction for future ages, and left it entirely unimproved? In what a different light, too, would this place the redemption of Israel from Egypt by the strong arm of the Lord! What an opportunity would this have afforded, for setting forth what the whole of the Old Testament history so constantly displays, the mercy and fidelity of the caller, in spite of all the guilt, corruption, and disobedience of the called ! How striking and comforting would have been the proof thus afforded at the outset of the history, that however frequently the chosen people might forsake the ways of God to tread their own ungodly ways, they were never forsaken by Jehovah, but were always chastised with the scourge of the Nemesis, that they might be brought back with the cords of mercy.

From what we have said above, we feel that we are forced to the conclusion, that if any such subjection of Egypt had taken place, it would assuredly have been mentioned in the biblical record ; and as there is no reference to it there, that it can never have occurred. But not only does the record contain no notice of any such occurrence ; it evidently precludes it. For if the author had been aware of the fact, and had passed it over in silence, he would have been guilty of dishonesty and partiality, inasmuch as so deceptive an omission would have indicated a desire on his part, to transfer to the innocent Egyptians the charge of guilt, which really belonged to his own people. If *Delitzsch's* view be correct, the Egyptians were fully justified in subjugating and oppressing the Israelites. They were only practising the right of retaliation. The scriptural record, however, not merely takes no notice of any such right ; but, on the contrary, charges the Egyptians with ingratitude and faithlessness (Ex. i. 8 ; *cf.* Deut. xxvi. 6 ; Ps. cv. 25, &c.). According to *Delitzsch*, the Egyptian oppression was a reaction against the previous ascendancy of the Israelites ; but according to the representations of the biblical record, it was the rapid increase of the Israelites, which first led the Egyptians to fear that they might obtain the ascendancy, and induced them to anticipate any such event, and endeavour to render it impossible by bringing Israel into bondage.

The biblical data adduced by *Delitzsch* in support of his hypothesis are of little weight. When it is stated in Ex. i. 7 that the Israelites had increased to such an extent that the land had become full of them, this does not mean that "the people had overspread the limits of their original dwelling place;" for by the land, which was full of them, we are certainly to understand the *land of Goshen*, which had been assigned them. When

Jacob's family (consisting of seventy souls) settled in the land of Goshen with a few thousand servants, they cannot possibly have filled so large a province; but after a short time, they increased so rapidly as to fill the whole of that land. And when again the king says in chap. i. 9, "behold the people are more and mightier than we," the purport of his words is such, that a little exaggeration seems quite in character. But even if we regard them as literally true, there is nothing in them to astonish us. According to chap. i. 8, the period had just arrived when a new dynasty arose, *i.e.* when the national dynasty threw off the yoke of the Hyksos and recovered the supremacy. It is true, the warlike dynasty had been driven over the frontier and compelled to leave the country. But many of the Hyksos settlers had undoubtedly been left behind; as we may gather from the Pentateuch itself, viz. from Ex. xii. 38 and Num. xiv. 4. And under these circumstances we may easily conceive, that the number of the Israelites was greater than that of the national Egyptians, who were then in power. When we read in 1 Chr. iv. 18 that an Israelite named *Mered* married a daughter of Pharaoh, named *Bitjah (cf.* § 15. 3), this does not favour *Delitzsch's* hypothesis, but tends rather to disaprove it; for it shows that the family of the Pharaohs and that of Jacob were not the same. And on the other hand, so long as the Hyksos dynasty, which was so friendly to the Israelites, held possession of the throne, there is nothing inconceivable in the supposition, that a distinguished Israelite may have married one of Pharaoh's daughters. Again the word הַמֹּלֶכֶת in 1 Chr. vii. 18 is a proper name, and therefore proves nothing. The military adventures, referred to in 1 Chr. iv. 22 and vii. 21 *(cf.* § 18), do not affect the hypothesis of *Delitzsch*, except so far as they actually seem to prove, that it was possible for strong warlike expeditions to be undertaken during the 430 years' sojourn in Egypt, without any reference being made to them in the *Pentateuch*. But, in the first place, neither of these events was of so much importance in the history of the *world*, as even *Delitzsch* ascribes to the supposed conquest of Egypt by the Israelites; and *secondly*, the author of the *Pentateuch* had no particular inducement to mention the former incidents, whereas there are a hundred places in the history of the Exodus in which the latter must have been called to mind. And we must also add, if the writer of the Chronicles had both an occasion and an inducement, to notice and describe at length those comparatively unimportant attacks upon Philistia and Moab ; he must certainly have felt a still stronger inducement to mention the much more magnificent and eventful subjugation of Egypt, with which he must have been quite as fully, if not more fully acquainted. In conclusion, we have one more objection to offer to this hypothesis : namely, that whilst at one time it raises the account of the Hyksos given by *Manetho*, into the position of a historical and trustworthy record, at another it is obliged to declare, that in its most essential points it is at variance with history. For according to the account contained in the Pentateuch, the Israelites remained more than a hundred years in Egypt after the rise of the new dynasty, and were so far from being driven away, that every exertion was made to retain them. And this discrepancy cannot be explained, on the supposition of the distortion of the actual fate of the Hyksos government for the purpose of pandering to the national vanity of the Egyptians. For if the national dynasty no sooner recovered the supremacy, than the Hyksos were humbled, enslaved, ill-treated, forced to render tributary service, and prevented from leaving the country as they desired, which the *Pentateuch* informs us that the Israelites were; there was undoubtedly much more to nourish and flatter the pride and national vanity of the Egyptians, in such an event as this, than in the supposed distortion of the facts of the case, which we find in the account of *Manetho*. No doubt the representation contained in the *Pentateuch*, in the history of the eventual deliverance of Israel, has in it an element that is greatly humiliating to the pride of the Egyptians. But the second extract from *Manetho*, concerning the expulsion of the rebellious lepers, shows that the national tradition of Egypt knew how to distort in its own way the liberation of Israel, which was the cause of its ruin, and yet at the same time to hand down the account of Israel's slavery, which was flattering to Egyptian pride.

(3). In order to overturn the *credibility of Manetho*, *Hengstenberg* in his " Egypt and the Books of Moses" (p. 236 sqq.), opposes the general opinion, that *Manetho* was the president of the priesthood at Heliopolis, and that he wrote his Egyptian history about the year 260 B.C., at the direction of the king, Ptolemy Philadelphus, and employed the archives in the temple. On the other hand, he endeavours to prove that the supposed *Manetho* was a "miserable subject, an intentional impostor, a confirmed liar, and a professional wind-bag belonging to the period of the Roman emperors." This unparalleled assertion, which stands in the most glaring contrast to the honour and esteem, in which the author of the *Aegyptiaca* has been generally held, by the ancients as well as by critics and students of a later period, is supported by arguments of so little value, that it is difficult to find words to express our amazement. *Hengstenberg* has studied the destructive critics of the *Pentateuch* to some purpose; he has learned from them how to treat an ancient author, whose good name is to be sacrificed at any cost in favour of certain preconceived opinions; in fact, so far as *Manetho* is concerned, he has really surpassed the critics referred to. Listen, however, to the chief arguments themselves: (1). "The supposed priest of Heliopolis betrays a striking ignorance of Egyptian mythology, and mixes up the names of Grecian and Egyptian gods in a singular manner." The latter is no doubt correct; but it ceases to be striking and singular, as soon as we understand the author's method and design. *Manetho* wrote in *Greek*, and therefore wrote for Greeks. In accordance with the syncretism prevalent in his day, he combined and identified, so far as it was possible, the names of the gods of Egypt and Greece; and just because he was writing for Greeks, and wished to make his work intelligible *to them*, he even substituted the latter for the former. He may certainly have been very unfortunate in these combinations and substitutions; but no one can call him either ignorant or an impostor on that account.—(2). "Just as striking is his ignorance of the geography of his own land, when, for example, he places the *Saitic nomos* to the *east* of the Bubastic arm of the Nile (c. Apion i. 14)." But may not the passage be corrupt? The ἀντίγραφα of the work of *Manetho*, which *Josephus* employed, was already so corrupt or full of interpolations, that in one the word *Hyksos* is interpreted shepherd-kings, whilst in the other it is said to mean captive shepherds. The conjecture that the text is corrupt, is rendered the more probable by the fact, that the reading in the Armenian translation is the *Methraitic nomos* instead of the *Saitic*, though even the former reading is probably spurious. *Bernardus* has made the proper alteration in the passage in question, and reads "the *Sethroitic nomos*."

The corruption was probably first occasioned by the circumstance, that the first king of the Hyksos, whom *Josephus* calls *Salatis*, was called *Saïtes* in other manuscripts. *Julius Africanus* appends to the name the following remark: *ἀφ' οὗ καὶ ὁ Σαΐτης νομὸς ἐκλήθη*. If we regard *Saïtes* as the original reading, which it probably is (see note 4), an ignorant copyist might easily be led to suppose that the city, which was built by Saïtes, must also have been situated in the *nomos*, that was called by his name. (3). "Pseudo-Manetho betrays entire ignorance of the Egyptian tongue, when he traces the first syllable of the word Hyksos to the sacred dialect, and the second to the vulgar. For there is nowhere else the slightest trace of the co-existence of a sacred and common dialect in Egypt. The author, in his thorough ignorance of Egyptian affairs, confounds the distinction between the sacred and common *dialect* with that between the sacred and common *writing*. Moreover, some suspicion is excited as to the author's acquaintance with the Egyptian language by the fact that *Hyk*, which, according to one account means *a king*, and is said in the other to mean a *prisoner* (an important difference), does not really occur in either of these senses." To this we reply, that with our present limited knowledge of the Egyptian language, the latter fact proves nothing. It is very unjust, however, that *Hengstenberg* should set down this difference in the explanations of the word Hyksos to *Manetho*'s account, since Josephus expressly says, that there were various readings in the codices which he possessed, and these must of course be traced to the copyists, and not to *Manetho* himself. *Lepsius*, moreover, has, in my opinion, fully proved that, at the time of *Josephus*, the genuine and complete work of *Manetho* had ceased to exist (probably it perished with the destruction of the library at Alexandria), and that nothing was left but the lists of the dynasties, and fragments of the history, contained in the other books. Again, to our mind, there is something truly astonishing in the statement that in Egypt there never was any distinction between the sacred and common dialect. For whilst the language of the monuments continued essentially the same; in the course of time, particularly during the Grecian rule, there grew up a marked distinction between the old Egyptian and the new Egyptian or Coptic (*i.e.* between the sacred and common speech) similar to that which we find in other languages. "Every sacred language," says *Bunsen* i. 310, "is in reality nothing but the popular dialect of an earlier date, which has been handed down in sacred books; *e.g.*, the Hebrew, in contrast with the so-called Chaldee; the ancient Greek in the Greek church, by the side of modern Greek; Latin, in contrast with the Roman dialects; and the early Sclavonic, in relation to the modern Sclavonic languages." The only question is, whether in *Manetho's* time the popular dialect (viz., the Coptic) was so distinct from the sacred language, or the language of the ancient documents and monuments, that they could be regarded as two different dialects. Now this was decidedly the case. By means of demotic MSS., we can trace the popular dialect to as early a period as the Psammetichs (*Bunsen* ii. 14). There is no objection, therefore, to the supposition, that the word Sôs was still in existence in the popular dialect, though the work Hyk had already disappeared, and *Manetho* had merely to seek his explanation in the various monuments and the documents in the temple. Nor is there anything very "serious" in the different explanations; for an obsolete word might easily be proved to have different significations. In any case, the very fact that it is so difficult to explain the word Hyksos, is an argument in favour of the age and historical character of the name, and therefore also of the persons represented by it. If that "miserable subject," that "professional windbag" (as *Hengstenberg* styles the author of the Aegyptiaca), had invented the name himself, he would certainly have based it upon some etymology that was intelligible at the time, or at any rate, to save himself from the appearance of ignorance, he would have given it an explanation that could be safely established.—(4). "In the work of *Manetho* concerning the period of the dog-star (Sothis), of which *Georgios Syncellus* has preserved some fragments, the author mentions the source from which his statements were taken, namely the accounts engraved by *Thoth*, the first Hermes, upon certain columns in the Seriadic land; they were written first in the sacred dialect, and with sacred characters, but after the flood their substance was translated into Greek and written in hieroglyphics by Agathodaemon, the son of the second Hermes, and the father of Tat, and placed among the sacred treasures of the temple; as if it could be necessary to make translations into Greek for the sake of the priests even in the most remote antiquity." But *Hengsten-*

berg, who declares that both of the works are equally the bungling performances of an impostor whose very name is assumed, is not warranted in drawing from the dishonesty of the one, conclusions prejudicial to the character of the other; for who can assure him that two different authors may not have made use of the venerable name of *Manetho*, for the purpose of helping their wretched productions to pass? Each of the two must be tested by itself. And the actual state of the case is this, that, whilst the Aegyptiaca is regarded by all competent critics as authentic, they are unanimous in pronouncing the book on the dog-star a forgery (*cf. Bunsen* i. 256 sqq., *Böckh* p. 15 sqq., *Lepsius* i. 413 sqq., &c.).—But, even assuming the genuineness of the Sothis, the case is far from being so bad as *Hengstenberg* supposes. *Zoëga* makes the highly probable suggestion, that the original reading in the Sothis may have been *εἰς τὴν κοινὴν* (instead of Ἑλληνίδα) *φωνήν*. Some copyist, or perhaps the Syncellus himself, may, either from a misapprehension or from hurry, have substituted the Greek *κοινή* for the common dialect (*κοινή*) of Egypt. (*cf. Böckh*, p. 16, and *Lepsius* i. 413 Anm. 2).—(5). "The hatred and hostility to the Jews, which gave rise to the second account of *Manetho*, had no existence before the age of the Roman emperors." But why must it have been this particular hostility which gave rise to the account? The disgrace and injury, which are said by the *Pentateuch* to have been inflicted by the Israelites upon the whole of Egypt, were surely enough to excite such bitter feelings in the minds of the ancient Egyptians, that we can very well imagine them to have lasted so long as to give rise to those *ἀδεσπότως μυθολογούμενα*, which were preserved till *Manetho's* days.—(6). "The statements made by *Manetho* do not receive anything like that confirmation from the monuments, which we should expect them to receive if *Manetho* were a trustworthy and honest enquirer." This argument leads to the very opposite conclusion, for, however great may be the differences, which we find on comparing the data obtained from the monuments and other ancient documents, with those given by *Manetho*, so far as names and numbers are concerned; the instances of agreement are so numerous, so strong, and so essential, that we are *forced* to the conclusion that *Manetho's* work must have been the result of the most careful research. The differences and discrepancies may certainly show, that his research sometimes led to misapprehensions and erroneous conclusions; though even these may be satisfactorily explained, either from the early loss of *Manetho's* book, and the faulty copies that were made of it, or from the arbitrary manner in which it was used and revised by the chronographers. The most *unfavourable* opinion, which a moderate, and, at the same time, keen criticism can express with regard to *Manetho*, is that of *Saalschütz*, who regards him as an honest, but somewhat uncritical, compiler.

(4). Hengstenberg (p. 247) thinks that he has proved "THAT THE HYKSOS WAS NO OTHER THAN THE ISRAELITES; that the account of *Manetho* is not founded upon any earlier native sources; but, on the contrary, has merely sprung from a transformation of the historical material preserved by the Jews, which is so altered *as to favour the national vanity of the Egyptians.*" With this *Hofmann* for the most part agrees; but he regards the transformation of the historical material, for the purpose of favouring the national vanity of the Egyptians, as having taken place at an early period in the history of Egypt, and therefore does not think it necessary to indulge in such unmeasured abuse of *Manetho*, as *Hengstenberg*. "The account of the Hyksos given by *Manetho*," says the latter, "presents such a striking resemblance to the history of the Israelites contained in the *Pentateuch*, and on the other hand, wherever it differs, is so evidently altered to favour the Egyptians, that we can have no doubt as to the identity of the Israelites and the Hyksos." This assertion is made with so much confidence, that we cannot abstain from a thorough and searching examination of the arguments by which it is supported. (1). "The Hyksos, like the Israelites, come *from the East*, and particular stress is laid upon the fact that, like the Israelites, the Hyksos were *shepherds*." Were the Israelites, then, the only shepherd-race in Asia? According to a tradition, quoted by *Manetho*, the Hyksos were *Arabs*. From an intimation, given by *Herodotus* (ii. 128), we might conjecture that they were *Philistians*. A Moslemite tradition (*cf.* Abulfedae hist. anteislam. ed. Fleischer p. 178) might lead us to suppose that they were *Amalekites*. And how many other known and unknown shepherd tribes were there in Asia at that time, who were strong enough to attack the favoured land of Egypt with the hope of conquering it? What is there to force us to think only of the

Israelites, who certainly came to Egypt without any such intention? The fact that Joseph's brethren state to Pharaoh that they are אַנְשֵׁי מִקְנֶה (Gen. xlvi. 34), and that Pharaoh wishes to appoint the best of them as שָׂרֵי מִקְנֶה, *i.e.*, as keepers of his own flocks (Gen. xlvii. 6), cannot certainly be regarded as a *proof* of the identity of the Hyksos and the Israelites; although *Delitzch* writes, as if these Hebrew expressions coincided with the name Ὑκσώς or ποιμένες βασιλεῖς. We should be much more inclined to discover an important resemblance in the words ποιμένες ἦσαν ἀδελφοὶ φοίνικες ξένοι βασιλεῖς, with which *Eusebius*, (in the Chronicon), and the Syncellus introduce the seventeenth dynasty. *Delitzsch* says: "Is not this a most striking description of the brotherly tribe of Jacob, which immigrated from Canaan?" But who can answer for it, that these are *Manetho's* own words? And they are worth nothing if they are not. On the contrary, when we consider that in his leading work *Manetho* expresses no decided opinion with regard to the origin of the Hyksos, as we may see from the extracts made by *Josephus*, it is very improbable that in the connected catalogue of dynasties he should have described them with perfect confidence as Phœnicians. *Delitzsch* also calls attention to the Semitic name of the Hyksos city Ἄβαρις, or city of the Hebrews, the north-easterly situation of which corresponds with that of the land of Goshen. (*Champollion* identifies this city with Heroopolis, *Lepsius* with Pelusium, *i.e.* Pelishtim, or city of the Philistines). But neither the situation nor the name proves anything in favour of the identity of the Hyksos and the Israelites. The name עִבְרִים was a very general one, descriptive of all the tribes whose original home was on the other side of the Euphrates (*cf.* Vol. i. § 46. 4). And this may have been the case with the Hyksos, without their being identical with the Israelites.

The *second* argument is this: "The first king of the Hyksos, who was elevated to this dignity from the midst of the people, was named *Salatis.* This unmistakeably Semitic name evidently sprang from Gen. xlii. 6, where we read that Joseph was the regent (הַשַּׁלִּיט) over the land." *Hofmann* (Zeitrechnung p. 22) denies that there is any such connexion, since the name, which we find in *Africanus* and *Eusebius*, is not *Salatis*, but

Saïtes. *Delitzsch*, however, is of opinion, that it cannot be disputed (Genesis p. 358, ed. 1). *Hofmann* was quite right, as we think, in giving up this argument; for the reading *Salatis* is all the more suspicious, on account of its being so serviceable to the purpose of *Josephus.* But even if this be the correct reading, instead of supporting *Hengstenberg's* hypothesis, it completely upsets it. For the name Salatis is either ancient and historical, or it is modern. If the former, then the Hyksos are also historical; but if it be the latter, and therefore, like the whole Hyksos fable, an invention of that "miserable subject," who assumed the name of Manetho, I would ask how this Hebrew name found its way into the Egyptian legendary lore, or how did Pseudo-Manetho get hold of it? Hebrew he certainly did not understand, nor is he likely to have read the *Pentateuch* in the original; and in the *Septuagint*, from which alone any acquaintance with Israelitish antiquities must have been obtained, the word השליט is not retained, but rendered ὁ ἄρχων τῆς γῆς.

Hengstenberg attaches still greater importance (3) to the statement contained in *Manetho's* account that Salatis went every year to *Avaris* at the time of harvest, τὰ μὲν σιτομετρῶν καὶ μισθοφορίαν παρεχόμενος, τὰ δὲ καὶ ταῖς ἐξοπλισίαις πρὸς φόβον τῶν ἔξωθεν ἐπιμελῶς γυμνάζων. Any one who reads the passage in its connexion would render it: He came to Avaris every year at the time of harvest, partly for the purpose of *provisioning* the place (as a border-fortress) and paying the garrison, and partly to strike terror into the minds of foreigners by exercising his troops. But *Hengstenberg* makes it mean, "Salatis occupied himself chiefly when there with *measuring corn,*" and then calls this a characteristic trait, in which it is impossible to mistake the reference to Joseph. Any Greek lexicon would tell him that σιτομετρεῖν means to *provision;* and in the present instance, this rendering is *imperatively* required by the context.

Hengstenberg also says (4), "the account of the oppression and harsh treatment, which the Egyptians suffered from Salatis and his successors, has its historical foundation in Gen. xlvii. 20, "and Joseph bought all the land of Egypt for Pharaoh," &c The distortions, to which this fact has been subjected, may be easily explained (?!!) from the endeavour to reverse the actual relation of the Egyptians and the Israelites, and thus to transfer the disgrace from the former to the latter. It was necessary that the charge of unjust oppression and cruel treatment, which the history attaches to the Egyptians, should be removed from them to the Israelites." With reference to any such distortion of the historical facts, to favour the national vanity of the Egyptians, all we can say is, that their national vanity must have been of a *very peculiar kind*, if this supposition be correct. That would surely be a rare and unparalleled description of national vanity, which would lead any one to represent his own people as oppressed, enslaved, down-trodden, and ill-used by a horde of men so despised as the Jews, when the very reverse had been actually the case, and on the other hand, to describe the Jews as a brave and victorious tribe, who were the rulers and oppressors of Egypt for several hundred years, whereas they were really timid, despondent, subjugated, and enslaved! A national vanity of this kind would be all the more rare and inconceivable, since it is well known that the views which prevailed in ancient times, with reference to the rights of slaves and helots, were not founded upon any very rigorous code of ethics. We have here, however, an actual specimen, and a very lucid one too, of the manner in which the national vanity of the Egyptians perverted the relation of the Israelites to the Egyptians. This specimen we find in the second extract from *Manetho*, which undoubtedly refers to the Israelites. The people, for whom the miracles wrought by their God had forcibly obtained permission to depart, are there represented as lepers and persons affected with the itch, as beggars and monsters, who were banished and hunted away, every kind of indignity being heaped upon them.

The most important argument in any case is (5) the one founded upon the statement, that after the Hyksos were banished from Egypt they went *through the desert to Syria*, and there built a city, which they called *Jerusalem*,—"a feature," says *Hengstenberg*, which ought in itself to be sufficient to convince our opponents of the error of their way." *Hofmann* (Studien und Kritiken, p. 409) also thought "the name Jerusalem of greater importance than all the rest." But first of all we would request attention to the fact that the Hyksos, who came from the countries πρὸς ἀνατόλην, would naturally return to the East when they were compelled to depart, and would also naturally go through the desert, probably towards Syria. Hence, neither of these elements is of any great importance. On the other hand, we must admit, that the account of the founding of Jerusalem still remains a most important point. Yet even this by no means proves the identity of the Hyksos and the Israelites; it simply proves (and this is still more clearly shown by the second extract from *Manetho)* that in the course of time the two legends, viz., that of the expulsion of the Hyksos, and that of the lepers, had been partially mixed up together. Whether there was any historical ground for this partial admixture, and if so, to what extent, are questions that we shall discuss in the next section (§ 45. 4).

Such, then, are the arguments, with which the attempt has been made to sustain the identity of the Hyksos and the Israelites; and we have seen how weak they are. Let us look now at the positive proofs, which may be offered, that such an assumption is inadmissible. And, *first of all*, we will examine the question from the *standpoint of the Pentateuch*, the statements of which we regard as indisputably historical. If now we compare the history of the Israelites, contained in the *Pentateuch*, with the account of the Hyksos given by *Mantheo*, it will soon appear that they are entirely different the one from the other. The discrepancies so thoroughly pervade the whole, that they appear in nearly every single feature, and in almost every word. If the Hyksos-legend was invented, for the sake of giving such an account of Israel in Egypt as would suit Egyptian tastes; the author has not introduced into his legion a single fibre of the historical truth, as we have it given in the *Pentateuch*. If, however, there is concealed in it a single element of historical truth, however small, it is impossible to think of an identification of the Israelites and the Hyksos. For in great things as well as small, in general statements as well as special details, we find on both sides nothing but mutually exclusive differences and contradictions. The Hyksos came in great numbers into Egypt; they came suddenly and unexpectedly; they came as enemies and conquerors; they murdered, plundered, devastated, and governed for five hundred and eleven years; and then they were overcome and compelled to depart. When the Israelites came, there were only seventy of them, with two or three thousand servants at the most; they first received permission to come; they came as suitors seeking protection; they lived peacefully among the Egyptians, but after a short time the latter oppressed, ill-treated, and enslaved them;

they then begged and entreated for permission to depart, but all in vain, &c.

Hofmann (Studien und Kritiken, p. 408) tries to persuade us, that we only need to forget a very little of the account of *Manetho*, in order to convince ourselves that all the rest harmonises very well with the history of the Israelites, as given in the *Pentateuch*. "If," he says, "for a moment we overlook the fact, that the Israelites did *not* enter Egypt by force of arms, and did *not* conquer the land" (and, *we* may add, a few other trifles, *e.g.*, the capture of the princes of the land, some of whom they slew, making slaves of the rest with their wives and children; the choice of a king from among themselves; his residing at Memphis; the fortification of the city of Avaris; the annual military exercises; the names of the successors of Salatis; the eventual appearance of a family belonging to the national dynasty; the tedious war of liberty; the siege of Avaris; and other things besides—if all this could be forgotten for a moment) "then the rest *applies to the Israelites very well.*" And what is the rest? To the Egyptians they certainty were *ἄνθρωποι τὸ γένος ἄσημοι;* they had come to Egypt without a conflict; Avaris was situated to the east of the Bubastic arm of the Nile, and that was also the situation of Goshen; and the fortification of the eastern frontier we might find in the building of the two arsenals Pithom and Raemses, though they were built by *forced labour* (this, of course, it would also be necessary to forget).

We will now, *secondly*, take as our starting-point, *the statements of Manetho* himself; that we may see whether we can thus arrive at the conclusion, that the Hyksos and the Israelites were the same. In addition to the account of the Hyksos, *Manetho* gives a description of the banishment of some leprous Egyptians, whose leader and lawgiver is called Moyses (§ 43. 2). As it cannot be questioned that this second account refers to the Israelites, who are described as banished Egyptian lepers, the two accounts are regarded by all who identify the Hyksos with the Israelites, as different versions of the same legend. But a comparison of the two will show that they are radically different; so different that it is impossible to discover any common ground, from which we may deduce the one primary legend that gave rise to the others. One thing which renders it impossible to establish any such connexion between the two accounts, is the fact that the second presupposes the first, and evinces a perfect consciousness of the difference between the Hyksos and the lepers, since the former are called to the assistance of the latter and unite with them. According to *Josephus*, *Manetho* places the expulsion of the lepers 518 years later than that of the Hyksos, and therefore must have been very far from supposing that they were identical. It is true that, for his own part, he has no great confidence in the credibility of the second account, but he has introduced it in its chronological order into his historical work. He must, therefore, have detected some historical germ, which rendered it possible for him to assign it a proper chronological position in his history; and his doubts as to the credibility of the narrative can only have had respect to its fanciful and fabulous dress. And the more his candid expressions of doubt, as to the perfect credibility of the second account, prove him to be a modest and sincere enquirer; the greater confidence shall we be able to place, not merely in the historical character of the first account, the trustworthiness of which he does not at all suspect, but also in the results of his enquiry, namely, that the two accounts refer to different persons, different events, and different times.

In judging of the statements of *Manetho*, we must carefully distinguish between the *Israelites of the Pentateuch* and the *Jews* of his age. It is only the former that are to be identified with *Manetho's* lepers. The Jews of later times he supposes to have originated in a combination of the Hyksos and the lepers. And we shall see presently that this view is not so thoroughly unfounded and unhistorical as we might at the first glance be led to suppose; but that, on the contrary, it receives a certain measure of support from a passing remark in the *Pentateuch* itself (*cf.* § 45. 4).

§ 45. There is a threefold difference, in the opinions entertained by those who agree that the Hyksos and Israelites were not the same, with regard to the relation in which they stood to each other. Some refuse to admit that they were connected in any way whatever, and assign the expulsion of the Hyksos to a period anterior to the days of Abraham and Joseph. This is the conclusion at which *Lepsius* has arrived, chiefly as the result of chronological calculations (1). According to another view, of which *Saalschütz* is the representative, the new king, who is said in Ex. i. 8 to have begun to oppress the Israelites, was the first king of the Hyksos dynasty (2). The third hypothesis, and the one which has met with the most general adoption in modern times, assumes that the Hyksos dynasty was in power when the Israelites went into Egypt, that it was by this dynasty that so much favour was shown, and that as soon as a national dynasty recovered the supremacy (Ex. i. 8), the Israelites were hated and oppressed as the friends and protégés of those who had been expelled (3). As an impartial examination of Egyptological researches, with their arbitrary methods of procedure and contradictory results, necessarily leads us to the conclusion, that no reliable means have yet been discovered of threading the labyrinth of Egyptian *chronology*, and that it is scarcely likely that any will be found; our safest plan will be to compare and combine the *actual* data in our possession. These are to be found, on the one hand, in the Pentateuch history, and on the other in the accounts given by *Manetho*. We are both warranted and constrained to make such a combination, by the general testimony of the earliest traditions and investigations up to the time of *Manetho*, to the effect that the Hyksos and Israelites were contemporaneous, and that there was some connexion between their histories. And when we compare the two, we find so much to support the third view referred to above, that we feel no hesitation in adopting it as our own (4).

(1.) The general features of the view entertained by LEPSIUS are the following: About the year 2100 B.C., during the period of the twelfth (the second Theban) dynasty, the Hyksos, a warlike pastoral tribe of Semitic origin, entered Egypt from the east, conquered the land without resistance, took possession of Memphis, adopted it as their own capital, and imposed tribute upon the upper and lower parts of the land. About 430 years afterwards, (in the year 1661), the native kings, who had maintained their independence partly in Upper Egypt and partly in Ethiopia, advanced from the south, and after a long (eighty years') war succeeded in expelling the Hyksos from their last stronghold, Avaris (the Pelusium of later times), and drove them back upon Syria, after they had been in Egypt 511 years. They now numbered some hundreds of thousands, and had probably received as much benefit from the culture they met with in this highly cultivated land of art and science, as the latter had received of obstruction and injury from them. Being expelled from Egypt, they were obliged to seek a new home in *Palestine*. This led to fresh expulsions and emigrations, and probably issued in the division of the Hyksos and their dispersion in different directions. The expulsion of the Hyksos took place under king Thummosis, *i.e.*, Thuthmosis III. Almost two hundred years elapsed after this, before the Israelites went down into Egypt; and both their immigration and their departure, after a sojourn of hardly a hundred years, occurred under the nineteenth dynasty. *Sethos I.* (1445—1394; called *Sesostris* by the Greeks), was *the* Pharaoh who was on the throne when Joseph was brought into Egypt; his son *Ramses II.*, Miamun the Great (1394—1328), was the king at whose court Moses was educated; and *his* son *Menephthes* (1328—1309), the Amenophis of Josephus, was the Pharaoh of the exodus. The exodus itself took place in the year 1314.—The second account of *Manetho* refers to the Israelites. The statement that they were lepers was not an Egyptian calumny, but a fact; leprosy being at that time a prevalent disease among the Israelites in Egypt. This is proved by the Mosaic laws with reference to leprosy, and by the history of Miriam (Num. xii. 14)!!

For the present, we will allow that the author is correct in his assertion, that the three great Pharaohs reigned at the periods assigned them. But he will never persuade us, to say nothing of convincing us, that their reigns coincided with those periods in the history of Israel to which he refers. So long as any confidence is placed in the credibility of the Pentateuch and the Old Testament history in general, the combinations which *Lepsius* has made will be rejected as baseless and visionary; but they will excite no less astonishment at the arbitrary nature of his criticism, and the recklessness with which the sacred records are handled, than at the extraordinary amount of learning and ingenuity displayed in the treatment of the subject. In support of his view *Lepsius* says (Realencyclopädie i. 145) : " The strongest confirmation is to be found in the fact, that there is one circumstance mentioned in the Mosaic narrative itself, which points in the most conclusive manner to the period to which we have assigned it. We refer to the building of the cities *Pithom* and *Raemses* by the Jews, under the predecessor of the Pharaoh of the exodus, and therefore under Ramses II. We know from other authorities, that this most powerful Pharaoh had many canals cut and new cities built, and particularly that he projected the canal connecting the Red Sea with the Nile, at the western end of which was Pithom and at the eastern Raemses (=Abu-Keishid). Among the ruins of this city there is still to be seen a group, consisting of two gods, with the deified Ramses II. on a throne between them." With regard to the last two circumstances, we refer the reader to § 41. 1, where we have pointed out the impossibility of the identity of Abu-Keishid and the ancient Raemses. We here make the additional remark, that as early as the time of Joseph there probably existed a city named Raemses (§ 1. 5 and § 41. 2) ; and, though the city is said to have been built either by or under a king Raemses, it is impossible to see why this may not have taken place just as well under an earlier king of that name.

Again, *Lepsius* lays great stress upon the fact, that in *Manetho's* second account *Amenophis* is mentioned as the king who expelled the lepers (*i.e.*, the Israelites). As this Amenophis (*Africanus* reads Amenophthis) is described as the son of a Ramses and the father of a Sethos, he cannot have been any other than the *Menephthes* of the nineteenth dynasty, whose father was Ramses II. and his son Sethos II. At first sight there is certainly something surprising in this coincidence. But when we consider, that *Manetho* himself describes the whole legend concerning the expulsion of the lepers as ἀδεσπότως μυθολογούμενα, and reckons 518 years from the expulsion of the Hyksos to the reign of this king Amenophis, it is very natural to suppose that he has made a mistake, in the position assigned to this uncertain legend in the history that he took from the sacred writings, and has placed it some centuries too late. In the eighteenth dynasty the name Amenophis (Amenophthis in *Africanus*) repeatedly occurs. We might be the more easily led to think of one of these kings, say of Amenophis III., or the Great, whom the Greeks called Memnon, since he lived about fifteen hundred years before Christ, and therefore at the period assigned to the exodus in the biblical chronology (480 years before the erection of Solomon's temple). *Manetho*, who was not acquainted with the biblical data, from which he might have obtained a safe criterion for fixing the true position of the exodus, could very easily be led into such a mistake.

On these grounds, then, though professedly from the greatest respect to the biblical records, the scriptural chronology and history are cut down and mutilated in the most miserable and arbitrary manner, to form a Procrustes' bed for the chronology of these three kings. The author proposes the question "whether the Old Testament accounts contradict the Egyptian (*i.e.*, as *he* has explained the latter), in such a manner that the Egyptian must necessarily be declared erroneous;" and to this he replies, "on the contrary, the Egyptian history, which is of so definite a character, apparently receives the most decided confirmation from the Hebrew records, provided we assume that there is an error in the reckoning of the interval between the exodus and the building of the temple, which reckoning, at all events (?!), was not introduced till a later age. According to 1 Kings vi. 1 this interval was 480 years, a period which neither tallies with the different numbers given in the Book of Judges, nor with the *Septuagint* reading, nor with the reckoning of the author of the Acts of the Apostles (chap. xiii. 20), and which was not even regarded by *Josephus* himself as correct (Ant. viii. 3. 1; *c.* Apion ii. 2). In most of these cases the number of years is said to have been *still greater;* but an unprejudiced (?!) examination and comparison of the genealogical tables, of which the Levitical deserves the greatest confidence, and a computation of the intervening numbers, lead to the conclusion that the interval was considerably shorter, and the number obtained is just such as we should expect, provided the Egyptian tradition with regard to the epoch of the exodus be correct."—The Levitical genealogies only mention three generations, from the time of the entrance into Egypt to that of the exodus (Levi, Kohath, Amram), and only ten or twelve from that period to Zadok (the high priest under Solomon). Now as *Lepsius*, quite arbitrarily and erroneously, gives only thirty years as the length of a generation; the Israelites, according to his opinion, can only have been ninety years in Egypt instead of 430, and the building of the temple must have taken place 300 years afterwards instead of 480 (*cf.* Chronologie i. 367 sqq.). And in the same way, since the period which elapsed between the entrance of Abraham into Canaan, and Jacob's going down to Egypt, only embraced three generations, this cannot have been longer than ninety years. To this we reply (1), that when the Old Testament speaks of a generation, it means something very different from modern statistics, the *Pentateuch*, according to Gen. xv. 13—16 (*cf.* Ex. i. 6), regarding it as embracing at least 100 years, instead of thirty, during the patriarchal and Mosaic period (*cf.* § 14. 1); —(2), that *four* generations, and not *three*, are mentioned as included in the period of the Israelitish sojourn in Egypt (Levi, Kohath, Amram, and Aaron), for Aaron was eighty-three years old at the time of the exodus;—(3), that during the same period *six* generations are named in Joseph's family, *seven* in Judah's, and as many as *ten* in Ephraim's (*cf.* § 14. 1). But *Lepsius* cannot make use of such facts as these, and therefore he declares that "they are evidently in a state of confusion, and lead to no result," *i.e.*, they do not square with our critic's premises, and lead to a result which he does not like. In our opinion, however, these different accounts prove with indisputable certainty, what we repeatedly find on other occasions in the biblical genealogies, that in some instances individual members are omitted, and in others several are linked together. The reason for such combinations we have already explained at § 14. 1.

Again, *Lepsius* points to the fact that "the correct view" (that is, his own), has been *retained (?!)* by the Rabbins. For example, according to the Rabbinical chronology, which was first invented by *Hillel ha Nassi* in the year 344 of the Christian era, and gradually met with general adoption, the exodus occurred in the year 2448, from the creation of the world, that is in the year 1314 B.C., according to the Christian mode of reckoning. The fact that this Jewish era was not heard of till the fourth century, and moreover, that nearly all the data, on which it is founded, are false, that is, at variance with the calculations of *Lepsius*, does not affect the question—the year 1314 B.C. suits his purpose, and therefore the Rabbins have *retained* the correct view! Now according to all the previous calculations with reference to the biblical chronology, viz., that of the *Septuagint* translators, of *Josephus*, of the chronological tradition (which was followed by *Stephen*), and of the Christian chronographers and others, 480 years was *far too short* a period for the events which occurred between the exodus and the building of the temple; and yet *Lepsius* finds *in the same fact* a support for his opinion, that this period of 480 years was *much too long* to be regarded as correct!!!

Hence, without further discussion, we must reject this reduction of the chronological data of the Bible from 430 and 480 years to 90 and 300; for the simple reason, that the history of the period referred to does not admit of any such reduction. We shall defer, to a later period, the proof that the same remark also applies to the interval between the exodus and the building of the temple. But with regard to the earlier period, it is so very conspicuous, that we can only wonder at the facility, with which it could be ignored or set aside. We will pass over the fact, that if there is any chronological datum of the Old Testament, which has all the probabilities in favour of its correctness, this is certainly the case with the history of the 430 years' sojourn in Egypt. Suffice it to say, that it would be difficult to persuade any one, that a family of seventy souls, with not more than 2000 servants, increased to two millions in ninety years. *Lepsius* will probably meet this objection with the favourite explanation of *Ewald* and others, that Jacob, Joseph, and his brethren are not to be regarded as the heads of single families, but as the representatives of whole tribes. If so, let him candidly affirm that the *Pentateuch* does not contain a *history*, but an unhistorical *myth*, and then we shall no longer have occasion to argue with him. Besides, it is not *merely* the *single* number 430, which has to be set aside. There are many other numerical statements in the books of Genesis and Exodus, which are most closely intertwined with the historical narrative, and must also be explained away; and of these *Herr Lepsius* will not be able to affirm that "they were at all events introduced at a later period." For example, according to Ex. vii. 7, Moses was eighty years old when he first entered upon the controversy with Pharaoh respecting the exodus. Hence he must have been born ten years after the Israelites went down to Egypt. How does this square with the other data contained in Genesis and Exodus? Joseph was seventeen years old, when he was taken to Egypt (Gen. xxxvii. 2), and he died there at the age of 110 years (Gen. l. 22, 26), after he had seen the sons of Ephraim of the *third* generation (Gen. l. 23). Moreover, Joseph was thirty years old when he was promoted by Pharaoh (Gen. xli. 46). Nine years afterwards his brethren came down to Egypt (Gen. xlv. 6). Hence Joseph lived seventy-one years after their entrance into the land. Now we read in Ex. i. 6 seq.: "And Joseph died, and all his brethren, even all that generation. And the children of Israel were fruitful and multiplied, and became very mighty, and the land was full of them." And it was not till after this, that the oppression and tributary service first began. But as these precautions did not suffice to restrain the extraordinary increase of the people (and certainly some decennia must have passed before this fact could be ascertained), the command was issued to murder all the new born boys, and *then* Moses was born. Who can read this and come to any other conclusion than that the period which elapsed between the entrance into Egypt and the birth of Moses, must have embraced at least a couple of hundred years? But according to the chronology of *Lepsius* there are only ten years left for the whole!—And how are we to understand the statement in Ex. i. 8: "And there arose a *new* king, who did not know Joseph?" We have already seen at § 14. 4, that there is every reason, both historical and philological, for regarding these words as the announcement of the rise of a new dynasty. Still we shall not insist upon this, as the arguments advanced there are certainly not absolutely conclusive. But with all the greater emphasis we enquire: how is it conceivable, that if *Joseph* filled a post of such extraordinary importance under the Pharaoh *Sethos I.*, as even *Lepsius* admits that he did, his son and successor, *Ramses II.*, should have known nothing whatever about him?

What we have already said will surely suffice to convince any one, who has the slightest confidence in the credibility of the *Pentateuch*, that the boasted discovery of the *Lepsius*-criticism is an untenable and baseless illusion, to which we can only exclaim, *transeat cum ceteris!* Still there is one more utterance of the critic, which is of so striking a nature, that it well deserves to be mentioned. In the *Real-lexicon*, p. 146, we read: "Thus it was under this king, the greatest of the nineteenth dynasty (Ramses the Great), that Moses, the great man of God, was born; and under his successor, whom *Herodotus* (2. 111) mentions by the name of Pherôs (Pharaoh), and describes as a haughty and wicked king who was punished with blindness on that account, Moses led his people away and founded the first Jewish theocracy by the law which was given at Sinai; just as we find that thirteen hundred years afterwards, when the revolution of the world's history was complete, Christ was born under Augustus, the greatest emperor of the Graeco-Roman world, and the second, or Christian theocracy, was established by his death, under Tiberius, the Roman Pherôs."—Does it not appear as though the author was desirous of conciliating and quieting the Christian readers of the *Real-encyklopädie*, whose sacred relic, the credibility of the *Pentateuch*, he had completely destroyed, and therefore offered them this historiosophical trinket, which he fancied would suit their taste?—For our own part, though by no means prejudiced against historiosophy, we must certainly beg to be excused from bartering so important a portion of the Bible and of history, for any historiosophical idea, however great its attractions may be.

Stark (ut sup.) differs from *Lepsius* in this respect, that he supposes the Hyksos to have been tribes of lower Egypt and therefore of genuine Hamitic descent, who had been once before in power as a Herakleopolitan dynasty, and who, after their expulsion, took possession of the coast immediately adjoining Egypt under the name of Philistines. Consult, on the other hand, *Lepsius* (in the theol. Real-encycl. i. 149).

(2). Saalschütz, in the work referred to above, has set up the clever and carefully developed hypothesis, that the new king, with whom, according to Ex. i. 8, the oppressions of the Israelites commenced, was the first king of the Hyksos-dynasty, that the destruction of Pharaoh in the Red Sea coincided with the overthrow of that dynasty, and lastly, that the next (national) king Sesosis (Sethos, Sesostris), who is described by *Josephus* as *Σεθῶν τὸν καὶ Ραμεσσήν*, was the same Ramesses, who is celebrated in the obelisk-inscription of Hermapion (cited by *Ammianus Marcellinus*) as: *ὅς ἐφύλαξεν Αἴγυπτον τοὺς ἀλλοεθνεῖς* νικήσας and πληρώσας τὸν νεῶν τοῦ Φοίνικος ἀγαθῶν. The armed supremacy of the Hyksos over a portion of Egypt lasted about eighty-one years, for it probably commenced shortly after the birth of Aaron, which took place before the command to kill the children had been issued. The second of *Manetho's* accounts is supposed by *Saalschütz* to be identical with the first (as everything is done by the Hyksos in both cases), but both of them are set down as equally confused and incredible.—According to *Saalschütz* (p. 95), the Hyksos were Philistines, or Gathites. In support of this he refers to 1 Chr. vii. 21, which passage he explains in a totally different manner from his former exposition (*vid.* § 18. 1). According to his present interpretation the Gathites, who were born in the land (*i.e.*, on Egyptian territory), had come down to Egypt to steal cattle, and it was on that occasion that they were massacred by the Ephraimites (p. 96).—We cannot subscribe to this view, for it neither appears to us to do justice to *Manetho's* first account, which we regard as essentially trustworthy, nor to be reconcileable with the statements of Scripture.

(3). Among the scholars of modern times, who suppose that the Israelites entered Egypt during the Hyksos-period, the first that we shall name is Bunsen. Whilst *Eusebius* and *Georgius Syncellus* recognise only *one* Hyksos dynasty, which they call the seventeenth, *Julius Africanus* speaks of *three* Hyksos-dynasties, the fifteenth, sixteenth, and seventeenth, the first of which continued 284 years, the second 518, and the third 151. *Bunsen* adopts most fully the statement of *Africanus*, and maintains that *Josephus* has arbitrarily selected the second number, 518 (511), and given that out as the sum total of all the Hyksos reigns, in order that he may be able to establish the identity of the Israelites (with their 430 years) and the Hyksos (with their not very different number of 511 years). After certain other critical operations he arrives eventually at the result, that the Hyksos supremacy lasted in all 929 years. Into farther discussions respecting the origin and history of the Hyksos, or their relation to the Israelites, he does not at present enter; but he so far anticipates the result of future investigations as to pronounce the Hyksos "Canaanitish tribes probably associated with *Bedouins* of northern Arabia." We must therefore wait for the rest, before we can enter into a thorough examination of his views,

Ewald (Gesch. i. 450 sqq.) describes the Hyksos as Hebrew tribes, related to the Israelites, who had forced an entrance into Egypt long before the Israelites wandered thither. Whilst he succeeds, on such an assumption, in making it very intelligible, why the Israelites met with so good a reception in Egypt, he resorts to fanciful conjurings in order to get rid of the difficulty, that, at the time of Joseph's promotion, everything about the court seems to have been of a thoroughly national, Egyptian character, and that the protégés of the Hyksos, the Israelites, were *not* banished along with their protectors by the returning national dynasty. For example, Joseph came to Egypt as the hero and leader of a smaller Hebrew tribe, some centuries after the more powerful tribe of the Hyksos, and at first under the protection of the latter. But, after a time, some disagreement probably arose between the tribe represented by Joseph and the more powerful ruling tribe of the Hyksos, which caused the former much distress. The only reference which is made to this in the book of Genesis is found in the account of the scene with Potiphar's wife, and the consequent imprisonment of Joseph. This led the smaller tribe of Joseph to attach itself to a native Egyptian ruler, to whom it rendered assistance, when the Theban and other kings of Egypt rose against the Hyksos; and, in particular, it made itself serviceable to Egypt by summoning the other and more powerful portion of Israel, to assist in defending the eastern boundary against any fresh invasion on the part of the Hyksos. But as this danger gradually diminished, the presence of so warlike and well-armed a people, as Israel was, began to be regarded by the Egyptians as in itself an evil. Hence the friendly relation, which had previously existed, was disturbed; collisions took place; the old hatred towards the Hyksos, who had been banished some centuries before, was now directed towards the Israelites who were their relations; and Egyptian kings at length commenced that oppression, of which we have a description in the book of Exodus, and which eventually stirred up the people to successful rebellion under the guidance of Moses and Aaron. Moreover, notwithstanding the great alterations which ages had made, the original bond of friendship between the Israelites and the rest of the Hyksos was renewed, and hence we find Moses entering into alliance with the princes of the tribe of Midian, which formed a part of the numerous collection of tribes included in the term Hyksos, which also embraced the Amalakites. *Sic !*

Far more simple and natural is the view, formerly taken by *Heeren*, and more recently advocated by *Bertheau*, *Lengerke*, *Knobel*, and others. They also regard the Hyksos as Hebrews in a wider sense (*Knobel* in his *Schrift über die Völkertafel* calls them Amalekites, *Bertheau* names them Terachites). The national, Egyptian physiognomy, which was characteristic of the court in Joseph's days, is accounted for on the supposition that the victorious Hyksos had by that time adopted the culture, the language, the customs, and the religion of the subjugated national Egyptians. The oppression of the Israelites they suppose to have commenced with the restoration of the national dynasty, after the expulsion of the Hyksos.

(4). *Our own view* is essentially the same as that of *Bertheau*, *Lengerke*, and *Knobel*. We should merely be inclined to substitute a different conjecture with regard to the *origin of the Hyksos*. We cannot believe them to have been a Terahite people, since according to the book of Genesis the formation and organisation of the Terahite tribes (Vol. i. § 46. 6), belonged to the time of Abraham, Isaac, and Jacob, whereas the Hyksos must have established their power in Egypt at a much earlier period. We should be more disposed to fix upon the Amalekites; provided, that is, they are not to be traced to Amalek, the Edomite mentioned in Gen. xxxvi. 16, but to be regarded, as *Ewald*, *Knobel*, and others suggest, as a much older Semitic race (*cf.* Vol. iii. § 4. 2).

However, the positive arguments, which could be adduced in support of this opinion, are not of much weight. *Delitzsch* has lately pointed out with perfect justice, that we cannot attach much value to the Moslemite testimony quoted by *Abulfeda* (hist. anteisl. ed. Fleisch. p. 178), according to which the Pharaohs of the time of Abraham, Isaac, and Jacob, belonged to the tribe of Amalika (see *Delitzsch* Genesis ii. 221) ; and the etymological attempt, to show that the meaning of the name Amalek coincides with that of the name Hyksos (עֲנָא or עָן = צאן, small cattle, and מֶלֶךְ, king, from which by the hardening of the כ into ק we get the name עֲמָלֵק, kings of flocks, or shepherd-kings) is not considered of much importance even by its originator, *Saalschütz* (p. 95). On the other hand there is great probability, that the Hyksos were either a Semitic tribe, or a tribe with a Semitic language, partly because they came from the districts, in which the formation of the Semitic tribes originally took place, and partly because the name of the city of *Avaris* or *Abaris* has undoubtedly a Semitic sound. (The name of the first Hyksos-king, *Salatis*, also called *Silites*, has less bearing upon this question, since its primary form was probably *Saïtes*, *cf.* § 44. 3, 4).

In opposition to the notion, "that the Hyksos were a *Semitic* tribe, though they were not Israelites," *Delitzsch* (Gen. ii. 75) urges the objection, from a scriptural point of view, that the people of Egypt, who oppressed the Israelites, appear throughout the whole of the Old Testament as a foreign nation, in no way related to Israel; that the very reason why the house of Israel was led into Egypt was that whilst growing into a nation, they might be far removed from the danger of intermarrying; and therefore that the idea of relationship is completely excluded. However, we cannot see that there is any great force in either of these objections. For to the *first* it is sufficient to reply, that the Egyptians, who oppressed Israel, were not the Hyksos, but the national rulers who had once more recovered the supremacy. And the *second* disappears quite as quickly, when we consider that the relationship may have been nothing more than a common Semitic origin, and that, on the other hand, the Hyksos had already adopted the speech, and manners, and religion of Egypt, at the time when the Israelites found a welcome reception there.

It is stated in *Manetho's* account, as given by *Josephus*, that there were some who regarded the Hyksos as *Arabs*, but the opinion is not cited as of much worth. According to *Eusebius* and *Georgius Syncellus*, *Manetho* himself appears to have thought them *Phœnicians*. And there are many who think, that there is some indication of this in the account given by *Herodotus* of the shepherd *Philitis*, who had led his sheep as far as to Memphis; inasmuch as the Phœnicians and Philistines were both included in the one common term *Palestinians*. But all these statements are so fluctuating and uncertain, that we cannot build much upon them. Yet when we add the fact, that after the expulsion of the Hyksos, they withdrew to Judea (= Palestine), probably to the country from which they had originally come, the opinion referred to becomes somewhat more plausible. And if we enquire further, whether there were other known migrations of any importance, in the pre-Abrahamic times, which might have occasioned their departure from Palestine; we are led at once to think of the forcible expulsion of the (Semitic) aborigines of Palestine by the Canaanitish tribes. But as there are other reasons, which render it probable that the immigration of the Canaanites into Palestine was of a friendly character (Vol. i. § 45. 1), and, therefore, could hardly have occasioned the departure of the original inhabitants, there is very little to sustain such an opinion.

On the other hand, we may be allowed to call attention to another feature of the case, which has hitherto received but little notice. *Manetho*, as quoted by *Josephus*, says that the Hyksos had strongly fortified the eastern boundary of Egypt, for the purpose of securing themselves against any attack from the *Assyrians*, who were at that time very powerful. This statement, to which commentators have not attached sufficient importance, appears to me to be of very great value, in assisting us to reply to the question now before us. If the Hyksos had such particular reason for fearing an invasion on the part of the Assyrians, they must already have stood in a hostile relation to each other, and on some previous occasion have been engaged in actual conflict. It is probable, therefore, that the Hyksos may have been dislodged from their possessions by the Assyrians; and if this were the case, it would naturally lead them to fear, that on the first favourable opportunity the latter would follow them to their new settlements, for the purpose of completing their subjugation. Let it not be said in objection to this, that it is improbable that the Hyksos, who had fled from the Assyrians, should be strong enough to conquer Egypt, which was then in its prime;—for *Manetho's* account informs us that Egypt was slumbering in peace and security, and submitted to the foreigners without the slightest resistance. Still this hypothesis can only be firmly established by testimony from other quarters, to the fact that such warlike expeditions were actually undertaken by the Assyrians, before the time of Abraham. And the book of Genesis appears to offer the testimony we want. There is nothing at variance with *Manetho's* account in the supposition, that at that time the Assyrians were the possessors of the cultivated lands on the Euphrates and Tigris. And the remarkable account contained in Gen. xiv., which all critics acknowledge to be trustworthy, indisputably proves, that before the time of Abraham expeditions had issued thence for the purpose of conquest, and had proceeded westwards as far as to Palestine. The rule of Chedorlaomer over the Pentapolis of the vale of Siddim was probably all that remained of more extensive conquests. If we duly consider these memorable circumstances, in connexion with the statements made by *Manetho;* the conjecture that the Hyksos were Canaanites,—either the Semitic aborigines of Canaan, or immigrants into Canaan who had adopted Semitic customs (Vol. i. § 45. 1, inaccurately termed Phœnicians or Philistines, *i.e.*, inhabitants of Palestine), but who subsequently yielded to the invasion of the Assyrians and sought out new settlements for themselves in Egypt,—may, perhaps, at least, deserve a place *by the side of* so many other conjectures, which certainly rest on no surer foundation. And there is the greater probability in this, since, as will presently appear, the journey of Abraham into Egypt (Gen. xii. 10) most likely took place but a very short time after the Hyksos had established themselves in the land.

But, whether we are correct in this conjecture or not, we must in any case adhere most firmly to the conclusion, that the immigration of the Israelites occurred during the period of the Hyksos supremacy, and that the restoration of the national dynasty was followed by their oppression. The objections to this conclusion (in *Hengstenberg* p. 160; *Lepsius*, Realencyclopädie i. 146; *Saalschütz* p. 56, and others), so far as they have any force, may be reduced to two: viz., (1), that at the time of Jacob and Joseph everything connected with the Egyptian court, language, customs, culture, and religion, was of a thorougly national, Egyptian character; and (2) that the Israelites were not banished along with the Hyksos when the national dynasty was restored, as we should expect them to have been if they were their protégés and friends. Our first remark, in reply to this, is that the two arguments cancel each other. For if the first be correct, and in general it must be admitted that it is so, the second necessarily loses its force. If, when the Israelites entered the land, the Hyksos had so thoroughly adopted the language and religion, the culture and the customs of Egypt, as the history of Joseph shows that they had; this would suffice at the very outset to establish such a wall of separation between the two, as to prevent any closer amalgamation. Moreover, the second argument has all the less weight, seeing that the banishment of the Hyksos was probably by no means universal, as we shall presently show.

Undoubtedly there appears to be an irreconcilable difference between the Hyksos rulers, as they are described in the extract from *Manetho* preserved by *Josephus*, and the court life of Egypt at the time of Jacob and Joseph, as it is represented in the book of Genesis. The Hyksos attack the national Egyptians with the fiercest cruelty, destroy the national temples and sacred relics, and maltreat the priests. On the other hand, in Joseph's time, the language and customs of Egypt prevailed at court; the king took the title of Pharaoh like the national rulers; and his courtiers had genuine Egyptian names (*e. g.* Potiphar); Joseph himself received an Egyptian name; the peculiar worship of Egypt was in full bloom; the Egyptian priests were highly esteemed, their privileges were recognised and increased, and the national dislike of the shepherd-life was undiminished. But, however glaring these differences may appear at first sight, they are by no means irreconcilable, if we take into account the difference in the periods referred to. All that *Manetho* says with reference to the cruelty, the harshness, and the spirit of destruction manifested by the Hyksos, applies merely to the time of the first invasion, and at most to the first six kings, whose names he gives. Besides, we are warranted in assuming that *Manetho*, as an Egyptian, or more probably still, the priestly sources to which he went for information, indulged their hatred of the foreign rule, by painting in the most glaring colours the injury inflicted upon their native land. But even granting that the whole is literally correct, it must still be admitted that the Hyksos cannot possibly have succeeded in completely exterminating the religion and culture, the language and customs of Egypt; for any people, and most of all, a people of such firmness and marked peculiarity as the Egyptians, would keep fast hold of these possessions under the pressure of the severest bondage. And even *Manetho* himself attests that such was not the case, since he states that the Hyksos subjugated the national rulers and made them tributary. This presupposes, as *Bunsen* (iii. 1 p. 33) correctly observes, that they not only left them alive, but allowed them to live in conformity with their previous habits. And if the customs of Egypt stood this first shock, no other result was possible, than that its manners and customs, which were most intimately related to its language and religion, should slowly and gradually, yet certainly and inevitably, exert their silent influence upon the rude and uncultivated conquerors. What has so often been repeated since then, in the history of the world, would be sure to occur in this instance; namely, that the barbarous conquerors of a cultivated people would very soon be conquered themselves, by the overwhelming mental power possessed by the nation they had vanquished. Thus did the barbarous conquerors of China adopt its religion, its speech, and its customs; and thus also did the Germanic tribes adopt those of the conquered provinces of Rome. If, then, we reflect that at the time when the Israelites entered Egypt, the Hyksos must have been some centuries in the land; there is nothing to astonish us in the fact that the language and customs of Egypt prevailed in the court of the Hyksos, especially as there are evident signs that these adopted manners were by no means assimilated, but rather resembled a coating of varnish that had been merely laid upon the surface.

When *Abraham* took refuge in Egypt on account of a famine (Gen. xii. 10 sqq.), the supremacy of the Hyksos had existed for some time, as certain facts and chronological calculations most clearly show. But at that time there was no trace of the hatred of shepherds, so conspicuous in Joseph's days. The Pharaoh of that day, as well as his court, did not hesitate to associate with them even in public. The king himself, who was desirous of doing honour to the Nomad Emir (as his future brother-in-law), sent him liberal presents of sheep, oxen, and asses, man-servants and maid-servants, she-asses and camels. Did that look like a national ruler? was it not much more appropriate for a shepherd-king? Two hundred years afterwards, when *Joseph* was in Egypt, the physiognomy of the court was completely changed. The language, customs, and religion of Egypt were then predominant at court; and a pastoral life was so far an offence, that it was unseemly for a courtier, as well as for the national Egyptians, to eat with shepherds (Gen. xliii. 32). But these were merely matters of expediency, to which the king and the court had seen fit to conform. Circumstances were repeatedly occurring, which proved that this was nothing more than an external adoption of the customs and notions of the country. That the ruling Pharaoh could venture to naturalise Joseph, the foreigner, the slave, and the shepherd's son, to place him in one of the highest posts of honour, and to give him a wife from the most distinguished priestly family, was a thing so thoroughly opposed to the national habits of the Egyptians, that we cannot conceive it to have been possible in the case of a native king, and are almost forced to assume the existence of a foreign and despotic government. Moreover, this Pharaoh was still the owner of large herds of cattle, for which he selected herdsmen from the immigrant Israelites. But such was the utter abhorrence in which a shepherd life was held by the Egyptians, that the wealth of a *national* ruler would have been much more likely to consist of landed property than of herds of cattle. The Hyksos rulers, however, most probably despised agriculture at first, just as much as the Egyptians despised a pastoral life; and hence it is very likely, that when they conquered Egypt they neglected to reserve a sufficient quantity of arable land. But the more thoroughly they entered into the habits of the Egyptians, the more sensible must they have become of the disadvantage under which they laboured, and Joseph very properly thought of providing a remedy (Gen. xlvii.). Just as irreconcilable with the idea of a *native* government, is the unhesitating readiness with which a pastoral tribe, like that of the Israelites, was welcomed into the land; especially if the Hyksos period was already past, as *Lepsius* believes. For in this case the recollection of the sufferings of that period would have been most vivid, the hatred of shepherds would have been at its height, and the danger of an offensive alliance between the immigrants and the banished Hyksos or other pastoral tribes of the East would immediately suggest itself. And yet these are the circumstances under which the Israelites are supposed to have been welcome (!), the best provinces being allotted to them, and even the gates and keys of the whole country being placed in their hands! Joseph advised his brethren to tell Pharaoh, without hesitation, that they were nomads: a poor recommendation, one would think, in the estimation of a national ruler. The aged patriarch, Jacob, took upon himself to bless the Egyptian king;—would a native ruler, with his national pride and his detestation of shepherds, have allowed such a thing to be offered to him by a despised shepherd-chieftain? And when Jacob died, the whole of the court and the elders of the land of Egypt formed a funeral procession, with chariots and horsemen, in honour of the unclean shepherd-chief, who was notwithstanding an abomination in their eyes! According to 1 Chr. iv. 18, one of Pharaoh's daughters, named Bitjah, was married to Mered, an Israelite. How is this conceivable, if the reigning house at that time had ceased to have any sympathy with shepherds?

Some centuries had elapsed since *Joseph's* time, when a new king arose, who knew nothing of Joseph (Ex. i. 8, *cf.* § 14. 4). The Hyksos had been banished, a native ruler (of the eighteenth dynasty) had recovered the throne, which had for centuries been occupied by usurpers. How intelligible and natural, that such a ruler should know nothing of Joseph, or rather should not wish to know anything of him! If the previous dynasty owed the maintenance of its supremacy to the wise measures devised by Joseph, and if the immigration of the Israelites, a people connected with them by their similar mode of life, and possibly also by descent, was a welcome event to them, the very opposite must have been the case with the new and native dynasty. From the very outset the Israelites, both as a pastoral race and also as the friends and protégés of the Hyksos, must have been an object of hatred and disgust. Moreover, this shepherd-race had grown to be a numerous people, probably even more numerous than the national Egyptians, who had just recovered their freedom and their independence; and they dwelt in that part of the land in which their presence would be most dangerous, if it should ever occur to them to enter into alliance with the enemies of Egypt outside (Ex. i. 10). How natural that the new dynasty should seek to oppress, to weaken, and to enslave a people which was so dangerous in its estimation! The Israelites were forced to render tributary service; they had to make bricks, and build fortresses. How thoroughly does this suit the character of the eighteenth dynasty! The Hyksos had destroyed so many ancient monuments, that the fresh dynasty determined to renew these objects of Egypt's pride; and it was actually under this dynasty, that the greatest number of buildings and the most magnificent were erected. *Josephus* says in his Antiquities that the Israelites were compelled to work at the pyramids. Whether this is founded upon historical tradition, or is an idea of his own; in any case, it is a conjecture which has every probability in its favour. The friends and protégés of the destroyers were compelled to restore what their protectors had demolished.

But why, it is asked, did not the national dynasty expel the Israelites at once along with the Hyksos, if their presence in the land was thought so dangerous and threatening? The answer to this question is so simple, that we can hardly understand how it could ever have been asked. The second book of the *Pentateuch* furnishes us with the reply. The policy of the Pharaohs rendered it more advisable to enslave the Israelites, and, by forcing them to perform tributary service, to render them harmless, than to drive away so many thousand men, who were actually needed for the accomplishment of their designs. In a state like Egypt (when governed by native rulers, and especially under the eighteenth dynasty), where the greatest glory was sought in the erection of colossal monuments, which must have required hundreds of thousands of hands; nothing could be more desirable, than to have a large population of helots in the land, who could without difficulty be forced to perform the hardest tasks. *Herodotus* (i. 108) and *Diodorus Siculus* (i. 56) both show how important this was to the Egyptian rulers. According to the former the great conqueror *Sesostris* brought back large crowds of people from the conquered lands, who were destined to render this hard tributary service; and the latter states that the same king (*i.e.* Sesoosis) did not employ a single Egyptian in the execution of his designs, but that the whole was performed by captives alone. Hence the inscription on all the temples: "no native has been employed in its erection."

Before leaving this subject, we must return once more to *Manetho's* accounts. Both of them contain some particulars, which need a somewhat closer examination. The first announces that the banished Hyksos went through the desert to Syria, settled in *Judea*, and built JERUSALEM. Is this statement historical or fictitious? What events had occured in Palestine, during the interval between the departure of the Israelites and the time of Moses, we cannot tell. Hence we are not in position to deny, without further investigation, that there was any historical foundation for this statement. If the Hyksos left Canaan on account of the Assyrian invasion, as we have already shown to be very probable; it is also probable, that when they were expelled from Egypt, they would first turn their steps in the same direction again. Whether they remained there, and settled in Judea, as *Manetho's* account affirms, or whether they were unable to find any settled abode in Canaan, and therefore wandered further and were lost among the tribes beyond the Jordan, the biblical history does not enable us to determine. We are not even in a position to reject altogether the statement that they built Jerusalem. In the time of the patriarchs, Jerusalem was merely called *Salem* (Vol. i. § 55. 1). In Joshua's days, and as late as the reign of David, the name of the city was *Jebus;* and it was not till David had conquered it, that it was called *Jerusalem.* This change of names is striking enough; and it is very possible, especially as *Jebus* was actually the name of a tribe, (Judges xix. 10, 11), that the Hyksos (or Jebusites) may have conquered Salem after their expulsion from Egypt, and called it by the name which formerly distinguished their tribe. On such a supposition, too, we might possibly obtain some clue to the striking fact, that the city was never completely taken by the Israelites till David's reign. Still we are more inclined to believe, that there is some error or confusion of names in *Manetho's* account, the cause of which is to be traced to a recollection of the close connexion which existed between the Israelites and the Hyksos.

To this conclusion we are more particularly led by the second of *Manetho's* statements, according to which the leprous Egyptians (or Israelites), led on by Osarsiph or Moyses, called the shepherds of Jerusalem to their assistance, and in alliance with them inflicted fresh calamities of every description upon the Egyptians. With regard to the latter, no one can overlook the fact that we have here an account of the plagues, though it is greatly altered to suit the interest of Egypt. And even the supposed alliance between the lepers and the Hyksos is not altogether imaginary. For we learn from the *Pentateuch*, that a number of *common people* joined the Israelites when they went away (§ 35. 7). Now as the Israelites allowed these Egyptian Pariahs to accompany them, they must have been closely connected with them; so far at least as to have suffered the same constraint and oppression. Hence our conjecture is that we have here the remains of the banished Hyksos. It is true that *Manetho's* first account says nothing about any of the Hyksos being left behind. But it is certainly not improbable that this was the case. All that the national dynasty wanted to do, was to overthrow the despotic rule of the Hyksos. For this it was sufficient to expel the king with his officers and soldiers. But it was far from being to the interest of the new dynasty to expel the Hyksos settlers, who were engaged in agriculture or rearing cattle, and had taken no direct part in the war; on the contrary they would be more disposed to do all they could to retain those from whom they had nothing to fear, that they might use them as slaves and helots. Common sufferings would then strengthen and knit more closely the connexion, which originally existed between the Israelites and the Hyksos; and we can very well imagine that the latter would eagerly avail themselves of the opportunity of attaching themselves to their fellow sufferers, who were about to depart, and thus escaping the oppressive yoke of the national Egyptians. And if the Egyptians continued to bear in mind the fact, that the Israelites and the Hyksos had left the country together; the traditions of a later age might easily confuse the whole affair, as *Manetho's* accounts have evidently done. The two fixed points, which had been handed down, were these: that the Hyksos had been banished long before the exodus of the Israelites, and also that numbers of the Hyksos had afterwards left the country along with the Israelites. But to those who lived at a later age, these two statements would appear to disagree, and either in the legends which existed before the time of *Manetho*, or by *Manetho* himself, they may have been reconciled and combined on the simple assumption, that Moses recalled the Hyksos, who had been previously expelled.

END OF VOL. II.

www.ingramcontent.com/pod-product-compliance
Lightning Source LLC
LaVergne TN
LVHW021140110826
845150LV00005B/1084

9781425548070